URBAN COMPLEXITY AND SPATIAL ST

Urban Complexity and Spatial Strategies develops important new relational and institutionalist approaches to policy analysis and planning. Addressing the challenge of metropolitan planning and development, this will be of relevance to all those with an interest in cities and urban areas.

Patsy Healey investigates the governance of urban places and is concerned with efforts that promote place qualities and which recognise that the spatial organisation of phenomena are important for quality of life, for distributive justice, environmental well-being and economic vitality.

Well-illustrated chapters weave together conceptual development, experience and implications for future practice. The book begins by opening out the challenge of the activity of spatial strategy-making for urban areas. This is followed by three in-depth, longitudinal cases of spatial strategy-making in the Amsterdam area, the Milan area and the Cambridge Sub-Region. These are then drawn upon to develop an understanding of the nature of spatial strategy-making in complex urban agglomerations.

Urban Complexity and Spatial Strategies offers concepts and detailed cases of interest to those involved in policy development and management, as well as providing a foundation of ideas and experiences, an account of the place-focused practices of governance, and an approach to the analysis of governance dynamics, useful for students and social scientists. Finally, for those in the planning field itself, it seeks to re-interpret, with a twenty-first-century relevance and grasp, the role of planning frameworks in linking spatial patterns to social dynamics.

Patsy Healey is Professor Emeritus in the School of Architecture Planning and Landscape at the University of Newcastle, where she was previously Head of Department, then Director of the Centre for Research in European Urban Environments. She is a specialist in planning theory, planning systems and planning practices, and has undertaken research on how planning strategies work out in practice and on partnership forms of neighbourhood regeneration experiences. She is Senior Editor of the journal, *Planning Theory and Practice*.

THE RTPI Library Series

Editors: Cliff Hague, Heriot Watt University, Edinburgh, Scotland
Tim Richardson, Sheffield University, UK
Robert Upton, RTPI, London, UK

Published by Routledge in conjunction with The Royal Town Planning Institute, this series of leading-edge texts looks at all aspects of spatial planning theory and practice from a comparative and international perspective.

Planning in Postmodern Times
Philip Allmendinger

The Making of the European Spatial Development Perspective
No Master Plan
Andreas Faludi and Bas Waterhout

Planning for Crime Prevention
Richard Schneider and Ted Kitchen

The Planning Polity
Mark Tewdwr-Jones

Shadows of Power
An Allegory of Prudence in Land-Use Planning
Jean Hillier

Urban Planning and Cultural Identity
William J.V. Neill

Place Identity, Participation and Planning
Edited by Cliff Hague and Paul Jenkins

Planning for Diversity
Policy and Planning in a World of Difference
Dory Reeves

Planning the Good Community
New Urbanism in Theory and Practice
Jill Grant

Planning, Law and Economics
The Rules We Make for Using Land
Barrie Needham

Indicators for Urban and Regional Planning
Cecilia Wong

Planning at the Landscape Scale
Paul Selman

Urban Structure Matters
Petter Naess

Urban Complexity and Spatial Strategies
Towards a Relational Planning for Our Times
Patsy Healey

The Visual Language of Spatial Planning
Exploring Cartographic Representations for Spatial Planning in Europe
Stefanie Dühr

URBAN COMPLEXITY AND SPATIAL STRATEGIES

TOWARDS A RELATIONAL PLANNING FOR OUR TIMES

PATSY HEALEY

LONDON AND NEW YORK

First published 2007
by Routledge
2 Park Square, Milton Park, Abingdon, Oxon OX14 4RN

Simultaneously published in the USA and Canada
by Routledge
270 Madison Ave, New York, NY 10016

Routledge is an imprint of the Taylor & Francis Group, an informa business

Typeset in 9.5/13.5 Akzidenz Grotesk by Wearset Ltd, Boldon, Tyne and Wear
Printed and bound in Great Britain by The Cromwell Press, Trowbridge, Wiltshire

British Library Cataloguing in Publication Data
A catalogue record for this book is available from the British Library

Library of Congress Cataloging in Publication Data
A catalog record for this book has been requested

ISBN10: 0-415-38034-0 (hbk)
ISBN10: 0-415-38035-9 (pbk)

ISBN13: 978-0-415-38034-8 (hbk)
ISBN13: 978-0-415-38035-5 (pbk)

CONTENTS

LIST OF ILLUSTRATIONS

FIGURES

TABLES

PREFACE

This book is written for all those with an interest in the present and future qualities of cities and urbanised areas. For those involved in policy development and management, it offers concepts and cases through which to reflect on the challenges they face and the contributions they are making. For students training to get involved in such governance work, it provides a foundation of ideas and experiences. For social scientists in urban geography, policy and politics, it offers not merely an account of the place-focused practices of governance. It also develops an approach to the analysis of governance dynamics which highlights efforts at transforming discourses and practices. Finally, for those in the planning field itself, it seeks to reinterpret, with a twenty-first-century relevance and grasp, the role of planning frameworks in linking spatial patterns to social dynamics, and in generating elements of both fixity and mobility in the complex evolving systems through which the material realities and imaginative possibilities of urban life are produced.

All contributors to academic literature come from particular trajectories. Authors develop their ideas in interaction with other authors, with colleagues and with their own life experiences. For those coming, as I do, from the field of urban and regional planning, such experiences may well encompass practising a particular professional craft. As I seek to show through this book, the interaction of intellectual development with knowledge of the challenges of practical endeavour is not some minor field of 'applied science', dependent on the 'real science' produced by theoretical abstraction from the messiness of the ongoing flow of life. It is the ground out of which, ultimately, come the focuses, purposes and resonances of intellectual inquiry. But our particular trajectories shape what we focus on, how we focus and the terms and languages we use to express our ideas and findings (Massey 2005).

My own trajectory has combined an early training in geography with experience as a planner, and an academic career in which I have been involved in the overall design of planning education programmes, in teaching planning theory, and in researching practices. These range from attempts at introducing ideas about planning in Latin American cities to the ways ideas about planning were used in the various local authorities in London, the practices of development plan-making in England, the institutional dynamics of land and property development processes, the practices of urban regeneration and community development partnerships, and most recently, initiatives in spatial strategy-making for urban areas in Western Europe, out of which this book arises. These experiences of empirical research and practice have led me to explore different social-science literatures to help develop an understanding of what I came to see in actual practices.

The result has sometimes been an eclectic mix of ideas, but has also involved a steady evolution, inspired by waves of ideas in social thought as well as practice development.

Critical of the simple normative and rationalist management models emphasised in the planning field in the late 1960s, I found inspiration in the 1970s in a mixture of Marxist-inspired urban political economy and phenomenology (Bailey 1975). Both my practice experience and my own sensibility made me question the high ground of 1970s 'structuralism', to find inspiration in the work of sociologist Anthony Giddens, whose concepts of how to overcome the opposition between theories of structural dynamics and theories of individual agency showed a way to understand the situated ethnography of practices and the complex energy generated by the multi-dimensional interaction of social constraint and human invention. This led me to explore in empirical work the interactions of structuring forces with the activities of those involved in particular practice arenas, a general perspective which is now widely used in policy analysis, planning, urban geography and urban studies, although with various inspirations, analytical concepts and vocabularies. I refer to the particular approach I use as 'sociological institutionalism', to give an idea of the trajectory I have followed.

But beyond a way of analysing practices, I am a planner and so I am interested in the interaction between ways of seeing the world and ways of acting in the world. So I have been interested in the implications of the phenomenological understanding that the reality of the world is always perceived by we humans in imperfect ways, structured by our limited perceptual capacity and the histories and positions from which we are perceiving. Cluttered with prior conceptions and imperfect sensibilities, we arrive at interpretations of the world out there, interpretations relevant to some purposes we have and tested by discussion with others as well as by encounters with the physical world. So we socially construct interpretations of what is going on, in interaction with others. It is this realisation that made me focus on the quality of the interactions through which planning work is performed, including their communicative dimensions. Pushed along by the inspiring work of John Forester and Judith Innes, and drawing on some of the ideas of philosopher Jorgen Habermas, I sought to bring together an 'institutionalist' account of what planning work involved and some normative ideas about how to evaluate the quality of the communicative work that planners engaged in. The result was a book on 'collaborative planning' around which there has been much debate in the planning theory field (Healey 1997/2006). In retrospect, this enterprise was part of a broader search in policy analysis to develop an interpretive, post-positivist approach to understanding and doing policy work in governance contexts (Fischer 2003; Hajer and Wagenaar 2003). This trajectory flows into the present book, where I develop the significance of a focus on interactions in terms of the time–place dynamics of the relations of interaction, while keeping in mind a normative concern with the kinds of worlds these interactions may lead towards and a commitment to the idea that, through collective action, it is possible to bring improvements to the experience of urban life.

In my intellectual trajectory, I have remained anchored in the planning field, with its disciplinary stories of the development of ideas and struggles in practice. For me, although full of frustrations and limitations, the core project of the planning field remains a fine endeavour. It centres on exploring what it takes to act in the world to pursue collective purposes which are seen to be inclusively beneficial in a particular social context, in ways which attend to the significance of the spatial dimension of all relations and to the particularities of place qualities as they evolve. Although many other disciplinary contributions provide inspiration and understanding to those engaged with this project, the planning field needs in particular to combine the fields of policy analysis and geography. This is what I have tried to do in this book.

The book, however, does not set out to be a work of planning theory. To the extent that it provides some theory, it is perhaps better understood as a contribution to social theory. Instead, it is an exploration of a key activity which has always been central to the planning project, that is, the development and deployment of a strategic imagination about the qualities of the places of urban areas, in particular the places which hover in imagination between the neighbourhood and the nation. I have called this a practice of 'spatial strategy-making for urban areas'. In the book, I combine empirical histories of evolving practices with conceptual development and with suggestions about the implications for those interested in urban governance or struggling with developing a strategic approach to urban area development. I hope in this way to provide both empirical accounts and analytical commentaries which will prove helpful not just to those students and academics seeking to understand the enterprise and practices associated with strategic spatial planning, but also to help those involved in strategy-making for urban areas to develop a richer, more situated and more practically effective understanding and ethically informed recognition of the potentialities and limitations of the practices they are engaged in.

ACKNOWLEDGEMENTS

I have been helped in the enterprise which has resulted in this book by very many people, and it is difficult to know how to thank you all enough. What follows barely does justice to the support and inspiration that many of you have given me over the past four years. So please forgive this rather dry listing of your contributions and any omissions which I may have inadvertently made. In the end, of course, I have to bear responsibility for the text, and for any distortions and misinterpretations of your ideas and critiques that I have made.

I must first thank the publishers, Routledge and the RTPI series, for consistent support since I sent the proposal to them in 2004. My thanks go especially to Caroline Mallinder and Georgina Johnson, their predecessor, Helen Ibbotson, and the two advisers for the RTPI series, Tim Richardson and Robert Upton, who both read the text carefully and offered good advice. In this formal category, I must also thank the Leverhulme Trust for a small grant to allow retired professors to complete their projects,[1] to Maarten Hajer and Alan Harding who supported my application, and to the School of Architecture, Planning and Landscape at Newcastle University for providing helpful facilities, including access to the Library, and the Thursday Seminar series!

I could not have produced the case studies without the help of very many people. For the Amsterdam case, I acknowledge the help of Luca Bertolini, Caroline Combè, Len de Klerk, Enrico Gualini, Maarten Hajer, Zef Hemel, Allard Jolles, Robert Kloosterman, Stan Majoor, Barrie Needham, Karin Pfeffer, Ralph Ploeger, Willem Salet, Max van der Berg, Lianne van Duinen, Eva van Kempen, Marijke van Schendelen, San Verschuuren, Johann Woltjer and Wil Zonneveld. My thanks especially to those who read and commented on drafts of what became Chapter 3, and to Allard Jolles for helping me to get copies of many of the illustrations. For the Milan case, I acknowledge the help of Alessandro Balducci, Alessandra de Cugis, Fausto Curti, Alessia Galimberti, Maria Cristina Gibelli, Sara Gonzalez, Enrico Gualini, Luigi Mazza, Silvia Mugnano, Giovanni Oggioni, Gabriele Pasqui, Filomena Pomilio, Paolo Riganti, Paolo Simonetti, Serena Vicari, Tommaso Vitale and secretarial and library staff at the Politecnico di Milano. Again, my especial thanks to those who commented on drafts of Chapter 4, and to Cristina Buttimelli at the offices of INU Edicioni, Rome, for help in finding good versions of Figures 4.2, 4.3, 4.4. For the Cambridge Sub-Region case, I acknowledge the help of Stephen Crow, Margaret Gough, Michael Hargreaves, Brian Human, Andrew Jonas, Michael Monk, Alan Moore, Nicola Morrison, Susan Owens, Simon Payne, John Pendlebury, Yvonne Rydin, Malcolm Sharpe, Peter Stoddert, Mark Vigor, Aidan While and Bill Wicksteed, as well as Melissa Wyatt at the RTPI Library. Thanks especially to those who commented on drafts of chapters and Mark Vigor (Cambridgeshire County Council) for help in getting my case figures sorted out.

Several other people made very helpful comments, as readers of draft chapters or through discussions about particular issues. My thanks go especially to Louis Albrechts, Ash Amin, Alessandro Balducci, Heather Campbell, Simin Davoudi, John Forester, Sara Gonzalez, Stephen Graham, Enrico Gualini, Zann Gunn, John Friedmann, John Hack, Jean Hillier, Judith Innes, Ali Madanipour, Tim Marshall, Luigi Mazza, Alain Motte, Frank Moulaert, Jon Pugh, Willem Salet and Geoff Vigar. I am also enormously indebted to four people who read the whole text: Robert Upton and Tim Richardson were required to do this for the publishers, but I really appreciated their detailed comments on content and structure; Urlan Wannop provided his friendly criticism from the point of view of a strategic planner and an academic; and Nairita Chakraborty reacted to the text in a careful and lively way from the point of view of a non-European masters student, with some planning experience. Thanks also to Derek Hawes and Paula Rutter at the University of Newcastle Print Service, for help with the illustrations, and to Kim McCartney at GURU, University of Newcastle, for much practical advice. Finally, Bernadette Williams has acted as a critical sub-editor, reading the whole text with careful attention to meaning as well as presentation and text quality, and who helped to break up my often contorted sentences. My thanks also to Georgina Johnson at Taylor and Francis, the publishers, and Hannah Dolan and colleagues, of Wearset, for all their work on the publication side.

Many of the illustrations have required copyright permission. My thanks to Bundesamt für Bauwesen und Raumordnung (BBR), Bonn, for the base map for Figure 2.1; Amsterdam City Council for permission to use the cover sketch and the maps for Figures 3.2, 3.4, 3.5, 3.6, 3.9 and 3.10; 3.7a and 3.7b (c) Physical Planning Department, City of Amsterdam; to the Netherlands National Planning Agency for Figure 3.8; to INU Edicioni, Rome for Figures 4.2, 4.3 and 4.4; to the Comune di Milano for Figure 4.5; to Cambridgeshire County Council for permission to use Map 33, from Holford, W. and Wright, H.M. (1950), *Cambridge Planning Proposals: a Report to the Town and Country Planning Committee of Cambridge County Council*, for Figure 5.2 and for the maps for Figures 5.5, 5.7 and 5.8; to the Office of the Deputy Prime Minister (ODPM), London, for Figures 5.9 and 5.10b; to the East of England Regional Assembly for Figure 5.10a; and to Pearson Education Ltd for permission to use Figure 1.1, p. 24, from Mintzberg, H. (1994) *The Rise and Fall of Strategic Planning*, for Figure 6.1.

Finally, as usual, I must thank family and friends who have tolerated the introverted existence of research and writing. I keep saying that, after the book is finished, I may be more sociable, but they and I know that addicted academics cannot really give up their obsessions. I hope in the future, as ever, for a better balance!

Patsy Healey, Wooler, Northumberland, March 2006

NOTE

1 Project No. EM/7/2004/0011.

CHAPTER 1

THE PROJECT OF STRATEGIC SPATIAL PLANNING FOR URBAN AREAS

> For [the citizen], evolution is most plainly, swiftly in progress, most manifest, yet most mysterious. Not a building of his [sic] city but is sounding with innumerable looms, each with its manifold warp of circumstance, its changeful weft of life. The patterns here seem simple, there intricate, often mazy beyond our unravelling, and all well-nigh changing, even day by day, as we watch. Nay, these very webs are themselves anew caught up to serve as threads again, within new and vaster combinations. Yet within this labyrinthine civicomplex there are no mere spectators. Blind or seeing, inventive or unthinking, joyous or unwilling – each has still to weave in, ill or well, and for worse if not for better, the whole thread of life (Patrick Geddes 1915/1968: 4–5).

> What matters within cities ... revolves around the fact that they are places of social interaction.... Cities are essentially dynamic.... Policy formulation must work with this; it must not think in terms of some final, formal plan, nor work with an assumption of a reachable permanent harmony of peace. The order of cities is a dynamic – and frequently conflictual – order. A new politics for cities must be equally fluid and processual (Amin *et al.* 2000: 8 and 10).

GOVERNANCE AND SPATIAL PLANNING

This book is about the governance of place in urban areas. It is concerned with governance efforts which recognise that both the qualities of the places of an urban area and the spatial organisation of phenomena are important for quality of life, for distributive justice, environmental well-being and economic vitality. It focuses on strategies that treat the territory of the urban not just as a container in which things happen, but as a complex mixture of nodes and networks, places and flows, in which multiple relations, activities and values co-exist, interact, combine, conflict, oppress and generate creative synergy. It centres around collective action, both in formal government arenas and in informal mobilisation efforts, which seeks to influence the socio-spatial relations of an urban area, for various purposes and in pursuit of various values. It is concerned with strategy-making which seeks to 'summon up' an idea of a city or urban region (Amin 2002), in order to do political work in mobilising resources and concepts of place identity.

There has been much discussion in recent academic and policy debates about the significance of the 'urban region' as a focus of governance and about the emergence of new forms of governance. In Western Europe, some strands of policy debate promote the significance of cities and urban regions as key actors in a new economic and political space of weakened and fragmenting nation states and stronger global economic forces. Some academic analysts relate this to the search for new modes of regulation resulting from changes in the dynamics of capitalist economies (Harvey 1989; Jessop 2000). Others emphasise the diversity of urban situations and experiences, and the uneven development of a capacity for city and urban region 'governance' (Bagnasco and Le Galès 2000a). It is widely recognised that the modes of governance that emerge in urban areas vary substantially in both their internal dynamics and the way responses are made to outside pressures. The promotion of an urban-region perspective in policy development is an example of a general idea attempting an organisation of this diversity and contingency. Experiences in developing spatial strategies with real power to influence urban development trajectories provide a rich laboratory for exploring the challenges and tensions of developing new arenas and forms of governance. This book is therefore a contribution to the debates on emerging governance forms and the potentiality of the 'urban region' (or 'city region' or 'metropolitan region') as a focus of political and policy attention (Lefèvre 1998; Salet *et al.* 2003).

It is also a contribution to a 'planning tradition', in that it emphasises the importance of attention to the qualities of places and to the material and imaginative ways through which people, goods and ideas flow around, into and beyond the many social worlds that co-inhabit urban areas. Over the past 100 years, this planning tradition, called variously town or city planning, urban and regional planning, spatial planning, territorial development and territorial management, has been concerned with the interrelation between fixity and mobility. In the traditional physical planning language, this was referred to as the relation between land uses and infrastructure channels (Chapin 1965). In the 1990s, the relation is more often conveyed in the 'network' language articulated by Manuel Castells (1996), as a tension between 'places' and 'flows'. This new network language not only emphasises the complex socio-spatial relations between physical spaces, places of meaning and the spatial patterning produced through dynamic social and economic networks; it also stresses the complex ways in which networks, or webs, overlay each other and reach out to others elsewhere in space and time. In the mid-twentieth century, it was thought that these networks were somehow integrated together in a coherent entity called a 'city'. But these days, as Mel Webber understood in the 1960s (Webber 1964), our experience tells us that our social worlds, even of daily interaction, may stretch well beyond the area of a particular city, and that the webs which matter to us may be quite different to those of our neighbour. As a result, the 'places' of cities and urban areas cannot be understood as integrated unities with a singular driving dynamic, contained within clearly defined spatial boundaries. They are instead complex constructions created by the interaction of actors in multiple networks who invest in material projects and who give meaning to qualities of places. These webs of relations escape analytical attempts to 'bound them'.[1] Efforts at strategy-making for urban regions are part of this material and imaginative effort to make some 'sense' of the complexity of urban life. The planning project, infused with this understanding of socio-spatial dynamics, becomes a governance project focused on managing the dilemmas of 'co-existence in shared spaces' (Healey 1997: 3).

The core of this planning 'project', as promoted by protagonists of a revival of the planning movement in the late twentieth century in Europe, centres around a particular concept of 'spatial planning' (Faludi and Waterhout 2002; RTPI 2001). This concept, inspired by the German *raumplanung*, has a fluid meaning and does not translate well into some languages (Williams 1996). In an earlier attempt to capture the range of these meanings in relation to my own perceptions of the nature of 'places' and spatiality, I suggested that in a general way, the term 'spatial planning' refers to:

> self-conscious collective efforts to re-imagine a city, urban region or wider territory and to translate the result into priorities for area investment, conservation measures, strategic infrastructure investments and principles of land use regulation. The term 'spatial' brings into focus the 'where of things', whether static or in movement; the protection of special 'places' and sites; the interrelations between different activities and networks in an area; and significant intersections and nodes in an area which are physically co-located (Healey 2004b: 46).

Most planning thought and practice of the past twenty-five years in Europe has moved beyond a simplified physical view of cities, in which place qualities and connectivities were understood through the physical form of buildings and urban structure. It is widely recognised that the development of urban areas, understood in socio-economic and environmental terms, cannot be 'planned' by government action in a linear way, from intention to plan, to action, to outcome as planned. Even where a government agency controls many of the resources for physical development and acts in an integrated and coordinated way, socio-economic and environmental activities make use of the physical fabric of urban areas in all kinds of ways that are often difficult to imagine in advance, let alone predict. What goes on in urban areas is just too dynamic, 'intricate and mazy' (Geddes 1915/1968).

Instead, those involved in spatial strategy-making are struggling to grasp the dynamic diversity of the complex co-location of multiple webs of relations that transect and intersect across an urban area, each with their own driving dynamics, history and geography, and each with highly diverse concerns about, and attachments to, the places and connectivities of an urban area.[2] This involves moving beyond an analysis of the spatial patterns of activities as organised in two-dimensional space, the space of a traditional map. Instead, it demands attention to the interplay of economic, socio-cultural, environmental and political/administrative dynamics as these evolve across and within an urban area. Within the sphere of governance activity, this means that planners from the 'planning' tradition, with its focus on place qualities, have to encounter analysts and policy-makers concerned with policy fields organised around other foci of attention, such as the competitiveness of the firm, or the economy as a whole, the health of individuals, or the operation of schools and systems of schools. In these encounters, clashes between conceptual frameworks and legitimising rationales are commonplace. Nevertheless, in this reaching out to, and joining up with, those working in many policy fields, efforts in spatial strategy-making are drawn into a widespread endeavour to re-think government and governance. This involves searching for new ways of 'doing government', driven in part by concerns for greater effectiveness in delivering policy programmes, but also for greater relevance and connection to the concerns and demands of citizens and organised stakeholders.

This search has led to all kinds of often contradictory initiatives. In one direction, 'partnership' governance modes have proliferated, between the different policy fields and levels within formal government and between formal government, economic and civil society organisations (Pierre and Peters 2000). In another direction, there are efforts to move the arenas for policy development and resource allocation from national levels towards more local levels, and/or to create new ways in which levels of government can interact. This has led to considerable analytical attention to what some call the 're-scaling' of governance attention (Brenner 1999) and to new forms of 'multi-level governance' (Hooghe 1996). In a further direction, there are initiatives to make government more responsive to the citizens who, in theory, it serves, through 'empowering' citizens, and through fostering a democratic 'public realm' of policy deliberation.[3] These initiatives take concrete form and often clash when evolved into specific programmes and interventions within specific urban areas. Typically, therefore, strategic spatial planning initiatives for urban regions involve working in, around and through complex tensions, struggles and conflicts. This book explores these struggles empirically through accounts of spatial strategy-making experiences in three dynamic and diverse urban areas in Western Europe.

The spatial planning tradition is not, however, the only policy domain with a spatial focus. In recent years, there has been a reawakened interest in the significance of the qualities of places and territories within the fields of economic policy and social policy, strongly supported by environmental considerations. Such policies embody, if sometimes only implicitly, certain principles of spatial organisation and ordering.[4] Policy-makers in these fields also increasingly recognise the positive and negative 'place effects' that influence the achievement of policy ambitions, such as improved health, better levels of education and more rapid structural adjustment to economic change. This new attention to place qualities and effects challenges the traditional organisation of government into 'sectors', focused around the delivery of specific functions: economic development, education, health, transport, social welfare, housing, environmental protection, etc. This is most obvious in the field of economic policy, where promoting urban assets as a contribution to 'regional economic competitiveness' has been a major preoccupation in recent decades at city, region, national and EU level. The 'competitiveness' agenda in Europe has recently widened to encompass considerations of environmental quality and social cohesion (CSD 1999). The challenge to functional/sectoral organisation, these days often called the 'silo mentality', generates a momentum to create more linkages between policy fields as they impact on the places and connectivities of urban areas, expressed as a search for 'policy integration' and 'joined-up government'.[5] But creating such linkages focused on particular urban areas is a challenging task. Intellectually, it involves imagining what to link, integrate and 'join up'. Politically, it involves developing coalitions with sufficient collective power to make the links and joins actually work. It involves building relations in the mind and in the social worlds of policy and politics. This book is about governance initiatives and practices that are struggling with ways of doing this, from different institutional positions and in more and less favourable circumstances.

THE GOVERNANCE OF PLACE

Urban areas have always had some form of place-governance, demanded by the challenge of the intensity and density of the interactions of urban life. Sometimes the focus has been on the internal organisation of cities, sometimes on their position in a wider geography. The resultant governance activities have been a variable mix of the regulation of economic activities, health and hygiene, provision of defensive considerations, protection from environmental hazard and the management of social relations, combined with periodic efforts at re-shaping the physical form of cities for welfare, wealth generation or symbolic and cultural purposes. All of these purposes have been important in the twentieth century, the era when large-scale urbanisation swept across the world. It is not surprising that it was in this century that land-use planning, territorial management, spatial ordering and town/city planning became an established part of government systems in most countries.

Yet different national cultures and governance practices provided a variable fertility for planning systems (Sanyal 2005). In the first part of the century, the idea of place governance and the management of land use and development in the 'public interest' conflicted with liberal concepts of individual property rights. In the second part, and particularly in North-west Europe, it conflicted with the organisation of the nation state into policy-delivery functions or sectors, linked especially to the delivery of welfare state services with their principles of universal access. A focus on place quality cuts across both a liberal reliance on individual initiative and market processes and a social-democratic reliance on the separate development of welfare services. Planning systems that aimed at an 'integrated' approach to developing and regulating the qualities of places have been pushed and pulled by the way these forces have interacted in the governance landscapes of individual countries. The result is substantial variety in the design and practice of planning systems, and in their ability to focus on place quality, as the cases in this book will show.

As with all policy systems, over time the institutional designs of one period become embedded in the practices of the next. Sometimes, this embedding creates valuable resources on which responses to new challenges and governance configurations can build. But it may also act as a resistance, apparently impeding adjustment and innovation. By the end of the twentieth century in Western Europe, planning practices were being attributed with both these potentials. 'Planning' was pilloried as part of the problem of governance adjustment to new conditions and promoted as part of the solution to the ever-increasing difficulty of managing co-existence in the shared spaces of dynamic urban areas. Some commentators present planning as a bureaucratic impediment to individual initiative and wealth generation. Others see planning systems and practices as a mechanism through which to manage the complex balancing of economic, social and environmental values in a coordinated and integrated way, and therefore a key activity of the governance of highly urbanised countries. In this latter view, an effective planning system is seen as part of the institutional infrastructure necessary for economically successful, liveable, environmentally considerate and socially just urban areas.

This second viewpoint received a surge of support in the late twentieth century in Western Europe. Economic, environmental and political arguments converged to emphasise the national and global significance of the qualities of sub-national territories, particularly cities and urban regions. Many reasons are given for this. Economic analysts have increasingly come to realise the power of 'place effects' to add and detract value from individual economic activities, particularly when firms operate transnationally and globally. This focuses attention on ways of creating and sustaining the positive place-based assets that add value to firms and hence to the overall economy.[6] Environmental analysts emphasise the importance of focusing on the interaction between natural resource systems, ecological systems and human systems as these play out in urban areas as well as globally (de Roo 2003; RCEP 2002). Other new social movements of the late twentieth century, and particularly those linked to feminism and to the recognition of socio-cultural diversity and difference, have brought into focus the difficulty experienced by marginalised social groups in negotiating the daily life environment in cities where the qualities of the locales and connectivities to which residents have access have been neglected. This puts the distribution of access to place quality and 'liveability' alongside access to income, education, health and socio-cultural facilities as a key arena of social differentiation, and therefore in need of governance attention if distributive justice is to be promoted (Amin *et al.* 2000). The concern with place quality is linked also to questions of identity and social cohesion as well as material welfare (Bagnasco and Le Galès 2000a). Attachment to place, and to diverse places within and around an urban area, may be an important dimension of people's well-being, part of their identity and ontology (Liggett and Perry 1995). The emotive feelings people have for place qualities lie behind many episodes of conflict between residents, developers and government. Finally, those concerned with the health of democratic politics have become increasingly aware that citizens are prepared to mobilise around threats to place quality and to stakes in places, whilst becoming increasingly disinterested in the mechanisms of formal party politics and representative democracy.[7]

These considerations have underpinned the attention given in many parts of Europe at the end of the twentieth century to urban areas as a focus of policy attention. The rising salience of this attention has influenced the discussion of the distribution of European Community funds aimed to reduce 'structural' territorial disparities in Europe (Faludi and Waterhout 2002). Municipalities seem too small to encompass significant interactions across an urban area, while the nation state is too large to manage how interactions between different webs of relations and spheres of governance activity work out to affect the experience of place quality. Urban areas come into focus as a governance level that seems to promise integration of different policy sectors as they interrelate in places and affect the daily life experience of place quality. Advocates of governance mobilisation focused on promoting urban 'region' development argue that such a policy and institutional focus has the capacity to bring together different government levels and sectors, as well as the array of special agencies and companies, not to mention the various partnerships that have grown up in recent years to deliver specific policies and projects. The urban region seems to offer a functional area within which the interactions of economic relations, environmental systems and daily life time–space patterns can be better understood than at a higher or lower level of government. It suggests an arena where diverse fragments of governance activity can come together, where key actors from different government levels and different segments of society can meet face-to-face and develop networks through which to identify priority areas where governance action is needed. Mobilising around such arenas may help to generate greater knowledgeability, more productive synergy and more appropriate conflict identification and consensus-formation.[8]

This new enthusiasm for 'regionalism' meanders around in academic discourse and in specific governance initiatives among all kinds of perceived areas between the nation state and the municipality. Many commentators try to find some way of aligning a vocabulary of levels of government with scales of functional activity, such as home–work relations or the supply-chain patterns of firms. This assumes some kind of hierarchical ordering, both in governance organisation and functional activity. Yet recent literature on both governance processes and on the patterning of social–spatial relations challenges the assumption of hierarchy, demanding more careful attention to the spatial reach of different networks or webs of relations as they weave across urban areas. This in turn problematises the notion of the aggregation of relations found in an urban area as having some kind of objective existence. In this book, I refer therefore to urban areas merely to call to mind the 'intricate and mazy' worlds of urban life. I use the term 'urban region' to refer to the conceived space of the urban, called to mind in analytical and governance initiatives of one kind or another. I use the term 'level' to refer to the institutional sites or arenas created for governance initiatives, where these are inserted in some kind of administrative hierarchy.[9]

However, the arguments for increased policy attention to urban areas do not go unchallenged (Lovering 1999). Some maintain that the nation state retains its strong integrating force, reducing the institutional space for sub-national scales of governance. Others suggest that, in the age of cyberspace and the global economy, all governance effort at national, regional and local levels is liable to fragmentation and disintegration (Amin 2002; Graham and Marvin 2001). I do not argue that an integrated, multidimensional strategic policy focus on the 'place' of an urban region is, in any general sense, either possible or desirable. Such a focus is not a recipe or formula that can be bolted onto an existing governance landscape. Instead, I explore how such a focus arises in particular situations and what can be learned from this about the potentialities and limitations of spatial strategy-making for urban areas as a governance enterprise. When and why does such an enterprise arise and gather momentum? How, and how far, do such initiatives get to affect material and imaginative realities? Studies undertaken so far highlight the difficulties experienced by initiatives to create governance capacity focused on urban 'regions', as they struggle to find leverage to expand in the well-structured institutional terrain of functionally organised government, in which the powerful arenas have been nation states and municipalities.[10] Breaking through this embedded power requires real efforts by many actors in all kinds of governance roles to imagine alternative ways of doing governance. It involves efforts in creating new relations with diverse people in different positions and networks in an urban area. It requires connecting understanding of the relations perceived to be important to economic actors, residents, other stakeholders and to non-human species, with the administrative jurisdictions of formal government through which to access public investment and regulatory power.

The project of spatial strategy-making focused on urban 'regions' is thus politically challenging. It is also intellectually challenging. Traditionally, planning strategies drew on a simple model of spatial integration. Cities were at the core of their hinterland regions, linked to smaller towns and settlements through a pattern of radial routes and a hierarchy of centres revolving in a centripetal fashion around the regional core. Place effects were experienced through the dimension of physical proximity. The closer were two phenomena in actual space, the greater their impact on each other. The city centre was seen as the site of greatest synergy, and the periphery the site of greatest isolation. In recent years, however, a new relational geography has developed to explore the dynamic complexity of the various relational webs which transect urban areas. Different webs have different space–time patterns of nodes and links. A place may be nodal in one relational web but peripheral in another. Synergetic dynamism may occur in all kinds of physical and institutional spaces, creating nodal place qualities. Isolation can occur in city centres as well as elsewhere in the urban fabric. By the end of the twentieth century, those involved in strategy development for urban areas were struggling with a recognition that the traditional spatial organisation of cities was being 'disintegrated', while the patterning of new, complex relational dynamics was very difficult to imagine and to grasp.

Policy attention focused on 'urban regions' therefore continues to be deeply ambiguous. It challenges the established institutional designs of formal government. It demands new geographical imaginations through which to understand and represent what the 'place' of an urban 'region' and the places within it are and might become. It involves rethinking how and where governance should be done and who should be involved in it. It involves mobilising social forces to create arenas for policy development and delivery. It is no surprise that the studies of recent planning initiatives with such a focus tend to conclude that these work best where there is already an institutional history which provides arenas for policy development focused on urban 'region' development (Albrechts *et al.* 2003; Salet *et al.* 2003).

Nevertheless, many efforts are being made in urban areas which lack such an inheritance. Are these efforts doomed to fail? Or are they precursors of transformations through which a new institutional history is in the making, in which an urban 'region' spatial strategy could, in time, have significant effects? If so, what external forces are sustaining them and how do these interact with local energy and mobilisation forces? In this book, I emphasise the nature of strategic spatial planning, both as a political project, which seeks to mobilise attention, change discourses and practices, and alter the way resources are allocated and regulatory powers exercised, and as an intellectual project, through which new understandings are generated and new concepts to frame policy interventions are created to sustain the political project. Overall, this political and intellectual project is about shaping, to some degree, the socio-spatial dynamics of urban areas, through explicit attention to spatial organisation and place qualities. I consider how these efforts may shape outcomes, understood in terms of material gains and losses, and also in terms of identity, knowledge frames and governance capacity. Overall, I am interested in how governance capacity with a focus on urban relations gets to develop the imagination and power to see and act differently, to innovate new governance practices and new socio-spatial imaginations.

A Perspective on Practices

This book is structured around empirical cases and conceptual discussion. The cases make up the first part of the book and provide narratives of the evolution of discourses and practices around spatial strategy-making in which attempts are made to view some kind of urban 'region' with a focus beyond that of a development project or neighbourhood management. In these accounts, I try to bring out how specific actors in organisational positions, policy communities and relational networks of various kinds interact in institutional arenas, both to produce strategic ideas and to insert such ideas into the flow of practices that affect the allocation of material resources. I aim to show how, in these interactions, structuring dynamics shape opportunities for mobilising governance attention in new ways and how active agents make use of these opportunities: in some cases through creatively enlarging them, in others through uncertainty on how to grasp them. I illustrate the multiplicity of relations that are drawn into such endeavours and the complexity of their evolution. I emphasise how the trajectories of the discourses and practices of governance activity in particular arenas evolve in interaction with their institutional settings, which are themselves relationally complex and dynamic. Thus those involved in spatial strategy-making for urban regions may imagine futures, but what evolves through time is continuously escaping their grasp and their power to define in advance.

This approach to 'telling stories of planning practices' is underpinned by a relational conception of social organisation and an institutionalist understanding of governance processes. In Chapter 2, I elaborate on this perspective in more detail. In the second part of the book, I draw on the experiences of the cases and on academic debates to explore what is involved in an approach to the activity of spatial strategy-making for urban regions which recognises the dynamic, indeterminate emergence of the place qualities of urban areas. I engage in conceptual development and make normative suggestions to help advance the political shrewdness and intellectual perceptions for addressing questions about when, why, where and in what way engaging in a spatial strategy-making could 'make a difference'. In this conceptual development, I explore four interrelated themes. The first looks at the way understandings are converted into actions through a focus on what emerges, implicitly or explicitly, as a *strategy*. The second focuses specifically on the *concepts of place and space* deployed in such episodes of place-focused governance. The third considers the sources of knowledge and creative probing through which *understandings and meanings* of place qualities are generated. This all builds to the fourth theme, which assesses the power of spatial strategy-making activities in shaping *governance capacities and landscapes* and the material and imaginative experience of urban life. I expand on each of these themes in Chapter 2.

A relational conception of social organisation emphasises that tellers of stories and academic analysts are not outside the worlds they explore, but are part of the dynamic, unfolding realities to which their work contributes. We are driven by insights and perceptions that are shaped by our own trajectories through which our understandings and our valuings have evolved. As authors, we cannot avoid being selective in what we present and normative about what we put forward as success or failure, as positive or negative developments. For those working within public policy fields, the pressure to make suggestions as to how to 'improve' governance discourses and practices is deeply felt, often making critical judgement difficult. In this book, I am concerned to shine an empirical and conceptual light on a governance activity, on how spatial strategy-making works, on how strategies get to be produced, on whether and how they produce effects and on the extent to which they develop a capacity to shape the multiple trajectories of urban life. In doing so, I seek to make explicit the bases for my analytical commentaries and normative suggestions. Chapter 2 outlines my analytical perspective and introduces the three cases.

But I am interested in these issues not in some abstract, observing capacity. I am concerned about how far, in specific situations, abstract notions of place quality are given concrete meaning and how this may affect the daily life experience of the socio-spatial relational worlds that co-exist in urban areas. I seek to show the possibilities and limits of bringing together the potentially conflicting values of distributive justice, environmental well-being and economic vitality, not as abstract principles but in their specific material and imaginative expression in concrete governance interventions that promote place qualities. I am interested, on the one hand, in the relation between such interventions and a 'public realm' of debate through which they are shaped, criticised, held to account and legitimated. On the other hand, I am interested in their effects, both materially, on who gets what where, and in terms of ontology and epistemology, on identities and understandings. Finally, I am interested in who get to become the critical actors in the processes I describe and examine. Who and where are the 'planners' in these developments? How do 'planning systems' fit into the governance landscapes of which they are a part? There are potentially several institutional sites within contemporary urban areas from which episodes in spatial strategy-making may be initiated. This implies that those trained formally as planners and working in planning systems, or in and around 'planning policy communities', may be only one amongst many of the players involved, and may not even play central roles. What potentialities and limitations do those with involvement in past spatial strategy-making activity carry forward to the challenge of developing approaches more relevant to the perception of the dynamic complexity of today's urban areas?

Overall, I seek to present a 'relational planning' situated within the evolving, complex, socio-spatial interactions through which life in urban areas is experienced. This relational understanding of the planning project has a double nature. As an activity of governance, it is concerned with how the relations of collective action create momentum to shape governance interventions. As itself constituted through an array of webs of relations within the intersecting complexity of the dynamics through which the futures of daily life experience of urban areas are produced, the relations of the planning project jostle and get jumbled up with all kinds of other relations. It is within the complexity of this jostling and jumbling in specific situations that governance interventions are both shaped and come to have effects.

NOTES

1 See Amin and Thrift 2002; Bridge and Watson 2000; Graham and Healey 1999; Massey 2005.
2 See Albrechts *et al.* 2001; Albrechts and Mandelbaum 2005; Healey 1997.

3 See Amin and Thrift 2002; Dryzek 2000; Friedmann 1992; Fung and Wright 2001.
4 This is brought out well in a recent study for the new Welsh National Assembly on the spatial dimensions of sectoral policy (Harris and Hooper 2004).
5 See, for example, the UK discussion, Wilkinson and Appelbee 1999; 6 *et al.* 2002; Tewdwr-Jones and Allmendinger 2006.
6 See Amin and Thrift 1994; Cooke 2002; Cooke and Morgan 1998; Morgan, K. 1997; Storper 1997.
7 See Crouch 2004; Fung and Wright 2001; Lascoumes and Le Galès 2003; Melucci 1989.
8 See Amin *et al.* 2000; Cooke and Morgan 1998; Le Galès 2002.
9 I avoid as far as possible the use of the term 'scale', for reasons I expand on in Chapter 7. See also Marston and Jones 2005.
10 Albrechts *et al.* 2001; Healey 1997; Lefèvre 1998; Motte 1995, 2001; Salet and Faludi 2000; Salet *et al.* 2003.

CHAPTER 2

URBAN 'REGIONS' AND THEIR GOVERNANCE

> The metropolitan arena is filled with public and private actors at manifold levels of spatial scale and they are active in all sectors of urban policy. In this multi-dimensional game many different coalitions and many conflicts may occur. . . . The main challenge for metropolitan governance is to find ways of organising the *connectivity* between the different spheres of action (Salet *et al.* 2003: 389).

DEVELOPING THE PERSPECTIVE

Strategy-making focused on urban areas involves creating some conception of an 'urban region' and forming institutional arenas in which to develop and maintain the strategic focus. It involves calling to mind significant relationships about urban dynamics and drawing together many actors and networks necessary for linking a strategic concept to the possibility of shaping how material resources and regulatory powers are used in urban development processes. Creating a spatial strategy focused on some idea of an urban 'region' adds another frame of reference into the mix of framing concepts and discourses through which ongoing investment and regulation processes in an urban area are being shaped. Such a frame creates an idea of an urban entity with particular place qualities (Amin 2004; Healey 2002). Explicitly or implicitly, it positions this entity within a wider geography and indicates how the places *in* an urban area relate to the conception *of* an urban area. For most people, Amsterdam, Milan, Newcastle, Barcelona or Gothenburg are places on some kind of map of cities in a country or in Western Europe. Each is also a collection of neighbourhoods and locales. Each is also a unity, an identity and an imagery, called to mind by the naming of an urban area. This naming involves a mixture of imagination and experience through which to 'see' such an urban area and to identify what interventions, if any, could and should be articulated to 'shape' the future trajectory of its development.

'Seeing the city', in terms of its socio-spatial dynamics, its spatial organisation, its urban form and its many identities has been at the heart of the planning tradition of the past 150 years, the epoch of massive urbanisation.[1] It often seems a messy, conflict-ridden and threatening enterprise because it seeks to 'integrate', to connect, different areas of knowledge and practice around a place-focus. An easier option has been to 'box up' policy attention to place qualities into a narrow agenda and range of influence, focused around localised impacts and rights to develop land and property. However, as introduced in Chapter 1, this brings its own tensions, as it drags against the momentum of the delivery of 'functional' policy programmes where these require sites and particular place qualities. The history of spatial planning efforts in the twentieth century can be read as a repeated cycle: bursting out from this narrow box with a new wave of place-focused strategic energy, followed by processes of routinisation and, often, narrowing (Faludi and van der Valk 1994; Healey 1998a). In the later twentieth century in Western Europe, a new wave of energy built up to break out of the box and develop once again a strategic approach to the place qualities of urban areas.[2] All three cases in this book were affected by this energy, but raise questions about the 'reach' of its influence and the way the place qualities of urban areas are called to attention.

In addressing such questions, some accounts of evolving spatial strategy-making practices focus on organisational elements – the difficulties of co-aligning administrative jurisdictions and formal government arenas with the realities of the social, environmental and economic relations of urban areas.[3] Others focus on the evolution of framing concepts and ideas, and on the competition between different discourses and priorities.[4] I am concerned with both of these dimensions, but I set them in the wider context of governance processes and cultures. I am interested in the interrelation between the processes through which framing discourses and practices are produced and the substance of the policies that are pursued.

As explained previously, as this book develops I make use of an approach to the understanding and development of spatial strategy-making practices that links two streams of academic thought. The first is interpretive policy analysis, as developed in the work of John Dryzek (1990), Frank Fischer (2003), John Forester (1993), Maarten Hajer (1995), Judith Innes (1990, 1992) and David Schlosberg (1999).[5] The second is relational geography, as developed in the work of Ash Amin (2002, 2004), Doreen Massey (2005) and Nigel Thrift (1996 and 2002, with Amin). Both focus on relations and interactions, and emphasise the social processes through which meaning is constructed. Both stress the complexity of the interactions that take place in specific social 'sites' (or arenas, or nodal sites in networks) and the way they are embedded in past trajectories and wider contexts. I link the two streams together through a 'sociological' variant of 'institutionalist' analysis (Hall and Taylor 1996).[6] This stresses the socially constructed work of creating policy meanings and frames, and the way in which such work is embedded in socially situated trajectories of experience and understanding. The 'sociological' term refers to the way that governance processes and policy meanings are produced through social relations in which potentially multiple frames of reference are constructed, mobilised and shaped into policy discourses which then interact with the various practices of governance. The 'institutionalist' term refers to the complex and evolving ensemble of formal and informal norms and practices through which governance processes and discourses are constructed, consolidated, challenged and transformed.

So far, there has been only limited intellectual interaction between this 'sociological institutionalist' analysis of governance processes, interpretive policy analysis and the development of a relational understanding of the geographies through which places and the spatial patterning of phenomena are produced. Geographers have become interested in the configuration of governance relations, particularly in the discussion of arguments about globalisation, regionalisation and localisation.[7] However, much of this work deals with broad generalisations about governance relations, despite emphasising that specific instances of governance are highly contingent on particular histories and geographies. The work of planners and policy analysts, in contrast, is accustomed to penetrating the relations and practices of governance, exploring their dynamics and how they are constituted. In this book, I seek to show how concepts of urban region 'geography' are produced, mobilised and become embedded in governance discourses and practices in specific instances. In particular, I am interested in the way the dynamic fluidity of evolving relational webs intersects with the 'fixes' that develop as certain ways of thinking and doing become consolidated into accepted practices, which then generate resistances to further transformations. Such consolidation is referred to in policy analysis as routinisation, or institutionalisation (Hajer 1995). In the regional economic geography literature, analysts refer to such processes as 'embedding' (Granovetter 1985). I am interested in processes of embedding and disembedding of policy discourses and practices, and in understanding the contingencies which make it appropriate to challenge fixities in one context and seek to stabilise fluidities in another. In the rest of this chapter, I develop this approach through an initial discussion of the four themes introduced in Chapter 1. I give particular attention to the issue of governance, as this is the overarching capacity to which the activity of spatial strategy-making seeks to make a contribution. I then move through the other three themes, from understandings and meanings, to concepts of place, space and of strategy. Finally, I introduce the three cases.

GOVERNANCE CAPACITY

URBAN GOVERNMENT AND GOVERNANCE LANDSCAPES

Thinking about governance and governance capacity involves venturing into broad debates about policy and administration, about politics and policy, about levels of government, about the state and citizens, about authority and legitimacy, and about what shapes cultures and processes of governance. In the mid-twentieth century, it was common to refer to government and the work of the 'public sector'. The public sector was seen as distinct and different from the 'private sector', the sphere of business and the economy. In democratic societies, the institutions of formal government – administrative law, political parties, executive government departments, the roles of elected politicians and appointed officials – were assumed to operate to realise the 'public interest', a general term used to mean the collective interest of the majority of citizens in a formal political and administrative jurisdiction, such as a nation, a region or a municipality.[8] Formal mechanisms of political representation and legal challenge were in place to check that government organisations acted legitimately and accountably, that is, within the law and responsively to citizens. An urban area was assumed to have a municipal level of government, perhaps within a larger political unit such as a county or a province, which was expected to pursue a coherent approach to the management and development of its 'territory'. Spatial strategy-making activity focused on urban areas could be neatly slotted into this government organisation, to provide a spatially articulated expression of a coherent development approach.

Half-a-century later, this orderly approach to identifying the activity of 'urban government' has been undermined both by the experience of governance activity in urban areas and by research and analysis of the performance of governance activity. It is as difficult to find a clear definition of what constitutes urban government as it is to find an objective definition of what an urban area is. Any urban area may have all kinds of governance relations threading through it, around it and over it. Some of these relations are attached to formal, hierarchically organised government organisations that provide a particular locus for an 'urban level'. Others are organised through coalitions of interest around particular issues or areas, which may or may not have any relation to the particularities of a specific urban area. There may also be other agencies focusing on the development of specific issues and areas within an urban area, or even partnerships and coalitions competing for authority with a formal municipal level of government. Sometimes, those promoting an initiative in strategy-making focused on urban-area development are seeking to bring some order into this confusion, to 'join up' diffuse efforts and programmes in some way.

In the above paragraphs, I have used the terms 'government' and 'governance'. In the mid-twentieth century, it was common in Western Europe to consider the sphere of government, often referred to as the state, as separate from business and civil society. Government got its authority and legitimacy from the politics of parties and from the citizen election of political representatives. Within government, the sphere of politics was imagined as separate from administration, itself entrusted in some political systems to legally trained bureaucrats, and in others to experts trained in various professions. A critical value in this separation was the desire to prevent 'corruption' of agreed political priorities by the interference of private interests in the delivery (or 'implementation') of government policies and programmes.

However, the reality of this conception of the separation of spheres never matched up to the model. In addition to the very real experiences of clear corruption, which we will encounter in the Milan case (Chapter 4), alliances between party groups and class or interest factions of society have been common. Thus, in the Netherlands, a long-standing style of democratic politics in the second part of the twentieth century brought together the main political party elites, key national business interests and representatives of trades unions to develop a relatively stable 'corporatist' consensus about a range of areas of policy. In relation to spatial organisation, a particular emphasis of this policy was on planned urbanisation, specifically to deliver low-cost, high-quality housing, while protecting landscape assets in the dense West Netherlands area (Faludi and van der

Valk 1994). By the 1970s, the close relationship between government and business interests in urban planning practices in Western Europe was being criticised by Marxist-inspired analysts who presented the state as an arm of the capitalist economy (Castells 1977). In the later years of the century, both practical experience and academic analysis came to emphasise the ways in which 'interest group' lobbies and 'single-issue' politics cut across the formal mechanisms of representative democracy.[9] Analysts of public policy who looked closely at the relations surrounding policy formation and implementation increasingly highlighted the existence of 'policy networks', 'policy communities' or 'advocacy coalitions'.[10]

Looked at in a relational way, these analysts perceived governance activity as driven by and performed through a nexus of complex interactions, linking the spheres of the state, the economy and civil society in diverse, if typically highly uneven, ways. These networks and 'communities' linked together, in different combinations, experts in particular fields, officials working in various levels of government, lobby groups and elected government ministers. The case accounts that follow show how 'planning policy communities' have formed, and how these challenge and are challenged by other policy communities and advocacy coalitions focused on different agendas. This focus on the relations through which governing activity is performed made it clear that such activity could not be confined to the domain of formal government organisation. The relations of governing linked state, economy and civil society in all kinds of ways, both in relation to policy formation and 'delivery'.

Interpretive policy analysts have helped to understand these emerging practices through a recognition that 'politics' has expanded out of the formal arenas of representative democracy into the complex interactive worlds through which policy formulation and delivery are accomplished (Gomart and Hajer 2003; Hajer 2003). This suggests that policy is made not necessarily in the cauldron of ideological politics, but in the evolution of knowledge and frames of interpretation that develop within policy communities. These policy discourses in turn shape the design of policy interventions – regulatory tools, programmes of investment and management, moral exhortations. They influence the evolving practices through which governance is performed. But these influences do not flow in a simple linear way. Old practices may resist new discourses. New policy discourses emerging in one policy community may be stalled by practices being shaped by developments in another. All three cases in this book illustrate an increasing instability in discourses and practices, as policy actors find themselves operating in arenas and practices that are increasingly challenged by developments around them.

The term 'government' is too narrow to encompass these governing practices. 'Governance' has come into use to refer to all 'collective action' promoted as for public purposes, wider than the purposes of individual agents. It is in this sense that I use the term 'governance' in this book. It signals a shift of intellectual attention from the description and evaluation of government activity in terms of formal competences and laws to a recognition that the spheres of the state, the economy and daily life overlap and interact in complex ways in the construction of politics and policy, and in the formation of policy agendas and practices. Understood in this way, the investigative lens has to widen from a narrow focus on what formal government does, to encompass the wider relations through which collective action is accomplished.[11]

By the 1990s, the assumptions that had supported the idea of a formal organisation of government separated from economic activity and the domain of civil society was also being challenged by political ideology. On the one hand, a neo-liberal agenda had developed, focused on reducing the activity of formal government in society and encouraging non-government agencies and individuals in the economy and civil society to take on more activities previously done by formal government. On the other, social democratic agendas promoted re-engaging citizens with democratic processes through initiatives to encourage participation, empowerment and political inclusion. Both developments in political thought have encouraged a proliferation of 'partnership' agencies, semi-public bodies and 'contracting' arrangements, in which government actors work together with representatives of business, communities, voluntary groups and interest associations to develop and implement policy initiatives.[12] These agendas have helped to create the diffuse urban governance landscapes of the late twentieth century, often referred to as 'fragmented'. The new organisational forms for governance activity raise difficult questions about how the accountability and legitimacy of such activity can be established and blur the boundary between the 'public' and 'private' sectors.

The new governance forms also emphasise the importance of looking closely at the webs of relations and institutional sites through which different groups are linked together as they weave through a diffused urban governance landscape. Where is a spatial strategy-making initiative located in such a landscape? Which relations are drawn into it and which excluded? Some analysts have argued that urban areas are paralleled by 'urban regimes', informal networks of social actors that build up an enduring coalition which commands the key institutional sites of governance in an urban area. Such regimes are reported as well-established in US urban areas, typically linking local business interests with political elites. In Western Europe, linkages between policy communities and political elites are stronger, and party networks may be more important in holding regimes together than business coalitions.[13] The three cases in this book illustrate different configurations in this respect. But the more important challenge to the concept that urban landscapes have urban regimes is that there may be no stable coalition of any kind holding governance activity in urban areas into some kind of focused coherence, or integrated, 'joined up' attention to urban region development. Tensions between different relational nexuses through which governance activity is performed in an urban area may be just too diffuse or riven by struggles for any kind of stable regime to develop. Initiatives in spatial strategy-making may be promoted both to destabilise and shift practices and discourses which some actors feel have gone past their sell-by date, as well as to create some coherence and stability in a dynamic, diffuse and tension-ridden urban governance context.

These shifts in political thinking about the work of governance have supported another meaning for the term 'governance' in the urban governance literature. Some analysts use the term to describe an actual shift in modes of collective action, in which the role of the formal state is reduced and the involvement of other societal relations is given greater scope in shaping collective action (Bagnasco and Le Galès 2000a). Jessop (1995), for example, uses the term to reflect a shift away from the 'welfare state' arrangements of the mid-twentieth century to a mode of governance in which non-state actors are much more explicitly involved. From the perspective of 'regulation theory', Jessop links this shift to a general search for a new 'mode of regulation' more appropriate for the 'mode of accumulation' of late capitalist economic organisation. But more detailed analyses of policy processes in the urban and regional context suggest that there is a considerable variety in 'modes of governance'.[14] The 'corporatist' mode, in which the state, large firms and trades unions shape government policy, has already been mentioned. The 'partnership form' could be seen as a looser and more flexible way of re-casting this mode. But such 'partnerships' also have echoes of the 'clientelism' that builds up when state actors develop patron–client relations with firms or citizens, dispersing funds and regulatory favours to individual supporters and friends. Alternative modes of governance have been developed to limit such clientelistic potentialities, including ideological politics, driven by core values; bureaucratic principles, driven by clear administrative rules; and technical expertise, driven by the legitimacy of scientific knowledge. It is these latter impulses that generated a 'policy-driven' mode of governance which permeated the landscape of governance across Western Europe in the second part of the twentieth century, providing an underpinning to the formation of multiple policy communities. The production of explicit strategies and their use to develop programmes of action, a key idea within the planning project, is in essence a form of a policy-driven mode of governance. But, as the cases show, such a mode of governance is always contested by other practices, and what starts out as a policy-driven mode may well be subverted by other modes. A governance landscape is thus likely to encompass several modes of doing governance work, some considered more, and some less, legitimate than others.

Urban areas thus vary enormously in the qualities and capacities of their governance landscapes. Practices of spatial strategy-making are therefore likely to be situated in very different conditions from one urban area to another. In some urban areas, such practices may have a central role in shaping urban governance landscapes. In others, they may be hidden away with impacts on a very narrow arena of action. The cases that follow provide illustrations of both possibilities. Analysts and practitioners of spatial strategy-making practices need some way to 'read' the dimensions and qualities of the governance landscapes in which a specific practice is situated in order to assess its influence and effects. Some analysts have attempted to track down this multiplicity into measurable 'factors', 'variables' and causal chains between dependent and independent variables in order to provide some kind of explanation of what interventions in which situations produce what kinds of outcomes. But such attempts fail to capture either the different speeds, scales and trajectories of these multiple relations or the complex ways they co-evolve and co-constitute each other (Fischer 2003). There is no easy way to classify and connect typologies of urban areas and typologies of urban governance landscapes, and the attempt to simplify the complex relational dynamics through which governance activity interrelates with the multiple relations weaving through an urban area is likely miss the struggles and synergies through which new potentialities are generated and new initiatives resisted. Interpretive policy analysts make use of the traditions of narrative analysis from history, biography, from cultural anthropology and qualitative social science to 'tell stories' about particular experiences (Fischer 2003). I follow this approach in presenting the cases that follow, in order to situate them in the dynamics of their particular institutional histories and geographies.

ANALYSING GOVERNANCE LANDSCAPES

In the discussion so far, I have already mentioned many terms and concepts that are commonly used currently in accounts of urban governance processes; for example, actors, arenas, networks, discourses, practices, structures, processes, cultures. In this section, I give some order to these many terms, through considering three issues. The first is the relation between structuring forces and individual agency. The second is the relation between the level of face-to-face social interaction and the deeper processes of routinised practices and cultural norms. The third relates to how power dynamics are treated. In this way, I aim to provide a conceptual vocabulary through which to assess the position and transformative power of the experiences of spatial strategy-making in the three cases that follow.

My general 'institutionalist' perspective stresses the significance of context, expressed in the way broader forces, through time, interact with the specific histories and geographies of social groups. I emphasise the socially situated and socially constructed nature of meaning, knowledge and value, and the complex relation between such situated ways of viewing the world and the capacity, through learning processes, to challenge and change these world views (Fischer 2003; Hajer and Wagenaar 2003). This leads to a concern with the relation between the shaping power of 'systems' or 'structuring forces', local particularities and the ability of individuals to imagine and to mobilise attention and action, discourses and practices, in ways that challenge and potentially change these structuring forces, as well as sustaining them. I therefore follow those sociological analysts who stress the interrelation of structure and agency, rather than giving a privileged position in analysis to 'structure' or to 'agency' (Giddens 1984). Individuals may play an important role in developing a spatial strategy and urban plan, as is evident in the cases that follow. These strategies, however, often then become part of the structuring parameters in which specific investments in physical developments and infrastructures are made, which, over time, shape the material opportunities and concepts of place which those in later generations experience and develop.

Giddens identifies three relations through which specific actions are shaped by structuring forces, and through which structuring forces are themselves produced. The first relates to allocative structures (the way material resources – finance, land, human labour – are allocated; for example, public investment in infrastructure or land and property investment processes). The second relates to authoritative structures (the constitution of norms, values, regulatory procedures for example, regulations over the use and development of land, or processes of environmental impact assessment). The third relates to systems of meaning (frames of reference, ideologies, rationalities, discourses).[15] As will become clear in the cases, spatial strategy-making initiatives are positioned in relation to particular configurations of resources for investment in urban development, of regulations governing urban development projects and programmes, and frames of reference and specific discourses about the qualities and appropriate development trajectories for an urban area. But the initiatives may also be motivated by the ambition of changing these configurations in order to achieve different material outcomes and identities for an urban 'region'.

How could this come about? Governance processes appear to be performed through routinised practices embedded in powerful social relations and cultural assumptions that seem to hold them in place despite energetic efforts to change them. Yet they do change. To penetrate into these transformative dynamics, it is helpful to separate out analytically three levels through which governance activity is performed (see Table 2.1). I do not refer here to the traditional hierarchical model of nested 'levels' of government authority. Instead, I refer to levels of conscious attention.[16] The first is the level of specific interactions played out in an episode of spatial strategy-making. Such episodes may occur over time, involving many actors in a range of arenas or institutional sites, each with a distinct setting, but they all involve direct interactions between people, developing and challenging agendas and concepts about urban region development in one way or

Table 2.1 Three levels of governance performance

Level	*Dimension*
Specific episodes	• Actors – roles, strategies, interests • Arenas – institutional sites
Governance processes	• Networks and coalitions • Discourses – language, metaphor, derived from frames of reference • Practices
Governance cultures	• Range of accepted modes of governance • Range of embedded cultural values • Formal and informal processes of critique through which governing processes are rendered legitimate

Source: adapted from Healey 2004a, page 93

another. The case accounts each provide a narrative about several episodes in the period from the mid-twentieth century to the present.

The second is the level of institutionalised governance processes, that is, the routinised practices and discourses of established agencies of formal government and the various informal communities and networks through which many governance activities are routinely performed. I use the term 'discourse' to refer to the policy language and metaphors mobilised in focusing, justifying and legitimating a policy programme or project. This vocabulary gives expression, implicitly or explicitly, to one or more frames of meaning, which shape how 'problems' and 'solutions' are perceived. By 'practice', I refer to the effects, meanings and values embodied in what those involved in governing activity actually do. Discourses and practices may be neatly co-aligned, so that what people do is what they say they do. But as we all know, they may well drift apart. This may be because of deliberate attempts to manipulate how governance activity is perceived. But it may also be because transformations in discourses proceed at a different speed and in a different direction to transformations in practices. The relation between discourses and practices is therefore better understood as in continual potential tension. If new ideas about priorities are to have effects, they will need to penetrate into the discourses and practices of those who have the authority over resources and regulatory powers to realise ideas. This will often mean challenging established networks and coalitions.

The third level refers to the cultural assumptions through which the rhetorics and practices of those involved in 'doing governance', in significant collective action, derive their meaning and legitimacy. These assumptions, about what values should be given priority and what modes of governance are appropriate, are activated in critical commentary on the work of those involved in governing. At this level, critics and monitors of governance activity, such as the media, pressure groups and protest movements mobilise norms of appropriate governance practice and ideas about urban 'region' qualities and trajectories, engaging in critical debate about governance initiatives and processes.

To have significant effects, spatial strategy-making initiatives focused on urban 'region' development need to accumulate sufficient power behind the idea of the 'place' of an urban area to shape resource allocations (particularly in relation to development and infrastructure) and regulatory practices (particularly in relation to environmental quality and how land is used and developed). This implies that such initiatives can develop framing concepts or policy discourses with the capacity to move beyond the 'episode', the institutional site of their articulation (maybe the efforts of an advocacy coalition, or a strategic planning office, or a consultancy exercise), to shape and transform the practices through which resources are allocated and regulatory procedures enacted. This requires substantial mobilising power at the level of an episode. But such energy may be encouraged or inhibited by the movements in the context of an episode. The dynamics of institutionalised discourses and practices, of governance processes, may be widening the moment of opportunity for a strategic initiative. Or the opportunity space may be very limited, leaving only cracks through which new ideas can seep into the wider context (Healey 1997; Tarrow 1994).

The wider governance culture may similarly be pushing forward or pulling away from the ideas promoted in a strategic episode. Thus each level may be evolving in a different way and along a different timescale, even as they interact. A spatial strategy-making initiative focused on an urban area may fail to penetrate into 'mainstream' governance processes at one period, but may yet have resonance with evolutions in a governance culture, which later may exert sufficient pressure on governance processes to effect significant transformations towards the discourses and practices promoted by the earlier strategy-makers. This means that the analysis of the impacts of strategies needs to be undertaken over a considerable timescale, with careful attention to the extent and manner in which the ideas generated in a discourse 'travel' to other institutional sites in a governance landscape, penetrate governance processes and sediment into governance cultures.

Unpacking governance activity into analytical levels in this way helps to highlight more clearly how struggles over authoritative, allocative and framing power are conducted and how, through the energy of specific agents, for example in a spatial strategy-making episode, structuring dynamics may themselves be changed. This implies that the power of structuring dynamics to shape agency possibilities is always limited, contingent on the way agents respond to the opportunities available to them. Skilled mobilising energy can challenge and change structuring power in specific circumstances. This suggests that every experience of spatial strategy-making in an urban area will be so different as to make comparison between experiences inappropriate. Yet, as the cases will show, similar influences often appear in many places. Can any generalisations be made about the broader driving forces that shape urban governance landscapes, how these may have changed from the mid-twentieth century to the present, and hence about the configuration of moments of opportunity to pursue urban 'region' spatial strategies, especially ones that seek to keep concerns with distributive justice and environmental well-being in conjunction with economic vitality?

The arguments of the regulation school referred to earlier claim that a coherent, integrated urban development strategy suited the logic of an 'industrial' mode of accumulation, providing sites, buildings and transport to make production more efficient and organising urban development to provide low-cost housing and welfare support for workers. With the break-up of this 'mode of accumulation', there is less need for an integrated approach to the management of urban development. Instead, more attention is given to providing appropriate spaces for new kinds of production, commercial, financial and consumption activity, and to fostering communications infrastructure so that individual firms and clusters can more easily stretch out to global markets. This leads, the regulationists argue, to a transformation dynamic that seeks to swing established governance processes 'locked in' to old integrated and 'managerial' modes of governance towards more 'entrepreneurial' approaches to developing the assets of urban areas.[17]

There are some signs of such a dynamic in all the cases that follow. But this thesis suggests that the economic imperative sweeps all before it. In such a context, integrated approaches to urban spatial strategy-making that aim to keep economic, social and environmental considerations in conjunction are doomed to be subverted into an economic dynamic. Yet economic forces are only one of the pressures to which urban governance processes respond. There are other movements in contemporary Western Europe – the concern for environmental qualities, shifts in lifestyle and cultural values, which also have significant leverage on governance processes and governance cultures, and have a potential to hold economic logics in check. The struggle over spatial strategy-making in urban areas is thus not only one between practices appropriate for an economic past and a different economic future, but over what kind of future to promote in the unfolding conditions of daily life existence in urban areas.

Transformations in governance landscapes thus involve struggles over materialities and meanings, over access to material resources and to regulatory authority, over creating frames of reference which shape governance attention and mould practices. Power, in this perspective, is much more than the formal authority of government agencies. It is more than the ability of powerful individuals or broad structuring forces to impose agendas on others. It is not a 'thing' to be possessed, but an energy that mobilises and suppresses attention, and thereby achieves control over others in some conditions but also generates the force to undertake projects, to infuse protest movements and concentrate effort towards collective projects. Power and governance capacity is not located only in the formally elected positions of government ministers. Instead, this energy has always been diffused through many arenas, some more and some less endowed with formal competences and legal authority. What differs from one period to another is the patterning of governance relations, in terms of who gets involved, in which institutional arenas, subject to what checks and balances and with what capacity to exert influence over others. Nor is power merely authoritative power, 'power over', in a command-and-control way. Power is also a generative force, expressed in potentialities – the energy to act, to do things, to mobilise, to imagine and to invent, 'power to' (Dyrberg 1997; Giddens 1984). Allen (2003) suggests a four-fold way of thinking about power as an immanent force in social relations: as the exercise of authority; as attempts to dominate; as manipulative and persuasive acts; and as attempts to 'seduce' or to attract others to a position or attitude. Initiatives in spatial strategy-making for urban 'regions', as the cases show, rely to a considerable extent on their power to mobilise attention through their persuasive and seductive qualities. But they also need to access the law-governed power of formal government administration, and are continually open to domination through framing discourses and practices which privilege, as often implicitly as explicitly, well-established social groups, as in the practices of 'corporatist' and 'clientelist' modes of governance, or the self-interested strategies of individuals or social groups.[18]

I have now outlined an approach to urban governance and governance capacity that will inform the case study accounts, though the accounts are presented as historical narratives rather than organised into analytical categories. I now fill out this understanding by looking at more specific dimensions of governance activity, those of particular importance to initiatives of spatial strategy-making. These form the other three themes I draw out from the case-study experiences. Each will be developed further in Chapters 6 to 8, so I only introduce them briefly here.

KNOWLEDGE AND MEANING

Spatial strategy-making focused on urban 'regions' involves generating framing ideas and organising concepts through which an urban region is 'summoned up' to become 'visible' in a governance context. It involves 'framing' and 'naming' the phenomena of an urban 'region' (Schon and Rein 1994; van Duinen 2004), converting a fluid and dynamic complex of diverse relations into some kind of conceptual entity. Such a frame, or way of 'seeing', is inevitably a simplified and selective viewpoint. But if sufficient actors buy into the frame and the discourses it generates, then the frame accumulates the power to flow from the institutional site of its formation to other arenas and practices and to generate consequences in its turn. The institutionalist analysis of governance processes emphasises the systems of meaning 'called up' in episodes of spatial strategy-making, how these interact, and how, through interactions, discourses are produced and diffused. Many analysts focus on the learning processes through which knowledge is accessed, interpreted and re-assembled in policy processes and organisational contexts.[19] But the processes of re-framing are more than this, as they involve shifting the parameters within which sense is generated from information and pieces of knowledge. Strategy-making that involves re-framing is about creative discovery as well as systematised learning. It involves knowledge creation as well as acquisition (Takeuchi 2001). It requires generating new ways of thinking about issues and about new priorities and pressures.

Frames are systems of meaning that organise what we 'know' (Schon and Rein 1994). Friedmann (1987) argued that the whole enterprise of planning was about the relation of knowledge to action. But what is encompassed by 'knowledge' and how do the spheres of 'imagining', 'knowing' and 'acting' relate to each other? The established idea in twentieth-century policy analysis and planning was that knowledge was primarily that of 'science', formalised through the routines of deductive logic and inductive inquiry from empirical evidence and experimentation. This knowledge provided the basis for identifying strategic parameters, which in turn could be expressed in a plan. This plan became the basis for action programmes that 'implementers' were expected to follow. Such a linear model emphasised research and analysis before 'plan'. The plan was then 'implemented'. However, in reality, many influential spatial strategies have been produced before anyone knew what to research. Concepts and priorities emerge, not just from the codified knowledge of science, but from experience, ideology, professional concepts and political fixes. Rather than being linear and logical, making the relation between knowledge and action in strategy formation is a complex, interactive, ongoing activity, in which diverse forms of knowledge are 'called up', generated and given meaning.

For analysts of urban governance processes, this means that attention is required to the range and types of knowledge mobilised in policy-making processes and to how imaginative conceptions of 'what could be' are confronted with diverse kinds of knowledge about 'what is going on'. What do these processes of discourse formation highlight and what do they push to the periphery of attention? Once produced, how well does a strategic frame move from one site to another? Does a new way of 'seeing' a set of issues survive as it travels from one arena to another, or does it get translated back into established systems? Such issues lie at the heart of assessing how and how far new policy discourses have the potential to diffuse and become institutionalised in governance practices (Fischer 2003; Hajer 1995). These issues move the discussion of 'policy learning' in episodes of spatial strategy-making into a complex understanding of the production of knowledge, of framing conceptions and specific discourses as processes of struggle between different perceptions and epistemologies; that is, of ways of knowing.

The established twentieth century view of knowledge as 'science' sought to set 'objective' knowledge apart from the power struggles of the formal political domain. The aim of what has become known as 'positivist' science was to uncover the principles governing phenomena in the natural world. Scientists sought causal laws validated by the techniques of experimental testing or statistical analysis (Fischer 2000; Lindblom 1990). This gave legitimacy to their conclusions and predictions, which then served to criticise, sustain and provide the grounds for changing political agendas and discourses. The legitimacy of the scientist was expanded to cover technical crafts – the experience of the doctor, the architect, the engineer and, later, the policy analyst and the planner. The view of the scientist and expert as above and apart from politics, holding a privileged knowledge, more legitimate than that of the politician or citizen, strongly influenced urban planners and politicians in the mid-twentieth century, as the cases will show. By the latter part of the century, however, this authority was widely challenged. Scientific knowledge itself was shown to be constructed through specific perspectives and paradigms, which structured and gave meaning to experiments and analyses (Barnes 1982; Latour 1987). The weaknesses in the formal knowledge base of experts were also exposed, as major development projects were challenged and discussed by protest groups and the media. Environmental pressure groups took a leading role in exposing the weaknesses in formalised, technical knowledge, and in highlighting the value of 'local' knowledge – what people know through daily life observation and experience and through cultural inheritances (Geertz 1983). Analyses of the knowledge used in the business field also highlighted the importance of experiential knowledge, and the tacit, craft knowledge embodied in practices as well as expressed knowledge.[20]

It is this view of knowledge that underpins the 'sociological' institutionalism outlined earlier. This view emphasises that all knowledge is constructed through social processes, which filter what is experienced, observed and imagined as it is arranged into systems of meaning. These meanings are shaped by contexts, by purposes, by values and by power relations. They are formed within social practices, not apart from them (Ingold 2005). It is within these practices – of the laboratory, the council chamber, the professional studio, the strategic planning team, the office meeting, the practice of negotiating financial grants for projects, etc. – that forms of knowledge encounter each other, are filtered, and arranged into arguments, justifications and concepts of cause and effect. In these social contexts, sometimes called 'communities of practice' (Wenger 1998), people engage in arguments about meanings and values, and make 'practical judgements' about significance, about validity, and about the integrity of the knowledge claims of others.[21] The politics of knowledge production and organisation lie in these social processes of filtering, meaning making, argumentation and practical judgement. As Sandercock (2003a: 73) states: 'There's nothing more political than epistemological struggles', struggles over meaning and which knowledge counts in governance processes. Struggles over spatial strategy-making are thus about the knowledge that is to count in the work of framing and legitimating a strategy.

Spatial strategy-making for urban areas has both to 'imagine' an idea about an urban region from the multiplicity of relational webs that transect the space of an area, and draw out resources of knowledge and understanding with which to explore, justify, develop and test the ideas that emerge. It has to tap into the 'distributed intelligence' (Innes and Booher 2001) in the array of relations and to justify the conceptions produced through this intelligence. The institutional arenas within which such strategies are formed are, in effect, sites for the social construction of framing concepts. Such institutional sites are places of encounter, learning, contestation and creative discovery, as will become clear in the cases that follow. In the dynamic, fluid mix of ideas and knowledge about what places are and could be, the work of strategy-making, of strategic frame formation, produces some kind of fixing of meanings, through which public policy interventions can be focused and shaped. Given that generating a policy focus around the place of an urban 'region' challenges the frames and meanings evolved in other policy sectors and, potentially, in the practices embedded in the relational webs of economic and social life, such urban region frames of meaning may often be fragile, or too weak to accumulate the power to 'travel' across a governance landscape and through time.

CONCEPTUALISING 'CITIES' AND 'URBAN REGIONS'

It is not just any kind of knowledge that is drawn into the processes of spatial strategy-making. A strategy with a place focus draws on and draws out conceptions of places, their qualities and their positioning as regards other places and their dynamics. The challenge for any episode in spatial strategy-making focused around urban areas is that an urban 'region' is not a 'thing', to which an analyst can approximate an 'objective' representation. It is an imagined phenomenon, a conception of a very complex set of overlapping and intersecting relations, understood in different ways by different people.[22] Such a conception may be driven by political–administrative logics, such as the definition of the boundaries of a city jurisdiction, or analytical logics, such as the demarcation through statistical measures of journey-to-work and labour-market areas. An idea of a city may be positioned in a map of a group of cities with similar characteristics, such as in classifications of 'global cities' (Taylor 2004b). Or it may arise from affective considerations, such as people's feelings of identification with a place where they grew up, or visited, or where they now live, or with a football team (Hillier 2000). The resonance of the 'imagined city' will then be linked to an ambience, or to some streets of buildings, which captures for people that particular feeling. In any area 'called into imagination' as a city, there are thus likely to be many potential 'imagined cities' (Healey 2002; Vigar *et al.* 2005).

The planning field has a rich and fascinating history of attempts to provide representations of the city and urban region. Underpinning this history is the effort to grasp the 'whole' city, in a 'comprehensive' way. Planners and urbanists have struggled to find ways to synthesise the complex socio-spatial dynamics of cities, with their multiple layers of relations operating at diverse scales and often in conflict with each other, and to shape some kind of 'integrated' conception, through which the city can be 'seen' and grasped. Planners have then proposed interventions to promote the material shaping of this imagined reality. Their representations have often become absorbed into a governance culture, as in the Dutch conception of the 'Randstad' (Faludi and van der Valk 1994), and the English notion of towns surrounded by a landscape of 'green belt' (Elson 1986; Hall *et al.* 1973).

Planners' 'cities of the imagination' have drawn on many different strands of inspiration.[23] One such has been the utopian tradition of imagined societies, complete with spatial and physical morphologies. Pursued in the context of novels, treatises and architectural imagery, these typically combine political, moral and aesthetic concerns with an attempt to link social dynamics with physical form. Ebenezer Howard's *Garden Cities of Tomorrow*, and Le Corbusier's *Cité Radieuse* provided a powerful imagery for campaigners for improved urban conditions in the twentieth century. A second tradition, which overlapped with the 'utopian' one in the first part of the twentieth century, believed the city could be found through empirical inquiry. Inspired by the work of Patrick Geddes (1915), it was argued that survey, research and analysis should provide the basis for developing a conception of an urban region. By the 1960s, planners in the UK were being offered a range of models of urban form from which to choose when undertaking an urban plan. Such ideas have re-surfaced in European planning discourses in concepts of compact cities, urban networks, gateways, nodes, concentrated deconcentration, polycentric development and development corridors.[24]

Since the 1960s, however, there has been an infusion of ideas into the planning field from the social sciences, and particularly geography, about the dynamics of urban spatial organisation. In Chapter 7, I show how a relational geography, developed since the 1980s, has challenged an older geography of physical proximity which focused on the integration of the relations transecting an urban area into a cohesive socio-physical unity expressed in a specific urban form. A relational geography focuses instead on the diversity of the relational webs that transect an urban region, each with its own scale, driving dynamics, organisation into centres, nodal points and flows, and spatial patterning. This geography presents space 'as a simultaneity of multiple trajectories' (Massey 2005, page 61). The relations of an urban region are not therefore necessarily 'integrated' with each other. They may be in tension or severe conflict, particularly over access to, and the value of, particular places. This conception emphasises the existence of multiple networks, of nodes where networks intersect, of urban areas as 'polycentric', as well as conceptions of the urban as comprising multiple flows of people, goods, water, energy, information and ideas.

This relational perspective on the spatiality of the city is not new in the planning field. Mel Webber sought to express such a perspective in the 1960s, focusing on the significance of increasing mobility and new ways of communicating (Webber 1964). Such a perspective also has resonance with personal experience. Companies working in their supplier–customer networks make choices about where to expand and when to relocate. Many people with family members and friends spread over a wide scale, who travel a lot and who intensively use mobile phones, email and the Internet to access their social lives and obtain knowledge and material goods, feel nearer to more distant others than to their immediate neighbours. In such a relational perspective, if place quality has a value, this must be sought in the way the experiential meanings of 'place' have 'presence' in these complex, shifting and conflicting relations through which the experience of daily life in urban areas is constituted.

In the case accounts that follow, I emphasise the resonances generated by conceptions of an urban area and their power to mobilise attention. This focuses investigation on the processes through which a conception of an 'urban region' is brought into policy attention, and its capacity to 'create' a powerful focus for action, knowledge formation and identity construction through which material effects come about and capacities for collective action around a 'place' focus are enhanced. The case accounts show how concepts of 'urban region' evolve over time, the power they accumulate and the institutional work that they perform.

STRATEGIC FOCUS AND SELECTIVITY

The focus of this book is on the practices of spatial strategy-making. But what is special about 'strategising' as an activity? What constitutes a relational and interpretive approach to strategy-making? I use the term 'strategising' to refer to the drawing out of a sense of potentialities and possibilities from multiple unfolding relations, within which to set actions that will intervene in these unfolding relations in the hope of furthering particular objectives and qualities. Such actions could be the shaping of a project, or the drawing up of a programme of action, or the creation of a persuasive 'vision' of an urban area that captures the attention of others and organises their actions. Strategising implies the 'calling up' of a frame of meaning, though this may not always be explicit.

Strategic thinking, as Mintzberg (1994/2000) and Bryson (2003) so persuasively argue, is not the same thing as having a strategy or plan. Strategic thinking involves a way of thought, in which events, episodes and possibilities are continuously interpreted in terms of their significance for an enterprise as it evolves over time in a specific and dynamic context. It encourages the continual shaping of actions in terms of new information and understanding of the resistances and potentials for an enterprise. It challenges practices that are justified in terms of 'following established procedures' or 'this is what we have always done'. In the context of collective action for the development of urban areas, strategic thinking involves selecting and focusing on key relationships through which such development is being shaped and on key interventions in these relations that could make a difference through time. Because collective action in the public domain is always likely to be challenged and needs to build authority and legitimacy, strategic thinking also needs to pay attention to the persuasive power of strategic ideas and their acceptability in a governance culture. The inherent selectivity of strategic thinking is thus deeply political. It highlights some issues and interests and 'lowlights' or ignores others. It synthesises some relations and linkages and neglects others. Its 'integrations' and 'joinings-up' are always to an extent partial, pulling some relations closer together, while 'disintegrating' others.

The concept of strategic planning is often used to elide strategic thinking, understood in this way, with the production and use of a formally-approved strategy or strategic plan. Recent writers in the management and planning fields have been careful to pull this elision apart. For Bryson (2003: 38):

> strategic planning may be defined as a disciplined effort to produce fundamental decisions and actions that shape and guide what an (entity) is, what it does and why it does it.

A strategy is likely to result from this 'disciplined' process, but this is not necessarily best expressed in a formal plan. Too much formalisation, as both Bryson and Mintzberg argue, may undermine the very properties that are associated with strategic thinking. A strategy is better understood as a discursive frame, which maintains 'in attention' critical understandings about relationships, qualities, values and priorities.

The understanding of the meaning and practice of strategic thinking in the fields of planning and public policy draws on several traditions of thought (Albrechts 2004). Until the later 1970s, a sharp distinction was made between the practices of formulating a strategy and those through which a strategy was 'implemented'. Drawing on military analogy and on the separation of the spheres of politics from those of administration, a hierarchical and linear relation was assumed between strategy and action. The development of strategy was differentiated from tactics, a differentiation also expressed in terms such as 'implementation', 'operationalisation' and 'detail'. In the 1960s in particular, it was proposed that strategies could be developed from scientific analyses and models of urban region dynamics through which appropriate interventions could be evaluated and selected. The resultant choices could be expressed in a 'plan' which could then be implemented. In this way, the intellectual work of strategy formation (the techniques of analysis, modelling, prediction and systematic evaluation) could be neatly separated from the political work of establishing general values through which choices between alternative 'directions' could be made and from the messy organisational work of actually making things happen.

This linear conception has since been widely challenged in the fields of management, policy analysis and planning (see Chapter 6). It has been replaced with notions of strategy-formation and use in the public domain as some form of collaboration among diverse actors, through a mixture of formal and informal interactive processes, drawing on diverse forms of knowledge. In this conception, intellectual and imaginative work is interpenetrated with political considerations and struggles. A strategy co-evolves with the knowledge, values and politics that will give it authority, legitimacy and framing power. Its formation is the product of a specific institutional setting which shapes what is imagined as strategic and yet which may come to have the capacity to challenge and transform that setting (Hajer 2005). Strategies, in the complex dynamics of urban areas, cannot be expected to 'control' emergent socio-spatial patterns. Instead, they are risky and experimental interventions, 'thrown in' to the ongoing dynamic flow of multiple relational webs, in the hope that some beneficial relations will be encouraged and other, potentially harmful, effects will be inhibited. This interactive and situated understanding of strategy formation processes in dynamic complex situations parallels the sociological–institutionalist understanding of governance dynamics developed in this book. It raises questions about the extent to which episodes of spatial strategy-making are actually 'strategic' in the sense outlined above, and about how far they carry, whether intentionally or not, the capacity to transform urban governance dynamics.

EXPLORING EXPERIENCES

In this chapter, I have fleshed out what I mean by a relational perspective, focused on spatial strategy-making for urban 'regions' as a governance activity. I have emphasised the importance of setting urban governance activity in an evolutionary context, which means giving careful attention to the way governance relations develop through the interaction of past dynamics with new driving forces and agency energy. I have suggested dimensions of governance capacity, the production and use of knowledge, ways of thinking about place and space, and the nature of strategy that will be used to provide a vocabulary within the narrative accounts of the three cases that follow. This vocabulary and perspective will be developed in each of the later chapters to draw out an understanding of the transformative potentialities and limits of spatial strategy-making for urban 'regions' and the possibility of a strategy-making that has the capacity to pursue a rich and diverse conception of an urban area which holds concerns about distributive justice, environmental well-being and economic vitality in critical conjunction, rather than a narrow focus around the objectives of a few actors and social groups.

The three accounts that follow provide narratives of the evolution of spatial strategy-making from the mid-twentieth century. Each concludes with a major episode undertaken in the late 1990s/early 2000s. My stories end in 2005, but the practices, of course, continue to unfold. Each account illustrates several explicit initiatives or 'episodes' in spatial strategy-making. An episode is defined as a period when a particular effort is being made to articulate a strategic response to urban area development, though start and end points are never precise for such endeavours. They tend to rise up and then fold back into an ongoing flow of governance activity. As noted earlier in this chapter, an episode involves many interactions, weaving through several arenas, in which diverse actors are drawn into encounters and activities through which strategies are formulated, consolidated and diffused.

The three cases chosen – the Amsterdam area, the Milan area and the Cambridge Sub-Region, are very diverse and should not be considered in any sense as a 'sample' or as exemplars of 'good practice'. They are merely examples of efforts at spatial strategy-making for cities and urban areas. All display complex dynamics of growth and decline, although their institutional contexts are very different. All face challenges in relation to investment in development and infrastructure, and as regards the regulation of development, but these challenges are resolved in different ways. All three areas are located in what is recognised in European spatial planning discourse as the dynamic 'growth zone' of northwest Europe (CSD 1999) (see Figure 2.1), and are significant to their national economies, but have very distinctive local sensibilities.

Amsterdam has a long history of strategic planning which is still celebrated today (Jolles *et al.* 2003). It is the capital city of the Netherlands, acknowledged as the largest and internationally most renowned of Dutch cities. It is not the administrative capital, but is a centre of finance, commerce, industry and tourism, with a striking heritage environment in its city core. The City Council has managed to acquire direct authority over a wide area, and has been very active in positioning its own territory in the wider scales of 'the Amsterdam region', the area of the West Netherlands, and in national policy for cities. It has used the powers and finances of the national state to maintain a direct role in physical development and, as a result, is a major land and property owner as well as a service provider and development regulator. It thus has a lot of experience of 'doing' physical development. Strategic spatial plans to guide this work have been produced for the Amsterdam area since the nineteenth century, with a key internationally recognised plan being produced in 1935. In 2003, the city planning office could celebrate 75 years of a continuous planning effort. The account in Chapter 3 locates the most recent experi-

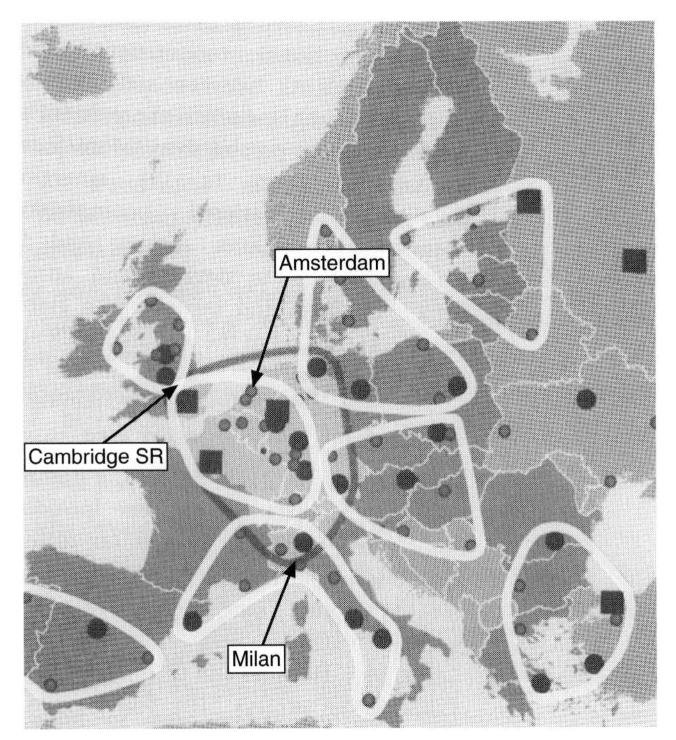

Figure 2.1 Three urban areas in Western Europe

Source: The base map is from Mehlbye 2000, with permission of BBR, Bonn. This shows an early version of a map of potential European zones of metropolitan co-operation, produced in the development of the ESPON project

ences in the post-war history of strategic planning for the city and region. The story concludes with the period during and after which the latest strategic plan was produced, the *Structuurplan* of 2003 (DRO 2003a).

Milan is one of the key cities of Northern Italy, and, like Amsterdam, is now recognised as Italy's de facto commercial and financial capital. It also has a tradition of both small, family-based industries and of very large heavy engineering plants, the latter badly affected by economic restructuring. A key economic nexus of the twentieth century has centred around fashion and design. There is a rich experience of producing urban development plans and projects, to be argued over as design ideas in the magazines read by cultural elites. Once the culturally hegemonic core of a rich region in the north Italian plain, with elite activities and residences clustered in the city centre, the physical area of the city and its economic and social relations have been expanding relentlessly across an ever-widening metropolitan area. Within Italy as a whole, there has been a wave of interest in strategic city planning. But the development of a collective actor capacity to support a city-wide strategy, still less one which is positioned in a sense of the wider region, is hindered in Milan by a largely disinterested city government, with a recent history of challenge to corrupt political practices that has broken many established practices and governance networks. In this difficult context, the account in Chapter 4 concludes with an attempt in the late 1990s to develop a different kind of approach to managing urban development, resulting in the production of a 'framework document' intended to provide a base for both the development of a strategy and principles to guide a new land-use regulation practice (Comune di Milano 2000).

The Cambridge Sub-Region in southern England has experienced dynamic economic growth in recent years, driven by the expansion of both new-technology industries and the London metropolitan region. This has transformed the area from a prestigious academic retreat, a heartland of identity and association for Britain's elite politicians and civil servants. Only 50 miles (80 km) from London, national, regional, county and district levels of government all emphasised conservation of the historic university environment and the surrounding landscape. But such a strategy became increasingly unsustainable as investment in motorways, railways and a new airport for London at Stansted, along with the expansion of the new 'knowledge economy' industries, drew the area into the ever widening globally-significant metropolitan region of 'London and the South-East'. The account in Chapter 5 illustrates the practices of a regulatory approach to managing urban development and the difficulties these have faced in switching from a growth-restraint strategy to a growth-oriented strategy. It exemplifies the wider struggle in southern England to develop an integrated approach to urban development in a highly centralised state with a strong cultural resistance to development in rural areas and a perception of urban areas as 'problem places', in need of regeneration rather than growth management.

These accounts are not written as 'good practice' examples. Instead, they aim to situate contemporary endeavours in their historical and geographical specificities and to show how, despite broader forces that lead to some commonalities in all the cases, these specificities really matter for the prospects of the enterprise of strategic spatial planning. The accounts are narratives, providing 'thick descriptions', guided by the themes outlined above. They are written largely in a narrative chronology, as institutional 'histories'.[25] There are just too many actors and too many events to be described in the form of a personalised diary or even to delve into the detailed way in which each stage has been experienced. For some actors, the events of an episode are the core of their working lives over several years. Some give it occasional attention from time to time. Others mobilise vigorously to mount a campaign or defend a position, but then lose interest and return to other preoccupations. It is often the rhythms of formal procedures and the activism and commitment of key actors that keep the process going, even where there are strong structural pressures to sustain an episode. In writing the accounts, I have tried to trace the interplay of ideas and practices, planning activity and wider governance processes, as these have evolved through time and continue to evolve. I provide more detail in an appendix about how I went about constructing each case account. The challenge has been to show the particularity and contingency of each strategic spatial-planning 'story', while providing material with which to explore what each account has to say about the themes discussed in this chapter.

NOTES

1 Boyer 1983; Hall 1988; Sandercock 2003a.
2 See Albrechts *et al.* 2001, 2003; Fedeli and Gastaldi 2004; Fuerst and Kneilung 2002; Healey *et al.* 1997; Motte 1995; Nigro and Bianchi 2003; Pugliese and Spaziente 2003; Salet and Faludi 2000; Salet *et al.* 2003; Tewdwr-Jones and Allmendinger 2006.
3 See Albrechts *et al.* 2001; Gualini 2004a; Lefèvre 1998; Salet *et al.* 2003.
4 See Dühr 2005; Jensen and Richardson 2004; Richardson 2006; Vigar *et al.* 2000.
5 For my own previous contributions, see Healey 1997, 1999, 2006a. See Fay 1996 for comments on the 'interpretive' label.
6 For a general discussion of institutionalist ideas, from different directions, see Hall and Taylor 1996; Hodgson 2004; Lowndes 2001; Peters 1999; and for a planning perspective, see Gonzalez and Healey 2005; Healey 1999; Verma 2006.
7 See Amin 2002; Brenner 1999; Macleod 2001; Macleod and Goodwin 1999.
8 The term 'public interest' as developed in Anglo-American policy literature needs careful translation into other languages.
9 See Fischer 2003; John 1998; Le Galès 2002.
10 See Fischer 2003; John 1998; Rhodes 1997; Sabatier and Jenkins-Smith 1993. Fischer (2003) provides a helpful review of the differences between these conceptions.
11 See Cars *et al.* 2002; Le Galès 2002; Stoker 2000.
12 See Imrie and Raco 2003; Le Galès 2002; Pierre 1998; Pierre and Peters 2000.
13 For discussion of urban regimes, see Fainstein and Fainstein 1986; Harding 1997; McGuirk 2003; Newman and Thornley 1996; Stoker 1995; Stone 1989, 2005.
14 A mode of governance refers to a way of carrying out the activity of governing. An alternative term in the policy literature is 'policy process form', see Healey 1990.
15 See Bryson and Crosby (1992), chapter 4, and Healey (1997), chapter 8 for applications in relation to urban and regional governance.
16 I draw here on Stephen Lukes' concept of three levels of power (1974) as re-worked with a Foucauldian inspiration by Dyrberg (1997). I develop these ideas further in a sequence of papers: Coaffee and Healey 2003; Healey 2004a, 2006b, d.
17 See the discussions in Amin 1994; Harvey 1989; Jessop 1997, 1998, 2000.
18 This is what Flyvbjerg (1998) refers to as 'strategic action' in his analysis of the planning politics of Aalborg, Denmark.
19 See Fischer 2003; Forester 1993; Sabatier and Jenkins-Smith 1993.
20 See Amin and Cohendet 2004; Blackler *et al.* 1999; Nonaka *et al.* 2001.
21 See Fischer 2003; Forester 1999; Flyvbjerg 2004.
22 See Amin and Thrift 2002; Bagnasco and Le Galès 2000a; Beauregard 1995; Healey 2002.
23 See Boyer 1983; Hall 1988; Liggett and Perry 1995; Ward 1994.
24 See CSD 1999; Dühr 2005; Faludi and Waterhout 2002; Jensen and Richardson 2000; Zonneveld 2000.
25 For discussions about story-telling in policy analysis and planning, see Eckstein and Throgmorton 2003; Fischer 2003; Flyvbjerg 2001; Sandercock 2003b; Throgmorton 1996.

CHAPTER 3

THE STRATEGIC SHAPING OF URBAN DEVELOPMENT IN AMSTERDAM

> Dutch strategic planning has reached a degree of maturity which seems unequalled (Faludi and van der Valk 1994: 122).

> Conscious physical planning has always been important in Amsterdam. . . . Unlike other cities, town planning in Amsterdam was always inextricably bound up with urban culture and urban politics (DRO 1994: 215).

> The Amsterdam approach leads to policy innovation through acting . . . 'in the spirit of the law', according to [an] opportunistic view of policy opportunites and constraints (de Roo 2003: 289).

INTRODUCTION

The Netherlands is admired internationally for its striking capacity to create and manage a built and natural environment through well-coordinated public investment, arising from political processes that have sought consensus among different segments of Dutch society. Within the Netherlands, the City of Amsterdam stands out for the strength and continuity of its urban-planning capacity. Amsterdam manages to be many cities and, at the same time, one city, with a strong sense of 'itself'. The many cities can be found in its daily life environments, its socio-cultural diversity, its industrial, commercial and financial activities, its transport nodes and its cultural, entertainment and sports activities. It is a city of very small neighbourhoods and of economic and cultural nodes that participate in global networks. It is the largest and most internationally engaged of all Dutch cities, the country's capital, though not the centre of national government. It is also a multifaceted tourist destination, with the attractions of its distinctive historic morphology combining with the lure of its liberal culture (Terhorst *et al.* 2003). It is somehow small-scale and large-scale at the same time, the ideal of those for whom the heart of urban place quality is an open and diverse cosmopolitan ambience (Amin *et al.* 2000). It has a lively and diverse civil society, in which conflict is endemic between the different activities and understandings of what 'Amsterdam' is, argued through in the media, in meetings, in demonstrations, in electoral politics and, from time to time, in direct action. Within this seemingly anarchic multiplicity, the *Gemeente Amsterdam* (Amsterdam City Council)[1] presides as the expression of the city's unity, a powerful voice in the region, the nation and in Europe, as well as a major presence in almost all the activities of the city – as strategist, regulator, funder, manager, landlord and land developer. This presence belongs not just to the City Council itself but is acknowledged, valued and vigorously criticised by its citizens.

Planned physical urban development has been a major activity for the City Council during the whole of the twentieth century. Much of the city's present urban morphology is the result of such strategies and the detailed land development projects nested within them. The City Council Planning Department[2] celebrated a 75-year history with a special exhibition in 2003 (Jolles *et al.* 2003), a history actively present in the memory of the city's spatial planners and its wider governance culture[3]. This is not just an example of extraordinary continuity in a professional planning tradition and its practice. It also reflects the iconic meaning of its distinctive geography to all those with a stake in Amsterdam, a city of canals and commerce, of heritage, domestic neighbourhoods and lively cultural engagement and political protest.

This account focuses on the evolution of Amsterdam's strategic planning tradition and practices as various key stakeholders struggle to adjust to pressures for change from different directions. As with the other cases in this book, national, regional and local actors were, by the 1990s, demanding strategic action to improve the 'competitiveness' of the locale of the city, to ensure that it remained one of Europe's core commercial locations. In parallel, national government revised the way funding was made available for major capital investment in infrastructure and urban development. In Amsterdam, the city government was coming to terms with the explosion of the urban area into a wider metropolitan region and a more diffused and polycentric urban morphology. By 2000, the City Council area had a population of over 730,000, growing again after a period of decline, in a wider metropolitan area of over 1.5 million. Many of the development opportunities in the urban region were located in this wider area.

This account locates present dilemmas within a timespan that starts in the early twentieth century. From mid-century, there have been several explicit episodes of spatial strategy-making for the city area, typically intelocked with national initiatives, including the production of Amsterdam's 2003 *Structuurplan* (DRO 2003a). This last period has been one of reduced confidence in the planning policy community in the Netherlands generally. But Amsterdam City Council has retained its large staff of professional planners and researchers. The 2003 *Structuurplan* may well be the last in a 70-year tradition of 'comprehensive' physical development plans, but it is unlikely to be the end of strategic thinking about city and urban region development in Amsterdam's governance processes.

URBAN EXTENSION IN A WELFARE-INDUSTRIAL SOCIETY: THE MID-TWENTIETH CENTURY

THE 1935 GENERAL EXTENSION PLAN

The story of twentieth-century spatial governance and planning in Amsterdam is a fascinating and important one for the history of Dutch and European planning. For many years, the City Council was at the leading edge of planning innovation, and Amsterdam's planners participated in the development of planning ideas not only in the Netherlands but in the wider planning movement developing in Europe at the start of the last century.[4] History and geography combine to create the context for a flourishing planning tradition and practice throughout the twentieth century. Two critical elements of this context have been the massive reclamation of wetlands and lakes for farmland and urban extension, and the Dutch political traditions of consensus agreement, developed in a bourgeois society divided in religion but anxious to avoid sectarian conflict. Always a centre of trade, finance, culture and industry, and situated as it is on the huge delta at the mouth of the river Rhine, Amsterdam has been a dominant location of the Dutch economic powerhouse in the second part of the twentieth century. Land-drainage and water-management activity helped to create a tradition of major state involvement in urban development. To the southwest of Amsterdam, the Haarlem lake was drained in the nineteenth century to provide 'polder' land on which the city and Schiphol Airport expanded from the 1950s. In 1876, the *Noordzeekanaal* was opened up, connecting Amsterdam by a short route to the sea, and providing opportunities along the shores of the river IJ and the canal for industry, transport and port-related developments. To the east, new polders were created in the IJssel lake. Since polder lands were often state lands, Amsterdam City Council could acquire them for its expansion purposes and extend its jurisdiction through annexation. In the process, it became a major land and property owner[5] (see Figure 3.1). Throughout the twentieth century, the City Council has undertaken substantial development projects, shaped by planning principles.

Until the 1960s, the emphasis in Amsterdam's urban development policy was on urban extension. Master plans were produced for individual extension areas. By the 1920s, however, following ideas on city planning developing internationally, more emphasis was given to a 'comprehensive' view of the city. City planner Van Eesteren, who in 1929 took over a newly-created Town Planning Section in the city's great *Dienst Publieke Werken*, saw not only the need to provide quality neighbourhood living environments but to connect the new parts of the urban area to each other and the city centre.[6] This led to a strong emphasis on promoting a good public transport network. The 1935 *General Extension Plan* proposed expansion of the city to the south, towards Amstelveen, to the west across the polder land of Sloten, and along the *Noordzeekanaal* primarily for industry. The result was a spatial form of fingers of development surrounded by lobes of green space, a metaphor still used by Amsterdam city planners today (see Figure 3.2).

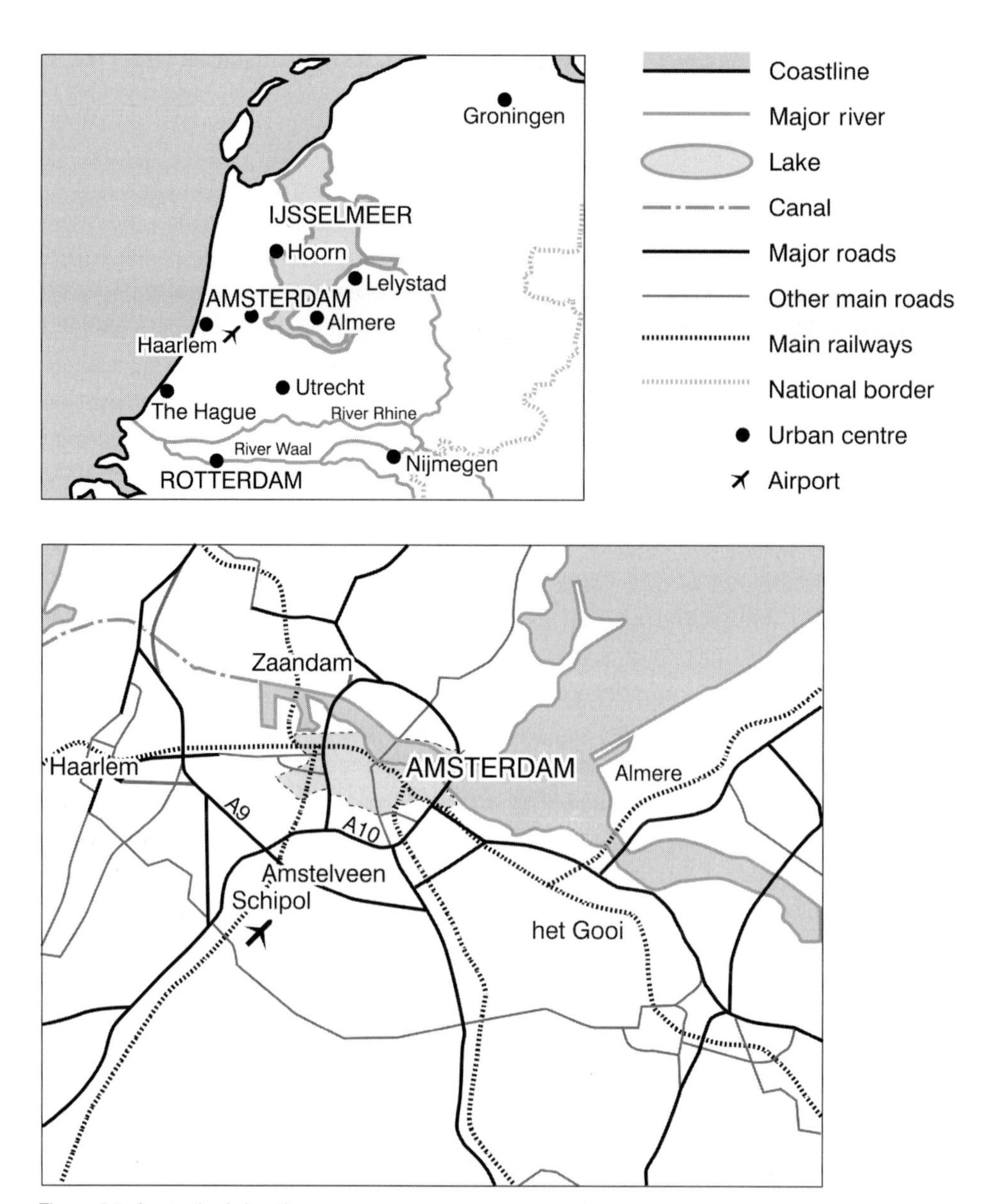

Figure 3.1 Amsterdam's location

Planning in Amsterdam at this time was a major city government priority. Underpinning this planning effort was a political context promoting good living conditions for the working classes. This encouraged substantial state intervention to promote better housing conditions and improved public transport.[7] The city was at the forefront of developing the infrastructure of a 'welfare society' which eventually underpinned the political nexus that dominated the Netherlands for most of the second part of the twentieth century in a 'co-sociationist' or 'corporatist' political relation between the unions, the state and industry

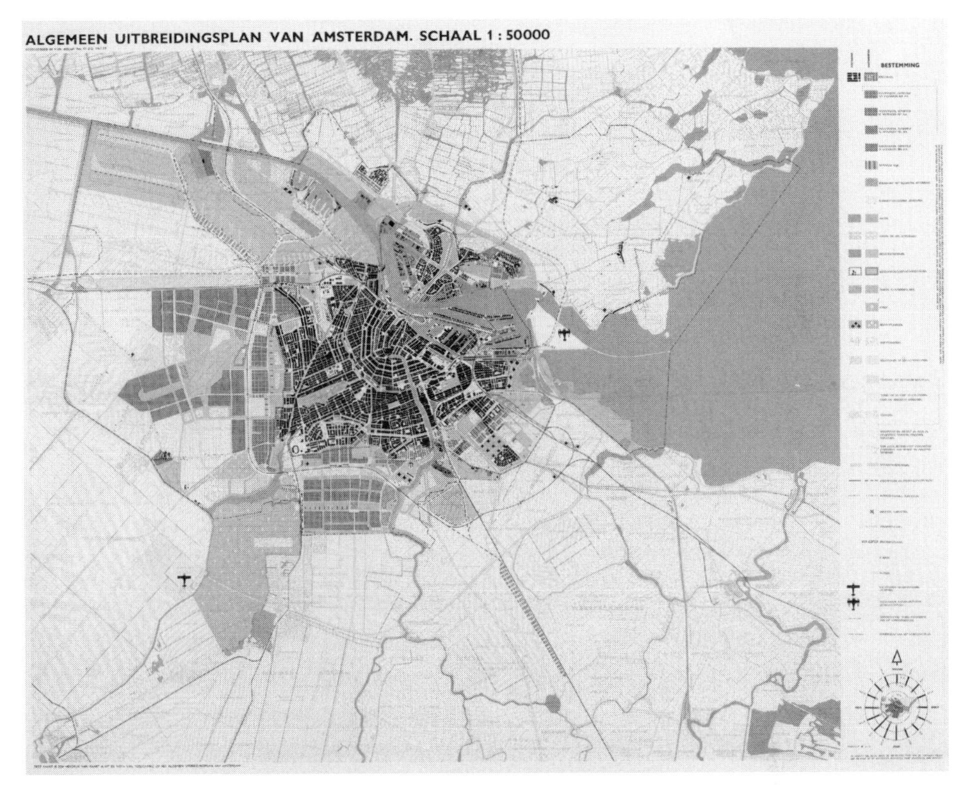

Figure 3.2 The 1935 *General Extension Plan*

Source: Amsterdam City Council, with permission

(Mak 2003; Wagenaar 2003).[8] Yet even in the 1930s, richer families were beginning to move to surrounding settlements, particularly areas of sand dune landscape to the west and the water landscapes of *het Gooi* to the east (Schmal 2003).

STATE-FUNDED URBAN DEVELOPMENT

The implementation of the 1935 *General Extension Plan* was largely left until after the Second World War.[9] Influential ideas about post-war national and urban spatial development were actively developed during the wartime period, as elsewhere in Europe. A national spatial plan was never produced as such, but the Ministry of Housing and Spatial Planning that emerged after the war became a major player in government, shaping investment in physical infrastructure and housing. This national planning effort injected important strategic ideas into Dutch planning, most notably the concept of the '*Randstad*'. Used in planning discourse from the 1940s (Faludi and van der Valk 1994), and celebrated in the 1960s by Peter Hall (1966), this morphological concept identified both the ring of cities in which Dutch urbanisation was concentrated, and the threat that continued urban extension presented to the traditional Dutch polder landscape in the areas within the ring, the so-called 'greenheart'. Protecting the greenheart and resisting

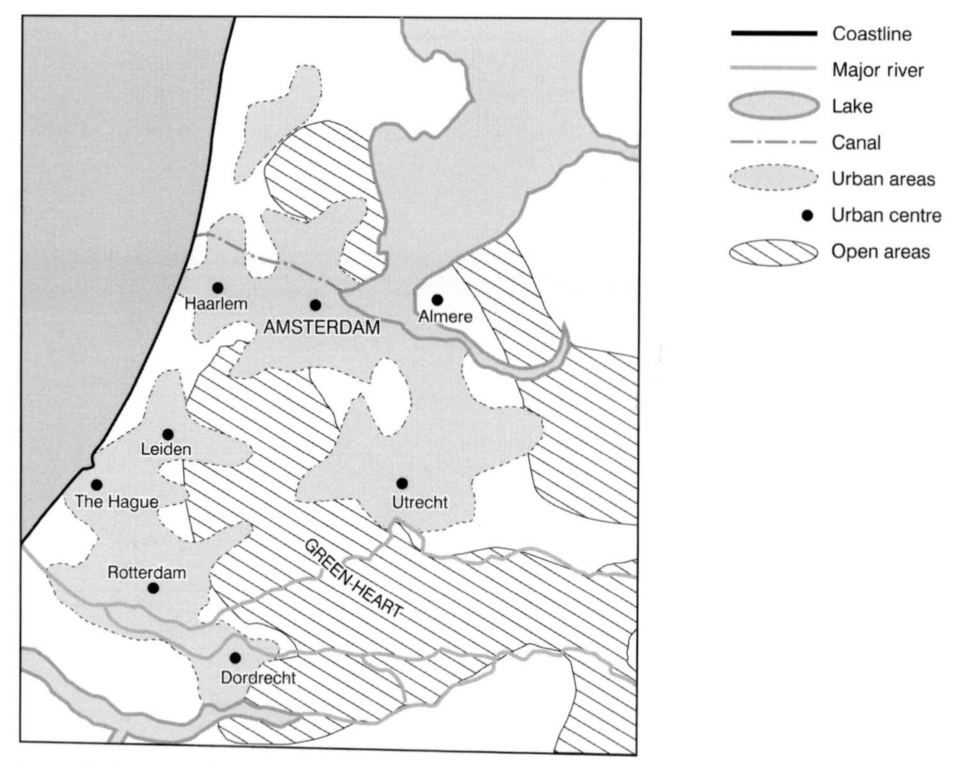

Figure 3.3 The Dutch *Randstad*

Source: adapted from Hall 1984: 113

sprawl began to take on the quality of a hegemonic policy discourse, or doctrine, as Andreas Faludi and Arnold Van der Valk call it (1994).

Amsterdam continued its extension plans in the 1950s, though with some delays due to the difficulty of funding and deciding locations for major river crossings. The national economy at this time was weak, finances were limited and the main emphasis was on building rental housing to meet a large and growing housing deficit. Amsterdam concentrated on providing rental housing areas, emphasising the amenities of local living environments. The main area of housing expansion was in the west (in the Slotermeer area), as envisaged in the 1935 *General Extension Plan*, and the city's boundaries were expanded to incorporate this area. But these developments slowly began to meet the expanding airport in the municipality of Haarlemmermeer, owned until 1958 by the Amsterdam City Council (Ploeger 2004). Because household size was falling, implying that larger numbers of new dwellings were needed than anticipated in the 1930s, the City Council sought to expand north of the river, with two tunnels completed in the 1960s,[10] and to the south. A *structuurplan* to guide urban extension to the north was agreed in 1958 (Jolles *et al.* 2003). Through these developments, Amsterdam became a major owner of subsidised rental properties. The main economic focus of both the City Council and the national government was on developing the city's industrial activity around its port area. Amsterdam's commercial and financial role was given much less attention (Ploeger 2004). However, the City Council produced a policy note on the city centre (the 1955 *Nota Binnenstad*). This affirmed a monocentric image of the city, with the city centre as an economic and cultural core. Most of the historic centre was to be retained, but with redevelopment of war-damaged areas in the eastern part of the city centre for business activities, including major road improvements. These investment projects would incite major protests in the following decade (Jolles *et al.* 2003; Ploeger 2004).

In all these developments, national funding for urban development played a critical role. The Netherlands is described as a 'unitary, decentralised state' (Needham *et al.* 1993). This means that it is legislatively and fiscally centralised (Terhorst and Van de Ven 1995), but that the other levels of government, the provinces and the municipalities, co-operate with the national level in developing policy and discussing how funding should flow.[11] In Amsterdam's case, its size and political weight has always given it a strong voice in national-level discussions. The province of *Noord-Holland*, in which Amsterdam is situated, has limited competences and financial resources of its own, and is forced to play a largely coordinative role. For its urban development activities, Amsterdam City Council, in contrast, was able to obtain substantial financial resources for major projects from the national government, justified by the policy of keeping housing costs, and hence labour costs, low.[12] The council could also raise funds through acting as the major developer of urban extension sites which, until the 1960s, could be incorporated into the City Council boundaries.

However, as the national economy began to pick up during the 1950s, there were national concerns about the concentration of development in the delta area of the western Netherlands. This led to a national strategy not only for 'deconcentrated concentration' of development pressure in discrete new towns rather than by continual urban extension, but also to a focus on dispersing development to the north and east of the country (Faludi and van der Valk 1994; Kreukels 2003). Amsterdam City Council resisted this policy for a while, preferring urban extension. It was then realised that more housing areas were needed to accommodate households displaced by inner-area redevelopment projects. Meanwhile, more affluent residents continued to move out. An Amsterdam sub-region was beginning to appear as a physical presence, as well as in economic and social networks.

URBAN EXTENSION REACHES ITS LIMITS: THE 1960S AND 1970S

A POWERFUL SPATIAL PLANNING REGIME

The 1960s was a period of major publicly promoted urban development projects across Amsterdam and in the surrounding region. This investment was both driven and demanded by the upturn in the Dutch economy, linked to the general growth dynamic of the European and global economies in the 1960s. This was a period in the Netherlands, as elsewhere in North-West Europe, of the institutionalisation of the delivery machinery of welfare states. Stable political regimes at national and city level provided the context within which the organisation of development activity was consolidated into large, professionalised government departments. The 'technical' knowledge of professionalised policy communities was in the ascendant. Urban development and spatial planning activity was centred in the Ministry of Housing and Spatial Planning (*Ministerie van Volkhuisvesting en Ruimtelijke Ordening, VRO*, later *VROM*).[13] The Ministry of Transport and Water Management (*Ministerie van Verkeer en Waterstaat*) managed major engineering infrastructures[14] and the Ministry of Agriculture dealt with the production of agricultural land and landscape planning in rural areas. Over time, each developed its own policy communities and cultures, and its own training and research institutions through links with particular universities.[15] *VRO/VROM*'s capacity to shape spatial development was always in tension with the transport and water-policy communities. The planning and agriculture ministries shared a common interest in the importance of controlling development. *VRO/VROM* focused on urban policy centred on housing delivery ('red' activities) and the Ministry of Agriculture on rural development ('green' activities).[16] Among planners, ideas of urban systems influenced analyses and concepts of 'rational process' began to shape strategic planning approaches.[17] In the language of regulation theory referred to in Chapter 2, the Netherlands at this time could be seen as a leading exemplar of a managed, welfarist mode of regulation. However, this was driven by cultural and political forces as much as by the demands of the economy.

In contrast to the UK, where spatial planning was always in tension with major sectoral policy programmes at national and local level (see Chapter 5), in the Netherlands it could get leverage over other departments through the strong linkage between planning and land development activity and a governance culture that continually sought co-alignment of programmes and projects, despite divisions between ministerial departments. Planners in government agencies made strategies, produced master plans and created urban and rural landscapes. The resultant policies and projects emerged from interactive policy processes involving all three levels of government (Dijkink 1995). As a result, strategic plans and planning concepts had substantial material effects on the physical environment. They helped to shape the locations of development and the routes of the infrastructures that other departments then brought to realisation. This was not just a result of formal legal requirements demanding conformity between the projects of other departments and spatial plans. It was also the result of co-alignment in government practices. Through intensive consultation processes, within and between the technical policy communities and among politicians,[18] strategic agreements were reached that focused attention and shaped agendas of projects, of funding and the criteria embodied in regulatory processes.

These agreements and the framing concepts that shaped them were consolidated in planning instruments at the three levels of government (see Table 3.1). At the national

Table 3.1 Chronology of formal plans and major policy notes: Netherlands, Noord-Holland and Amsterdam

	Netherlands	*Noord-Holland Province*	*Amsterdam City Council*
1950s	1958 Advisory Report on the *Randstad*	*Streekplannen* for various parts of the Province	1955: *Nota Binnenstad* 1958: *Structuurplan* Amsterdam-Noord
1960s	1960 First *Nota* on spatial planning 1966 *Tweede Nota Ruimtelijke Ordening*	1968 *Streekplan* for area around *Noordzeekanaal* (excluding Amsterdam)	1965: *Structuurplan* Amsterdam Zuid and Zuid-Ooost 1968: *Nota Binnenstad*
1970s	1973 *Derde Nota, et al.*	1979 First *Streekplan* for area around *Noordzeekanaal*	1974: *Structuurplan*: parts A and B
1980s	1983/85: *Structuurshets Stedelijke Gebeiden* 1988: *Vierde Nota*	1987: Amsterdam – *Noordzee Kanaal Streekplan*	1981: *Structuurplan* Part C 1985: *Structuurplan* (*De stad centraal*)
1990s	1992: *Vierde Nota Extra* (*Vinex*) 1999: new *Vijfde Nota* initiated	1991 *Streekplan* 1995 *Streekplan*	1996: *Structuurplan* (*Amsterdam Open Stad*)
2000s	2000: *Vijfde Nota* approved by Cabinet 2002: *Vijfde Nota* withdrawn 2005: *Nota Ruimte* approved	2003: *Noord-Holland Zuid Streekplan* (Feb.)	2003: *Structuurplan* (*Kiezen voor Stedelijkheid*) (April)

level were national reports, called *Nota.*[19] First produced in 1960, the *Tweede Nota Ruimtelijke Ordening* (*Second Report on Spatial Planning*) of 1966 emphasised accommodating housing growth, while avoiding sprawl through 'deconcentrated concentration', and the dispersal of growth pressures away from the congested western core zone of the country (Faludi and van der Valk 1994).

PROJECTS AND PLANS IN THE 1960S

Amsterdam City Council's development activity at this time was centred in the large and influential *Dienst Publieke Werken* (Department of Public Works), which combined managing the city's large land and property stock, with development projects, strategic planning and land-use regulation. Housing and industrial development in the west and north proceeded largely according to plans of the 1930s and 1950s. During the 1960s, the focus of attention for expansion shifted to the south-east, where a new large finger of development projected into the Bijlmermeer area. The site of the proposed southern ring road, the A10, was also defined as it went through these areas (see Figure 3.4). Amsterdam City thus continued with its urban extension strategy. It also continued to promote city-centre redevelopment, but this became increasingly controversial. Within the Social Democratic party on the council, there was a split between those advocating the commercial role of the city centre and those emphasising liveability. Meanwhile, in the context of significant expansion in the commercial economy, firms were finding it difficult to acquire city-centre property and there were signs of a move to accessible sites on the periphery. The 1968 *Nota Binnenstad* (City Centre Policy Note) represented something of a compromise between these two views, with a rapid transit underground project serving to increase the disruption generated by redevelopment in the eastern part of the city centre (Ploeger 2004).

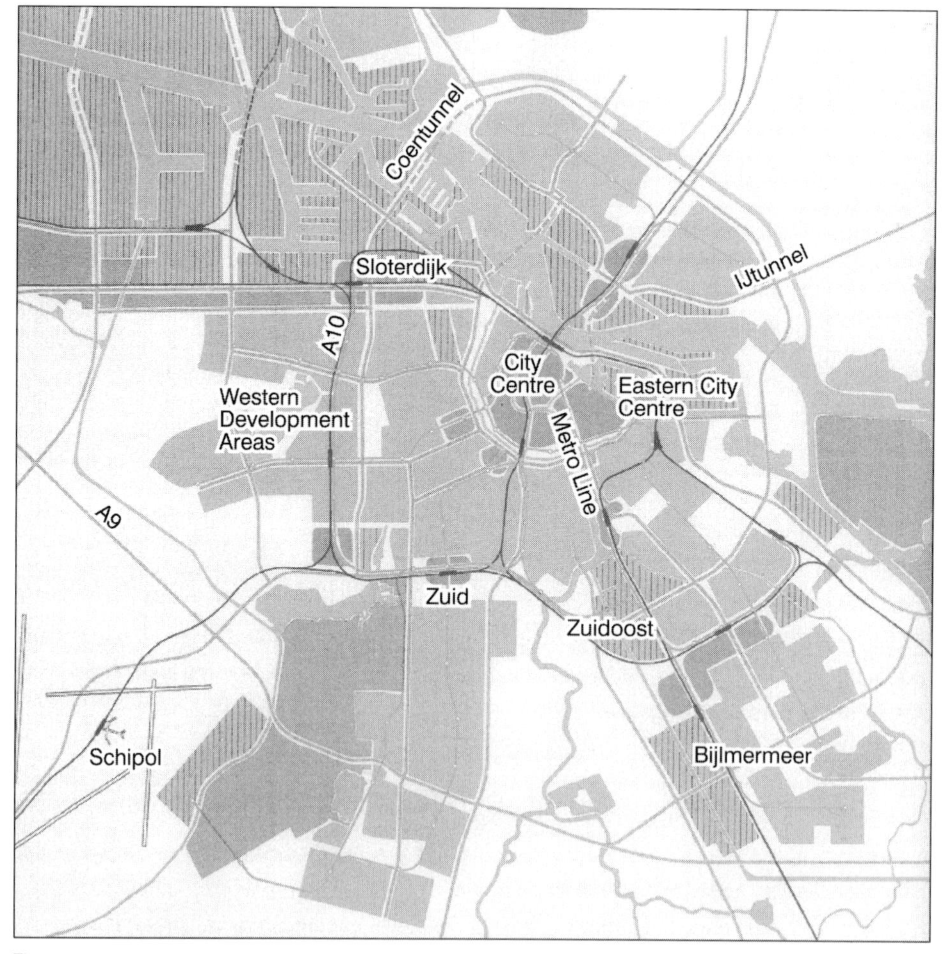

Figure 3.4 Major urban developments and infrastructure projects in the 1960s and 1970s

Source: The base sketch is from Jolles *et al.* 2003: 120, with permission of Amsterdam City Council

But the national funding for housing development was increasingly targeted on the new town growth centres outside the big cities (Needham *et al.* 1993, page 36). The new towns were promoted in the *Tweede Nota* (1966) and revised in the *Derde Nota*, produced in parts from 1973.[20] Within the City Council, the major shift in policy was from proposed redevelopment to an emphasis on housing renewal and public transport improvements, particularly the metro line through the east of the city centre. Overall, the strategic planning frame in Amsterdam continued to portray a centralised city, revolving around its central core, but with the recognition that some subsidiary centres would evolve. The most dramatic such centre, Schiphol airport, gets little mention in Amsterdam policy statements because it lay beyond the municipal boundary.[21] Yet, by the 1970s, it had become a major regional employment node. Thus a new, polycentric spatial structure for the Amsterdam area was beginning to emerge, both in daily life movements and in policy concepts. But it took time to articulate this in terms of a spatial strategy.

In 1965, a new national planning act consolidated the formal tools of the planning system: advisory *Streekplans* (strategic plans) at provincial level, and advisory *Structuurplans* at municipal level, for the whole or part of a municipality, implemented through legally binding *Bestemmingsplannen* ('destination' plans) which allocated land and property development 'destinations' and rights. *Bestemmingsplans* are formal zoning instruments, required for any development on non-urban land and are often also used for remodelling projects within the urban fabric (Needham *et al.* 1993). At the national level, legally binding Key Planning Decisions were introduced in the 1970s. These formal instruments were supplemented by the less-formalised mechanism of 'policy statements', the *Nota*, which were often as important as the formal plans and 'key decisions'. The influence of both formal and informal statements was underpinned by the vertical and horizontal co-alignment processes through which they evolved. These enabled a close link between proposals for development projects that emerged in the plans and the allocation of resources, primarily from national government. The statements in essence carried forward the framing ideas and agreements reached in these processes into mechanisms for coordinating and integrating project development and delivery.

Amsterdam City Council continued as a major actor in these urban development processes and in the arenas that shaped national and provincial development policy. The focus on urban extension continued into the 1960s, with a *structuurplan* for the *Zuidoost* (*Bijlmermeer*) area agreed in 1965. However, expansion directly to the south met opposition from the neighbouring municipality of Amstelveen, which, with national government support, resisted any proposal for annexation. Within its own jurisdiction, City Council authority suffered some serious political shocks during the 1960s. As a socialist, welfare city, the dominant politicians within the social democratic party had assumed that delivering low cost, primarily rented, housing and well-designed neighbourhoods was well-supported by citizens. Urban renewal had been envisaged as an attempt to upgrade city-centre neighbourhoods to similar standards. But this, along with schemes to accommodate road traffic, and the feared disruption resulting from the east metro line, provided the fuel for local conflict. This soon linked to wider urban social movement resources and led to substantial protest.[22] Amsterdam became one of the iconic sites of late 1960s/early 1970s European urban protest (Mayer 2000). The memory of protest and direct action such as squatting remains in the 2000s.

This protest partly challenged the conception of the city centre as a commercial core. In contrast, neighbourhood protest centred on an agenda that would become the planning fashion of the 1990s – mixed-use neighbourhoods, reduction in road space, retention of the historic fabric, and encouraging space for diverse lifestyles and 'scenes'. The protest was carried into the City Council through elections that strengthened the 'new left' faction within the Social Democratic Party. This led to two significant shifts in urban development policy. First, the council turned away from large-scale development, towards smaller-scale, urban renewal projects. In this it was encouraged by similar changes in national policy (Jolles *et al.* 2003; Ploeger 2004). Second, the council increasingly realised that the city's economy was based more on commercial and financial services than on industrial activity.[23] The concept of a 'multifunctional and varied historic inner city' was emphasised in the parts of a new city structure plan approved in 1974. But this shift disturbed the Chamber of Commerce which demanded more attention to the space needs of the country's national commercial and financial centre. Market processes were already generating a ring of subcentres on the peripheral transport ring. Some of these centres were indicated for the first time in the 1974 structure plan allocations. This emerging 'polynuclear' form for the city was strengthened by other City Council projects from the mid-1970s onwards. The *structuurplannen* of the 1980s and 1990s, however, were slow to give status to these new centres, incorporating them initially into a concept of a hierarchy of centres, focused on the core city centre (Ploeger 2004).

URBAN PROTEST AND THE RESPONSE

The protests, led by local activists including many young 'intellectuals' living in some of the older parts of the city, also challenged the priorities and practices of the City Council, and political and technical paternalism in particular. Following a further strengthening of 'new left' influences in the City Council in 1978, three innovations were launched. The first was the introduction of a 'process protocol', called the *plaberum*, a form of local co-alignment process. This was a seven-step procedure, still in use today, which defines the process through which any planning policy or project has to proceed before City Council approval. It specifies which parties need to be involved at different stages of the policy and project formation. The parties include other Amsterdam City Council Departments as well as residents and other stakeholders, providing a 'clear-cut, traceable process' for all concerned (Jolles *et al.* 2003). The second innovation was to break up the council area into districts. This started slowly, with the area north of the IJ river becoming the most significant example. Finally, the powerful City Public Works

Department was abolished, and divided into three parts: *Dienst Grondbedrijf* (real estate); *Dienst Openbare Werken* (Department of Public Works); and *Dienst Ruimtelijke Ordening* (Department of Physical Planning). Despite the local co-alignment mechanisms, this division created potential coordination problems that emerged later in the century.[24]

The comprehensive approach to city spatial planning faltered during this period. Influenced by ideas of planning as a process, and in parallel with the example of the *Derde Nota* (the third National Policy Report), a series of thematic sections of what was intended as an integrated plan were produced instead, though the series was never completed (Faludi and van der Valk 1994; Jolles *et al.* 2003). By the 1970s, in any case, the economic and social dynamics of the Amsterdam area were changing. As an industrial complex, the city was in decline, especially in the employment sector. Industrial restructuring processes across Europe were beginning to be felt, along with technological developments in water transport. The big industries along the waterfronts of the IJ and the *Noordzeekanaal* were hit. Meanwhile, the city's old commercial and financial role resurfaced as the major economic nexus of the city, together with cultural activities and tourism. In addition, the public sector had also expanded dramatically.

This sustained the relative economic prosperity of the city and its attraction for immigrants, particularly those from Surinam, Turkey and Morocco. It also led to a shifting social composition of the city in other ways. The more affluent middle-class families continued a steady dispersal from the city, in search of home-ownership and a more suburban lifestyle. The population of the city fell, while the relative numbers of unemployed and of recent immigrants increased. The latter found housing in some of the new housing schemes of the 1960s. Intended as socially 'balanced' neighbourhoods, some of these became home to concentrations of the poorest households (Cortie 2003). By the mid-1970s, the Bijlmermeer housing project initiated in the 1960s had acquired a label as a locale of the poor and deprived, while poorer families and the homeless were also concentrated in the inner-city core and, to an extent, in the newly built western suburbs. Overall, the social mix within the Amsterdam City Council area was increasingly differentiated from the areas around it. Meanwhile, new centres of business expansion were appearing not only at Schiphol, but also in the other nodes in and beyond the city boundaries, especially along the ring roads to the south and west. The economic and social webs of relations transecting the city were thus changing the area's internal geography and external linkages. The confident city builders of the 1960s were faced with the prospect of the city's urban core being no longer the unchallenged centre of the city but instead a focus of 'urban problems'.

The City Council could no longer respond to this new context by urban extension and encountered conflicts if it sought to increase the density of activities in the inner core (Terhorst and Van de Ven 1995). Its land opportunities for expansion were increasingly limited – by the growth of the airport and its ever-tightening 'noise contour' regulations to the west; the protection of wetland landscapes to the north and the 'Green

Heart' to the south; the increasingly stringent regulations related to hazardous installations in its remaining harbour areas, and by the opposition of affluent municipalities to the south. To the east, the new town of Almere was developing a confident sense of itself, while the area to the north of the river remained rather inaccessible to the rest of the city. In any case, peripheral expansion threatened still further the conditions in the city centre. The situation was made even more difficult by funding limitations, as the national exchequer suffered in the European recession of the late 1970s/early 1980s. As a result of these changing political relations, City Council strategists began to look both inward, to remodelling the city, and outward, to building support among surrounding municipalities for a collective approach to the 'Amsterdam region'. In 1972, the *Informele Agglomeratieoverleg* had been formed, involving 25 municipalities in the province of *Noord-Holland* and in neighbouring *Flevoland*, where Almere was located. Within the city boundaries, several redevelopment locales and new development nodes, including the idea of land reclamation in the IJ lake, began to appear in policy debates and statements, along with ideas for neighbourhood improvement. Heading into the 1980s, the City Council was fighting to ensure that urban renewal and urban issues retained a strong position in national government policy and funding, while its development strategies were shifting to reflect new ways of thinking about the spatial organisation of the city, new relations between the state and citizens, and, by the 1990s, new relations between the public and private sectors.

THE EMERGENCE OF AMSTERDAM AS A POLYCENTRIC URBAN REGION: 1980–1996

REINFORCING CENTRALITY

By 1980, the Dutch co-sociational/corporatist model of multi-level governance was well-established. National government played a key role in urban development through funding for land development, for housing provision and for physical infrastructures. Since the 1960s, the objective had been the 'universal' development of the whole national territory, and hence the dispersal of development impulses towards the less dynamic north and east of the country (Faludi and van der Valk 1994). By 1980, these expenditures were producing substantial financial strain on the national economy. This crisis was thrown into sharp relief by the economic recession (Terhorst and Van de Ven 1995). Industrial employment fell rapidly, leading to rising unemployment among the workforce of the four Dutch 'big cities' (Amsterdam, Rotterdam, The Hague and Utrecht).

The increasing problems of the 'big cities' (the *Grotesteden*), and of the economy generally, challenged earlier national spatial concepts of equal development of the national territory and dispersal of economic development and urbanisation away from the core western Netherlands. Instead, economic interests promoted the dynamic western area as the key to national economic strength. In turn this led to particular attention to the economic role of the Netherlands with respect to international movements in logistics, commerce and finance, in contrast to the earlier emphasis on manufacturing industry. This economic impetus encouraged national strategies to develop the country's 'mainports' – Rotterdam harbour and Schiphol airport in particular. In parallel, the 'big cities', faced with increasing social and environmental problems in the inner cores, combined to lobby for more attention for their special difficulties. Not surprisingly, Amsterdam City Council was a major voice in the promotion and maintenance of these policies. National government had provided funding for urban renewal since the 1970s, but in 1985 legislation was introduced that allowed funds to be transferred directly to municipalities and provinces. Most of this funding went to the big cities, along with 85 other municipalities. This encouraged cities to focus their attention more strategically on renewing cities, rather than merely renewing neighbourhoods (Needham *et al.* 1993). Nationally, however, it meant a significant break with the idea of spreading development across the country, as pursued in previous decades. Within the big cities, attention shifted from a focus on building new housing areas on city margins to reviving inner cores.

In Amsterdam, with a new, more radical council elected in 1978, the emphasis was on building more rented housing, improving conditions in the inner neighbourhoods and resisting expansion beyond the city's boundaries.[25] Older industrial areas along waterfronts were gradually re-imagined as residential areas, while new industrial estates were located in peripheral areas. Spatial development policy asserted a new model of a more 'compact' city, focused around the central core (de Roo 2003; DRO 1994), a concept strongly backed by left-wing alderman, Van der Vlis, in charge of the spatial planning portfolio from 1978–1988 (Ploeger 2004). Despite the emerging polycentric organisation of the urban area, city-planning strategy focused on restoring the city centre's centrality, as well as continuing with housing extensions in the western suburbs. The 1985 *Structuurplan* (DRO 1985), called '*Stad centraal*' ('the city in the centre') (Ploeger 2004) drew on the theme plans of the 1970s, but attempted a new, integrated expression of the city's spatial order (see Figure 3.5). The primary focus was on increasing housing supply and on neighbourhood improvement. New housing opportunities could still be provided in the western suburbs, so long as the 'environmental contours' around Schiphol airport could be maintained. Regeneration in the city centre was another opportunity for increasing housing supply. A third proposal, developing since the 1970s, was to extend the city centre eastwards onto former harbour lands and reclaimed land to the north of the city centre along the southern banks of the IJ river. The result was a conception of the city as focused around its traditional core, but with an east–west development axis stretching through the city centre. This became known by the end of the decade as the 'IJ axis'. The city centre was to be conserved in its existing fabric, with additional space for commercial development to be found in development lobes, interspersed with green lobes stretching out to the 'greenheart' of the *Randstad*. The connectivities between the parts of the city were to be supported by a tram network

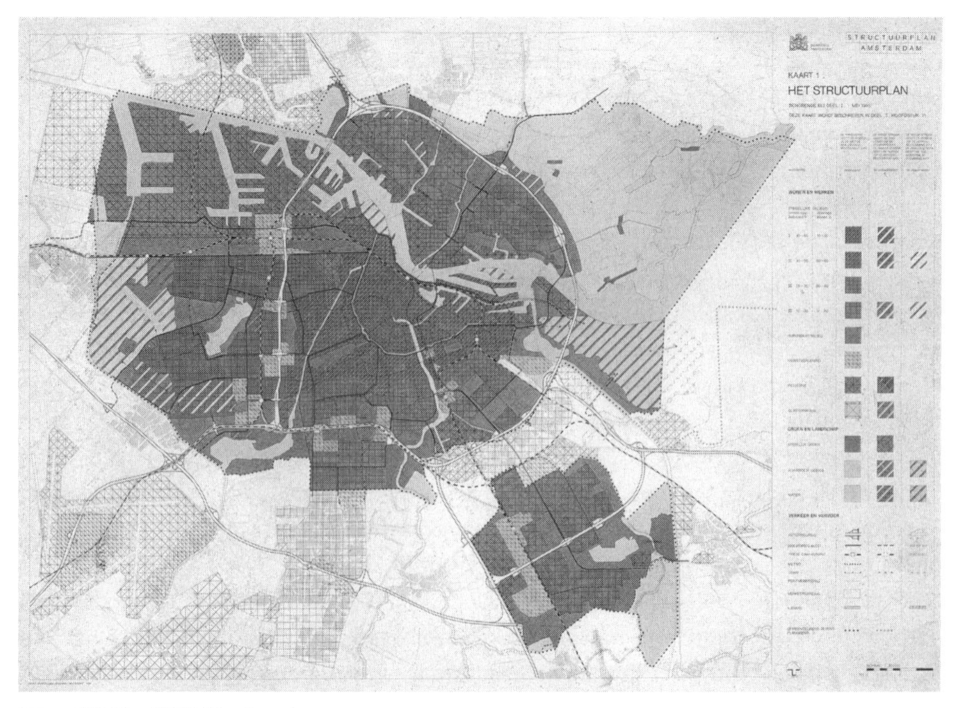

Figure 3.5 The 1985 *Structuurplan*

Source: From Jolles *et al.* 2003: 96/97, with permission of Amsterdam City Council

centred on the city centre (Jolles *et al.* 2003; Ploeger 2004). The proposal for an east metro line was dropped. As in all previous plans and policy statements, there was a strong emphasis on the quality of new development in neighbourhoods, together with the provision of local amenities and green spaces. Considerations of environmental quality and sustainability added new dimensions to conceptions of urban neighbourhood quality.

This plan thus played down the emerging polycentric reality of the urban region, and was criticised for this by academics and some business and labour interests (Terhorst and Van de Ven 1995). However, remaining city-centre business and property-owning lobbies supported the approach, a main purpose of which was to lever in national investment for urbanisation projects. The plan also played to residents' concerns about local liveability (Terhorst and Van de Ven 1995). The *structuurplan* was also well-aligned within the emerging spatial planning discourse at national level in the *Vierde Nota* (1988/9), in Province *Streekplanennen*, and in the Amsterdam *Structuurplan* of 1985. The emphasis shifted definitively, from dispersal to 'urban compactness', to stop

> the forces of the city draining away by encouraging them to remain in the metropolitan districts. The idea was to situate housing construction and industries within the city and along the edge. The outlying areas were cordoned off with 'green buffers' and urban expansion was prohibited beyond these limits (Salet 2003: 180).

The policy shift embodied in the 1985 plan responded to Amsterdam City's concern to resist the undermining of its social and economic resources by development in surrounding municipalities, and to acknowledge the increasing salience of environmental concerns in national and local politics.

At national level, the *Vierde Nota* reinforced the *Randstad* concept as a spatial-organising idea, despite some criticisms (Dieleman and Musterd 1992; Faludi and van der Valk 1994). By this time, national policy emphasised the importance of the Randstad area to the 'competitive position' of the Netherlands economy (Ploeger 2004; van Engelsdorp Gastelaars 2003). The *Vierde Nota* called for redevelopment at higher densities in the major cities, and encouraged public–private partnerships in urban development projects, particularly urban revitalisation schemes and major infrastructure projects. Amsterdam took the lead in pioneering such arrangements. This policy stance was accompanied by the increasing decentralisation of national investment funds, through the system of development agreements signed with specific municipal departments (de Roo 2003; Kickert 2003). In the spatial planning area, these agreements were linked to the specification of major projects in the relevant *streekplannen* and *structuurplannen*. Special arrangements (*ROM*) were set up for major development areas of national significance, particularly the 'mainports' of Schiphol and Rotterdam (de Roo 2003). The *Vierde Nota* also signalled a major shift in housing policy, with stronger encouragement for more diversity in housing tenure, and a big expansion of owner-occupation. This challenged Amsterdam's emphasis on social rented housing. By 1987, 50 per cent of the city's stock was in such tenure.[26] The city's policy was already under pressure as more affluent citizens continued to leave. By the late 1980s, City Council policy shifted to encourage more privately-developed housing for owner-occupation in the city centre. This 'privatisation' implied, over time, a weakening of the City Council's direct role as a property owner and developer, though it still retains its land-owning role even today.

RECOGNISING REGIONAL DEPENDENCIES

There can be no doubt of the power of the 'big city' lobby in shaping national planning and development policy in the 1980s through the focus on urban renewal. But the policy also generated tensions, given the 'fiscal crisis' faced by the Dutch exchequer. Demands for public expenditure were rising exponentially without the prospects of tax returns to support them. One response, as elsewhere in Europe, was to get the private sector more involved in development funding through public–private agreements and partnerships, complementing the embedded practice of multi-level government agreements.[27] In Amsterdam, this meant that subsidies for housing development were linked to targets for private housing provision rather than just rental housing alone.[28] It also meant that the city could raise funds by leasing sites it owned for big redevelopment projects (Terhorst and Van de Ven 1995). But city officials had only limited experience of working with private-sector partners and were not accustomed to understanding the city's development in terms of market values and dynamics.

Another response was to create new formal arenas for intermunicipal co-operation in metropolitan areas (Dijkink, 1995). The political forces in the major cities combined to produce a new strategy for the promotion of the *Grotesteden*, the 'big cities'. This led to the Montijn Commission, which argued in 1989 for the importance of the cities in developing the country's international profile (Salet 2003).[29] Municipalities in metropolitan areas were encouraged to co-operate in apportioning national funds for urban development and renewal through the formal 'agreement' system. The *Informele Agglomeratieoverleg* already existed in the Amsterdam area as an informal 'platform' for interregional co-operation. In 1986, this was consolidated into the *Regional Overleg/Orgaan Amsterdam* (*ROA*), consisting initially of 23 municipalities in *Noord-Holland* and Almere in *Flevoland*.[30] In 1987, the Province of *Noord-Holland* and the City of Amsterdam made an agreement that gave the city more power, particularly in the preparation and approval of *structuurplannen* for Amsterdam.

The Montijn Commission proposed the creation of a formal, directly elected regional organisation to take over strategic planning, transport, housing allocation, economic and environmental policy from the municipalities in metropolitan areas, and take over the strategic coordination role of the province[31] (Alexander 2002). Amsterdam City Council envisaged re-scaling itself into this *Regional Orgaan*, while at the same time decentralising many delivery functions to the districts within the city. These would then have the same status as the other municipalities in the region. The city centre would remain, however, as the responsibility of the re-scaled Amsterdam *Regional Overleg/Orgaan*. In this way, it was hoped to pursue policies of 'equalisation' between the richer and poorer areas of the expanding metropolitan area and at the same time encourage 'collaboration' in building the capacity to compete in national and international arenas.

However, this strategic transformation was never completed. The system of districts envisaged in 1980 slowly became city-wide, though it took time for these to establish significant powers and political identities (Dijkink and Mamadouh 2003). The surrounding municipalities were ambiguous in their support for the *Regional Orgaan*, some withdrawing from the arena, including the major development area of Almere.[32] Amsterdam's mayor became disenchanted with the weakened proposal that emerged. Finally, in 1995, the citizens of Amsterdam were asked to vote on the proposals and rejected them, in a defence of the significance of *Gemeente Amsterdam* as a collective voice and identity for the city. The *Regional Orgaan Amsterdam* survived as a forum, but from this time on, a metropolitan region approach had to be pursued through less formal channels (Salet and Gualini 2003).

From the late 1980s, there was a rapid evolution of both spatial organising ideas and mechanisms for decentralised investment coordination. In relation to spatial organising ideas, the notion of a physically 'compact' city was being challenged by concepts of networks and flows, and by recognition of the 'polycentrality' of urban agglomerations. In 1991, the City Council produced a revised *Structuurplan*. This largely followed the 1985 plan, but with an increasing emphasis on green spaces and landscape. The metro line proposal from the city centre was revived, on an alignment further to the west, creating a new focus of proposed development activity along the route.[33] However, housing expansion to the west on greenfield sites had to be reduced, due to the widening of the Schiphol airport noise contours. This removed from the City Council's real estate portfolio sites that were easy to develop and hence able to generate finance for the city's other development activities. Ambitious ideas for the further development of the IJ river banks were promoted, but, since agreement on ways forward had not been reached, these do not feature in the revised plan (DRO 2003a). At national level, the economic emphasis of *Vierde Nota* was overtaken by a new environmentalist philosophy driven by national politicians and the planning community.[34] This led in 1992 to the *Vierde Nota Extra* (*VINEX*), which is recognised in the Netherlands as the high point of the influence of environmentally focused spatial planning concepts. This confirmed 11 special planning areas (*ROM*) and other key national projects, linked to investment funding agreements, as well as maintaining the system of programme-funding agreements with the provinces and municipalities (Needham and Zwanniken 1997). The *VINEX* also introduced the famous 'ABC' classification of employment centres in terms of transport accessibility requirements.[35] Meanwhile, the preparation of a *Streekplan* for the metropolitan area began, formally by the Province, but informally under the aegis of the *Regional Orgaan Amsterdam*. Finalised in 1995, this focused on transport issues and housing locations, aiming to arrive at priority locations and projects which would then attract investment through the 'agreement' system.

Thus, by the early 1990s, urban development investment was being pursued through processes emphasising horizontal coordination and co-alignment, rather than vertical co-alignment. There was also more attention to involving the private sector in development projects. But these arrangements still operated within networks of the well-established planning policy community,[36] and there were significant tensions between the different policy communities (Needham 2005; van Duinen 2004). Amsterdam continued to have a good reputation for its consultation and co-alignment processes, but coordination problems developed as urban regeneration activities were departmentally separated from physical planning and Districts began to assert their powers over the local *bestemmingsplannen*. The City Council proceeded to prepare a further *Structuurplan*. The instrument of the plan, rather than shaping investment, was taking on a function as a rolling policy mechanism, consolidating and legitimating project proposals. New private housing projects were now emerging in the city centre, but the development of the IJ river banks was encountering difficulties. Citizens had rejected a flamboyant Rem Koolhaas scheme for the waterfront, development costs were proving higher than expected, and the private sector showed little interest in locating commercial activities there (de Roo 2003). Budget losses were predicted on several other development sites in the city. In contrast, the emerging 'southern axis' (*Zuid As*) along the A10 was attracting major commercial investments (notably the ABM/AMRO bank at the Zuid station[37]).

The City Council saw that some city-owned sites there could generate substantial financial gains if developed for commercial purposes. Thus two development axes began to emerge as spatial conceptions. By the early 1990s, the city's Economic Development Department was strongly promoting the idea of a southern development axis. In the 1996 *Structuurplan*, entitled '*Open Stad*' ('Open City') (DRO 1996), the axes were pulled back into a concept of major nodes around the city, with city-centre qualities of mixed development, defined as a category of '*perifeer centrummilieu*' ('peripheral city centre space') (Figure 3.6). Although city planners argued that the 1996 *Structuurplan* was merely an update of the 1985 plan (Jolles *et al.* 2003), the emergent polynuclear reality of the city was now clearly acknowledged (Ploeger 2004).

THE PLANNING TRADITION UNDER CHALLENGE

As national investment in urban development and infrastructure shifted from long-term, large-scale programmes to more specific targeted programmes and projects, with more substantial private-sector input, so the traditionally tight nexus between planning strategies, land and property development projects and infrastructure investments began to unravel. Rather than shaping the locations, forms and agreements through which development took place, the formal *structuurplannen* and *streekplannen* appeared to follow rather than shape the investments. The formal plan had become a conclusion of a negotiative

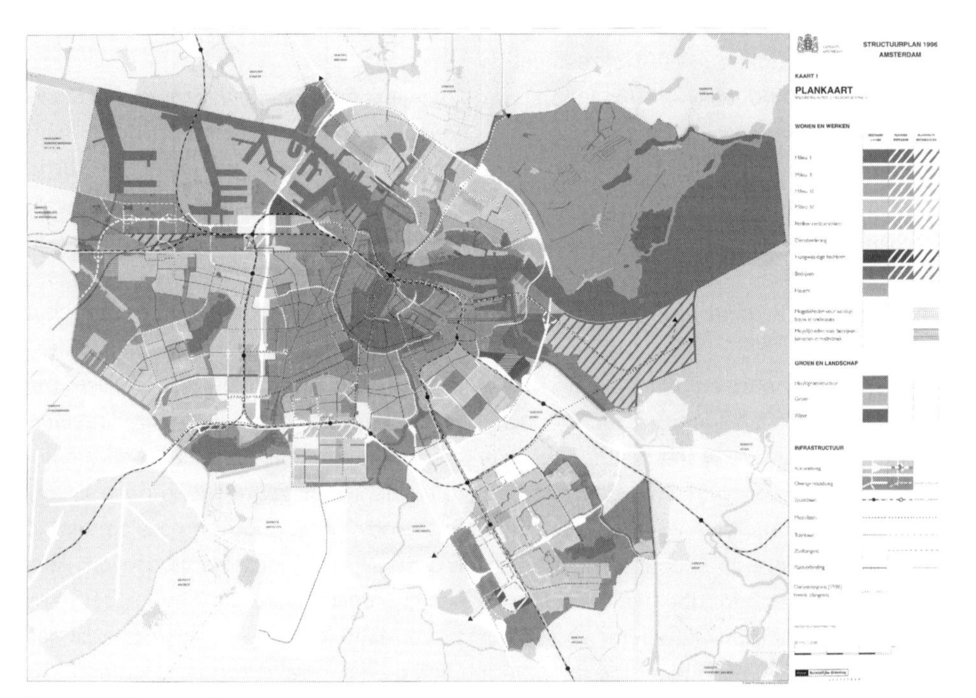

Figure 3.6 The 1996 *Structuurplan*

Source: From Jolles *et al.* 2003: 102/103 with permission of Amsterdam City Council

process rather than shaping of development strategies. The energy in planning had moved from planning itself to the implementation and management of major development projects. The Dutch planning tools and their use in practice had been designed primarily to manage a process of urban extension, especially in the Amsterdam area with its surrounding resource of reclaimed polder building land. Now urban areas faced challenges of reconfiguring already urbanised territory and reorganising the spaces/places within the city. Spatial organising concepts also had to be revised and adjusted to reflect both new emerging patterns in the urban structure and changing attitudes to environmental and place qualities. By the end of the 1990s, some were arguing that the initiative in shaping the spatial evolution of national territory in the Dutch multi-level planning policy community, centred on urban development and housing provision, had shifted to infrastructure development initiatives.[38] In addition, new ideas on how to think about space and place were filtering into the planning policy community. These challenges became more focused as the 1990s progressed. But they were challenges not just to the traditions of the Dutch planning community. They were also being worked through in a situation of economic difficulty and of undercurrents of disaffection with the Dutch way of 'doing government'. The combination erupted in the national election of May 2002 which displaced the social democratic/liberal hegemony of the previous decade.

In effect, by the end of the twentieth century the Dutch were in a process of critical reflection on the traditions of social democratic, co-sociational governance that had evolved in the second part of the century. For half-a-century, the planning function, coordinated through the political and technical consensus-building processes of multi-level governance, had shaped, developed and managed the precious resources of land in a crowded country. Strategic spatial organising ideas, such as the *Randstad*, the dispersal of development, deconcentrated concentration and compact development, commanded attention among other sectors of government because they were linked to resource allocation as well as to regulatory power, because they were respected as being based on high-quality expertise and appropriate knowledge, and because they seemed to reflect how Dutch people wanted their territory to be. But, by the 1990s, the links between general spatial organisation principles and public investment in infrastructure were being uncoupled, and the planners' influence over resource allocation reduced. Politically, the consensus over spatial planning was breaking up, with social democrats increasingly associated with the promotion of cities as development nodes, and the liberals with the fortunes of suburbs and less-urbanised areas.

For city councils such as Amsterdam, this uncoupling and break-up of consensus politics was particularly acute, as the council's resources of easily developable land were limited now as they could no longer extend their jurisdictions onto surrounding territories and thereby capture urban development value. Meanwhile, the city area had become increasingly diverse socially, with a mixture of buoyant economic activities, intense tourist pressures in the city centre, and severe social problems concentrated in various parts of the city. The city area had become a more cosmopolitan place, attractive to some more affluent households who were moving back into the city centre, but also more different and diverse than the surrounding metropolitan region. Its diversity was reflected in the fine-grained variety of its locales, particularly in the city centre. So what was once a great social welfare city in a leading European social democratic state found itself in a new resource context and with a changing socio-cultural base and governance landscape. Within this scenario, the city's Planning Department, the *Dienst Ruimtelijke Ordening*, struggled to maintain a strategic conception of the city in its regional context with sufficient persuasive power to act as a framing force. This was needed to shape the policies and programmes of the multiplicity of stakeholders through whose actions and perceptions the 'spatial ordering' of the Amsterdam area is now being produced.

THE STRUGGLE TO INNOVATE: STRATEGIC PLANNING IN THE AMSTERDAM AREA, 1996–2005

RETHINKING AMSTERDAM'S FUTURE

By the mid-1990s, Amsterdam City Council was evolving a new conception of the city. While still emphasising the quality of life for ordinary residents, the open, cosmopolitan and multicultural qualities of the city were increasingly acclaimed. This was combined with an assertion of special status for the city, as both a different kind of urban place within a metropolitan region, and as the country's most important city in a European and global context. The 1996 *structuurplan* had celebrated the city's liberal openness to the world in its title *Open Stad*, but the city's planners began to search for a conception of the special nature of the municipality of Amsterdam as a city of European and global significance within a wider metropolitan region, but also a liveable city of lively, safe and secure neighbourhoods. In the language of the powerful policy discourse of 'economic competitiveness' being promoted vigorously across Europe, Amsterdam was perceived as a 'top location' in the Netherlands and in Europe. This discourse highlighted the importance of promoting 'top locations' within the city, notably those along the A10 'south axis' (*Zuidas*).

The politicians were still focused on the city centre and the inner neighbourhoods around it, the heartlands of political protest in the 1970s. Although new residential neighbourhoods were being created on the IJ river banks, the primary concern in the older parts of the city, and in the newer neighbourhoods of rented housing in the western suburbs and the Bijlmermeer, was regeneration and neighbourhood management. In this activity, issues of safety and security were as significant as urban development itself. Further strategic planning work at the city scale was a low priority. The failed attempt at creating a metropolitan region and the difficulties of focusing commercial development attention on the IJ axis had discouraged an interest in strategic planning. The new alderman for spatial planning, Duco Stadig, who took office in 1996, was not initially keen on another strategic initiative. The DRO planners had also expected the

1994 *Structuurplan* to be the last, anticipating that the next plan would be for the metropolitan region. Nevertheless, some kind of formal plan was now needed to underpin the agreements over national and regional resource allocations for specific projects, including the spending of the *VINEX* funding allocations.

Meanwhile, the position of the spatial planning function within the wider municipal and governance landscape was becoming ever more complex. The multi-level government relations of the planning policy community were weakened as the power of the national Ministry for Spatial Planning (*VROM*) itself declined. Co-aligning development and investment became more difficult, with intense lobbying at national and province/regional levels around strategic ideas that could mobilise interest in specific investment projects. In this context, the DRO strategic planners, by this time with the support of Alderman Stadig, sought to mobilise attention within the City Council and the wider public realm onto the city's spatial development potentialities. They commissioned studies on future possibilities and promoted debates in the city about its nature and future, making use of the rich array of discussion arenas and media to which they were well-connected as officials and as citizens. In addition to issues about the distinctive identity of the city, they also sought to focus attention on the implications of the emergence within the region of other major nodes with city-centre qualities (notably in Almere and around Schiphol airport), and by the promotion of spatial concepts based on development axes rather than nodes.

These initiatives, coupled with the start of work at national level to prepare a new spatial development policy report (the *Vijfde Nota*), led to political agreement in 1999 to start preparation of a new *structuurplan* for Amsterdam, in parallel to the preparation of a new *streekplan* for the wider metropolitan area.[39] The production of the plan followed the formally established steps for consultation fairly smoothly and speedily, with an initial briefing paper, a draft strategy and public debate, a revised draft and finally approval by the City Council in April 2003. The resultant *structuurplan* did not in itself present a new approach to understanding the city. Rather, it consolidated, connected and sought legitimacy for agendas arising in regional arenas among municipal planners and aldermen and in the previous studies and debates. However, the debates on urban region futures involved in its preparation introduced ideas that strengthened later in subsequent discussion on strategy for the wider urban region. As the DRO planners commented, plan preparation involved a process of careful 'reading, listening and learning', searching for what is emerging and seeking for meaning, in a process which is '95% talking'. The formal preparation process mobilised data and information from the city's research and statistics department, while the specific consultation processes focused citizen attention on threats posed by development in the southern axis to the existing sports grounds, allotments and waterways.

There were also challenges to the DRO's process management. The main area of contention was over whether the planners should have presented one scenario or several for public discussion. The DRO planners were in a dilemma. Citizens were interested in debating alternatives, but the City Council and the DRO were campaigning in regional and national arenas for recognition of a specific agenda in the strategic co-alignment processes. Amsterdam City sought a long-term strategy with major project agreements extending to 2030, including the Almere link, along with revenue-sharing of the returns from the development process, so that the financial returns from greenfield development in the wider urban region could be used in part for development within the city.[40] But the Province, the national government and the municipalities could not reach agreement beyond a horizon of 2010. Although both the *structuurplan* and the *streekplan* were formally approved around the same time (Spring 2003), the cost was a shortening of horizons and the omission in the formal plan of a sketch of the perspective to 2030.

The resultant *streekplan* and *structuurplan* make few innovations in approach or presentation. The *structuurplan* rolls forward many of the ideas from the 1985 plan (Figure 3.7), and incorporates developments already underway into the framework of the plan. In terms of spatial organisation, it makes some significant breaks with established strategy and calls up a new idea about the city's identity around the idea of 'urbanity'. Drawing on studies by University of Amsterdam academics, DRO planners saw urbanity as expressed in the idea of the city as both a 'market place' and a 'meeting place':

> The city as a market place emphasises specialisation and the exchange of goods, products, services, knowledge and information. Necessary conditions are a differentiated working population, a varied range of housing and businesses, diversity of facilities and the best possible accessibility and volume of facilities. The city as a meeting place covers the variation, demonstration, happening and interaction opportunities for population groups, lifestyles, cultures/subcultures and opinions. The conditions for these include character, optimum accessibility and freedom of choice (Gieling and de Laat 2004: 316–317).

In particular, the *structuurplan* recognises the powerful dynamic behind the 'southern axis' of development, increasingly seen by national political and business groups as a key locus of national economic strength and a major development corridor in the North Wing (*Noordvleugel*) of the *Randstad*. *Zuidas*, now focused around the Zuid station, becomes a 'city region core area', a key development node. Other key development nodes are the city centre, and a major sport, leisure and retail complex built across the railway in the Bijlmermeer around the ArenaA football stadium. A special agency was set up in 1995 to promote *Zuidas* and campaign for national and regional attention to its development needs. In terms of spatial organisation, the *structuurplan* shifts definitively away from the 1985 idea of a centred and 'compact' city. Instead, the city is presented as a collection of different locales and development nodes, connected by axes of development which are in turn separated by lobes of green space.

The DRO planners were also very conscious that the 2003 *structuurplan* was an interim statement towards a more coherent expression of an Amsterdam metropolitan region. Amsterdam municipality is conceived, within this wider region, as a special kind

(a)

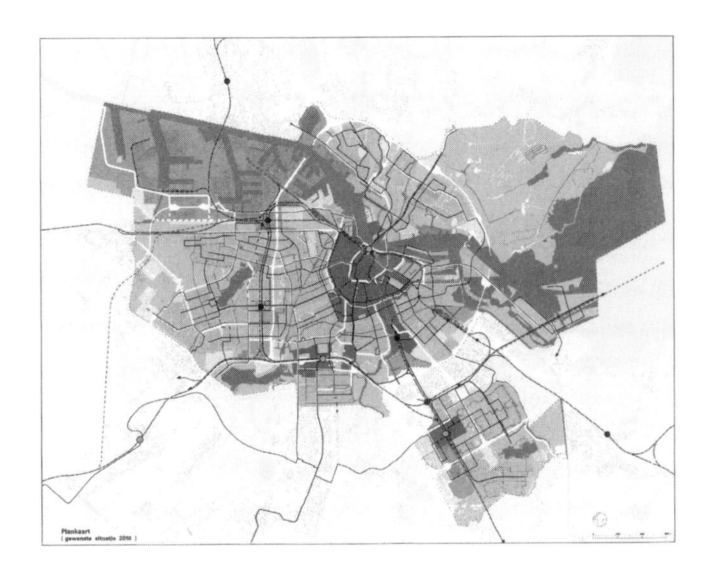

(b)

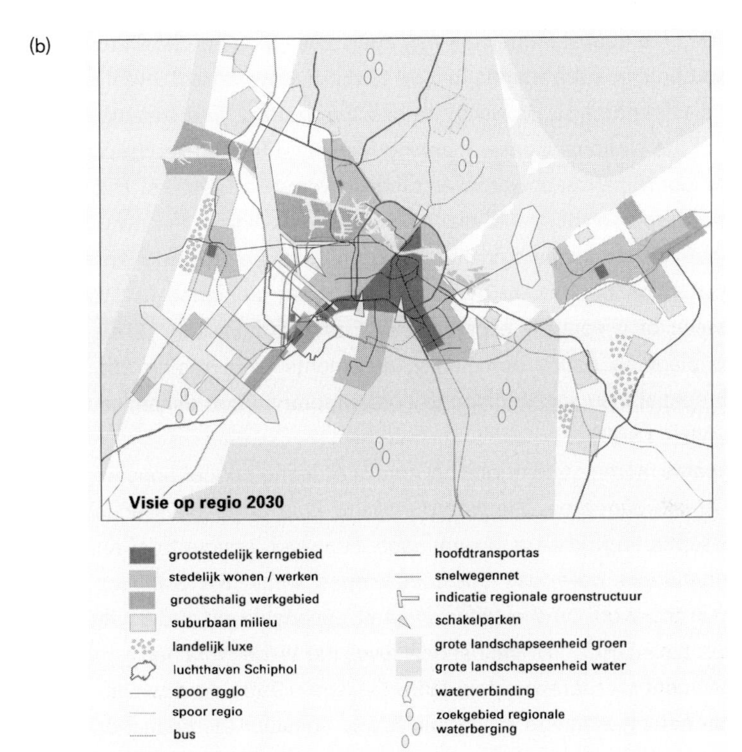

Figure 3.7 Amsterdam's 2003 *Structuurplan* and the wider regional 'vision' (a) The 2003 *Structuurplan*; (b) the Amsterdam area in 2030

Sources: (a) DRO 2003: 26, (b) Jolles *et al.* 2003: 104, both with permission of Amsterdam City Council

of urban area; it combines diversity, a cosmopolitan ambience and a liberal atmosphere, with a longstanding emphasis on liveability and connectivity through public transport, bicycle movement and walking. In developing a notion of 'urbanity', the city planners drew on both specially commissioned academic studies and on the ongoing debates within the city. These ideas reflected an experiential reality of an increasingly multicultural city, a business location with multiple linkages to other business centres internationally, and a busy tourist destination, adjacent to a dynamically expanding internationally oriented business and logistics node centred around a major international hub airport. This conception, however, co-existed uneasily with the image of the city held by many politicians and city residents, for whom the identity of the city was still associated with its old core and with the City Council's role in housing development and neighbourhood management. In the next two sub-sections, I examine in more detail the spatial concepts and the institutional context that shaped the production of this plan, and the continuing evolution of ideas about the city, the urban region and the shaping of their future trajectories.

EVOLVING IDEAS ABOUT SPATIAL ORGANISATION

Expressing a spatial order in a spatial framework has been a key focus of the Dutch planning tradition. From mid-century, such frames helped to focus government investment in urban development and infrastructure provision across the country. The framing ideas, attached to the power of the National Spatial Planning Ministry, *VROM*, exercised strong 'persuasive' and hence coordinative power over other government sectors. But if the coordinative and integrative power of spatial planning was weakening in the overall governance landscape, what role do spatial framing concepts continue to have? And if the focus of urban policy attention centres on qualities such as an ambience of 'urbanity', what is the role and value of a spatially expressed city strategy? By the 1990s, both within and around the planning policy community, arguments were raging about the relevance of the Dutch spatial planning tradition to contemporary urban dynamics (Hajer and Zonneveld 2000; WRR 1999).

Despite these debates, some quite traditional spatial ordering conceptions were still being deployed in the Amsterdam area. The *Noord-Holland Zuid Streekplan* focuses primarily on defining and separating urban and rural uses. However, concepts of relational dynamics and activity networks were also being developed within the planning policy community, particularly in the work on the *Vijfde Nota* at national level, and by the strategic planners in Amsterdam's DRO.[41] At the national level, the *Vijfde Nota* had a complex evolution. An initial document was prepared in 1999 (*VROM* 1999). At this time, there was a strong orientation among planners in the Ministry to consider the spatial development of the Netherlands in the context of a wider European perspective (Zonneveld 2005a).[42] The concept of networks is mobilised uncertainly, in terms of physical infrastructure routes and of flows of people, goods and information. Cities are seen to exist within 'clusters', as 'network cities on a regional scale' (*VROM* 1999: 12). Amsterdam is identified as one of the three primary such 'network cities', within the wider *Randstad* area. Corridors, development nodes and compact city concepts co-exist uneasily. By this time, the concept of corridors as a spatial organising principle was meeting considerable criticism and the planning community was identifying corridor development with 'sprawl' (de Vries and Zonneveld 2001; van Duinen 2004). Corridor development should focus around multi-modal 'infrastructure bundles' (*VROM* 1999: 17), with urban development concentrated at nodal points. Green spaces were to be used around the corridors to connect urban nodes to 'emeralds' of international and national importance for 'wildlife, recreation, water supply . . . and (national) cultural and historical heritage significance' (*VROM* 1999: 18).

These ideas mixed together two spatial concepts emerging among Dutch spatial analysts. The first imagined spatial patterns as created through 'layers' of activities. The 'layer' concept was promoted among the community of landscape and water-management planners. The second imagined urban areas as constituted through multiple networks, influenced by the emerging relational urban geography. Both concepts generated criticism, as those arguing for tighter control over sprawling development around cities sought a strengthening of compact city ideas, while those promoting economic competitiveness emphasised the need to promote the new kinds of locations emerging along the arterial 'corridors' (Zonneveld 2005a).

Struggles between competing spatial concepts and their consequences for the allocation of major development locations and government subsidies continued in the subsequent evolution of the full draft of the *Vijfde Nota*. By December 2000, it finally reached Cabinet ratification, but then failed to achieve parliamentary approval before national elections and a change of government in May 2002. The *Vijfde Nota* makes much use of concepts of 'layers' or 'strata', and of urban networks. Despite attempts to insert more relational thinking into the latter concept, well-established notions of the compact and 'centred' city, grounded in assumptions of the power of physical proximity, continually reinterpreted the network notion back into a physical conception. The 'layers' concept claimed to refer to flows, with geological structures, water flows, ecological flows and human flows in urban nodes and infrastructure networks layering over each other. It sought to introduce a richer understanding of the dynamics of spatial patterning than that centred on analysis of the built environment alone. But it derives also from traditions of landscape analysis that took a more physicalist position, linked to traditional ideas of separating town and country into green and rural areas versus red and urban areas, through 'contours' that could be policed by land-use regulation (Priemus and Zonneveld 2004; Zonneveld 2005a). Behind this reassertion of a traditional geography was a struggle by rural and suburban interests to protect their local landscapes and by other government departments to maintain their own approaches and spatial ordering concepts. There were also struggles over the interpretation of urban networks. In part, these were seen as the 'layer' of daily living in an urban region, but the *Vijfde Nota* also maintained the idea of networks of cities at different scales, which created a hierarchy with potential funding implications. Only some networks could break the restrictions of firm 'red' contours maintaining the compactness of urban development. As a result, city councils argued over their particular designation in network concepts (de Vries and Zonneveld 2001; Zonneveld 2005a).

Amsterdam, in the Cabinet-approved version of the *Vijfde Nota* (*VROM* 2000), was positioned in a new concept of the *Randstad*, called 'Delta Metropolis'. This spatial concept, promoted by a small group of spatial planners and urban designers linked to urban designer Dirk Frieling, had been taken up by the 'big city' mayors to push for a national emphasis on metropolitan development issues (Salet and Gualini 2003). An informal 'platform' of political and economic actors was formed in 2000 to promote the concept and, for a while, it proved very persuasive (Salet 2003; van Duinen 2004). The *Vijfde Nota* translates the Delta Metropolis idea into the vocabulary of 'layers' as constructed through a 'green–blue network', an 'infrastructure network' and an 'urban network'. All these various ideas are brought together in the *Vijfde Nota* map that was expected to be part of the Key Planning Decision (KPD) on national spatial policy (Figure 3.8).

For the Amsterdam area, the Draft KPD situates the city as a key node in the national urban network, indicates a 'rapid link to the north' which would provide a bridge to Almere, emphasises the national importance of the Schiphol 'mainport' and designates the city, along with five others, as a location for 'new key projects'. Almere was now seen in national policy as a major housing growth location for the Amsterdam metropolitan region. But the controversy over national spatial policy remained unresolved when the draft *Vijfde Nota* was withdrawn following the 2002 elections, with a new draft not available until 2004 (the *Nota Ruimte, VROM* 2005). This drops the Delta Metropolis concept in favour of a division of the *Randstad* into a north and south wing (*Noordvleugel* and *Zuidvleugel*). Two informal 'platforms' around these divisions had already been formed where spatial planners and aldermen met to discuss development issues. Some argued that these 'wings' related more clearly to the actual daily life networks of the western Netherlands than the *Randstad* or the Delta Metropolis ideas. *Nota Ruimte* has a strong emphasis on economic competitiveness and emphasises six key economic development zones, of which the Haarlemmermeer–Schiphol–Zuidas–Almere axis is the 'top location'. This indicates that national funding will flow to this axis, but the *Nota* has little to say about which nodes in the axis will benefit most.

Despite the difficulties experienced by the *Vijfde Nota*, Amsterdam city planners made use of some of its spatial concepts. The 2003 *structuurplan* is presented as built up through three 'layers', as in the national spatial policy, the main green and water structures, the infrastructure of 'roads, rail and cabling', and the built environment. Each is linked to specific actions, with an emphasis on intensification and mixed use in the built environment. The planners refer to a 'fourth layer' which they seek to treat in a non-physicalist way:

> There is, moreover, a fourth layer that can be distinguished in the city, namely the living culture. This layer plays an important role in consultation with the population

Figure 3.8 Draft Key Planning Decision on national spatial policy 2000
Source: VROM 2000, with permission of the Ministerie van VROM

about the desired development of the city. It is, however, difficult to reach hard and fast conclusions about this layer in a structure plan, since the living environment is only partly a function of spatial factors and can change more rapidly than the other three (DRO 2003b: 31).

The concept of layers was attractive to the city planners because it offered an integrated way of conceptualising the city, an alternative to the traditional division into sectors, activities and uses. It helped to break away from the sectoral emphases of the past (housing, agriculture, economic development, etc.), and provided a way to express the connections between environmental and socio-economic dimensions of space formation. The city planners used the 'layers' concept to discuss different development options and found that different groups and government departments could locate their concerns and how these related to other issues more easily than in the past (Gieling and van Loenen 2001). Now they wanted to emphasise mixtures of uses, and the complex interactions of the flows represented in the different layers. In graphics produced after the *structuurplan*,[43] the DRO planners expressed their thinking about the layers more clearly. The idea of a 'network region' is presented as the product of a ground layer (water and landscape), an infrastructure layer, a layer of nodes (from the global to the local) and a layer of new housing areas. Schiphol is designated as the world centre, *Zuidas* as a 'continental' centre, and the city centre as a regional centre with a global inflection (See Figure 3.9).

The final *structuurplan* remains traditional in its map form (Figure 3.7a). The plan re-affirms *Zuidas*, the Arena area and the city centre as major 'central city' locales. It emphasises that the other peripheral centres should develop as mixed-use locations, proposes four 'first-order' rail stations in the Amsterdam area and stresses the preservation of green and water areas. There is a strong emphasis on the relation of Amsterdam to other parts of the metropolitan region. But because of difficulties in regional co-alignment, ambitions for river crossings from the IJ river banks to Almere, which were in the draft plan as proposals beyond 2010, could not be included.

Compared to the *Vijfde Nota*, however, the Amsterdam *structuurplan* has less of an explicit emphasis on its international position. This is conveyed in the taken-for-granted role of the 'central city' locations. The focus is much more on regional linkages within which the city was struggling to get agreements over infrastructure projects and revenue sharing. In this context, a new spatial development concept was emerging, centred on transport axes. These axes are only weakly expressed in the *structuurplan*, but became much clearer in a paper produced later in 2003 which resulted from discussions on spatial strategy in the informal arena created around the *Noordvleugel* interests (Figure 3.10). This strongly asserts the economic development axes driving through the metropolitan region, but also illustrates how a conception of a broad development axis is being used to identify major locations for housing development projects (Mansuur and van der Plas 2003). For the City Council, the main concerns were the major strategic projects, not their strategic plan. It is these projects, rather than the *structuurplan*,[44] which are profiled on the council's website as the 'Big Seven', reinforcing the City Council's continued ambition to be seen as playing a major role in 'building the city'. The focus on key projects as the 'carriers' of strategy was thus evident nationally and in Amsterdam.

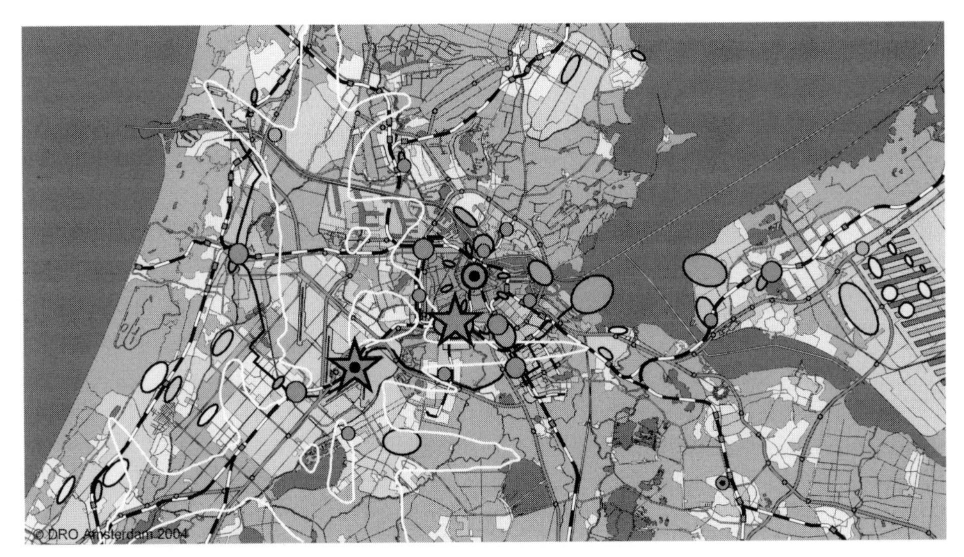

Figure 3.9 The network 'region'

Source: Amsterdam City Council, '*network regio*' pamphlet, 2004, with permission of ACC

THE COMPLEXITY OF URBAN GOVERNANCE

By the early 2000s, the *Gemeente Amsterdam*, the Amsterdam City Council, with its long tradition as a major force in shaping the fortunes of the urban area, the region and the nation, found itself in a much more complex governance landscape than in the mid-twentieth century. Then, the council's jurisdiction encompassed the daily life movement patterns of most of its citizens. The City Council was a major force in building an expanding city, following the path set by its pre-war strategic plan. It was a major land and property owner and garnered substantial public investment funds from national government. Both nation and City Council were committed to an agenda that stressed providing good-quality living and working places for its citizens. Its spatial planning department, combined with the city's land ownership and investment strength resulting from its development activities, was a key function guiding this development effort, setting strategies, shaping investments, and expressing and creating identities – for the city as a whole and for the places within it. But 50 years later, the council's situation was very different. The urban area was important to the nation, primarily for its economic role as a centre of commerce, finance, tourism and logistics, rather than as an industrial hub or a cultural centre. The city's jurisdictional area no longer coincided with the linkages of the wider urban region, which extended not only to the area encompassed by the informal *Regional Orgaan Amsterdam* (*ROA*), but to the wider 'northern wing' of the *Randstad*, the *Noordvleugel*. In theory, the City Council acquired stronger powers as a result of the national decentralisation of investment resources, other than for key national projects, to the Province and municipalities. In practice, this decentralisation represented a reduction in overall resources, and left it to the municipalities to fight over resources among themselves. And,

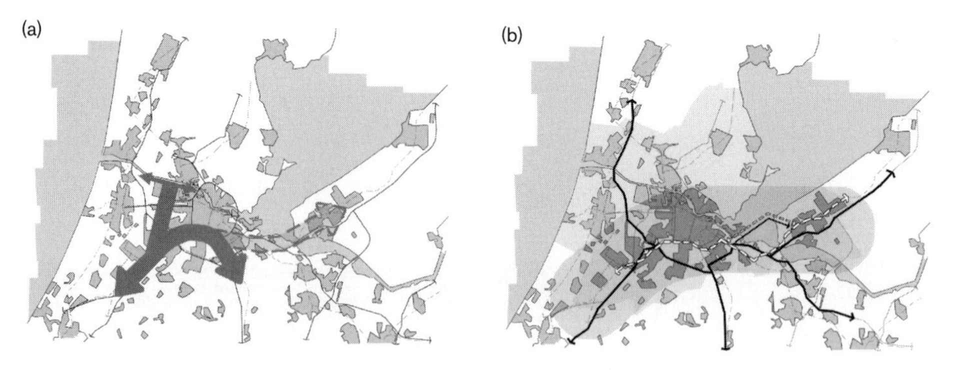

Figure 3.10 The Haarlemmermeer–Almere development axis (a) Economic development locations; (b) housing and employment locations

Source: Mansuur and Van der Plas 2003: 7 (a) and p.10 (b), with permission of Amsterdam City Council

although the council remained a major landowner, the spatial patterning of the city was increasingly being driven by the decisions of businesses and more affluent households, following, and also shaping, the demand for transport infrastructures.

Faced with limitations on the resources that could be acquired from central government, the City Council increasingly shifted from viewing its land assets in terms of their 'use value' in relation to particular city goals, to emphasising the market value of assets, as a way of filling the city exchequer, and mobilising private-sector development investment. In this context, the long-established social democratic agenda, with its concerns for liveability and neighbourhood conditions, had to co-exist both with the agendas of environmental sustainability developed in the 1980s and the focus on developing Amsterdam's 'world city' qualities in the context of the global competitiveness of cities.[45] The city's capacity to act as the dominant, integrative shaper of the trajectory of the Amsterdam urban region was thus under pressure.

In this context, the City Council had to re-think its governance relations and its role, particularly as regards the neighbouring municipalities and the Boroughs it had created during the 1980s and 1990s. For many years, Amsterdam had been the strong force in the Province, particularly for the surrounding municipalities. These tended to see the City Council as 'the enemy', while the City Council itself tended to reinforce this perception by acting as the most powerful and knowledgeable actor. It was in this political landscape that the city's efforts to create a formal Amsterdam metropolitan region were initiated, and then undermined as other municipalities, notably Almere, challenged the city's proposals for revenue sharing and its agenda of major projects. The second major development in the region, the growth of Schiphol Airport as a major urban node, was in any case outside the municipal ambit, as a key national development project, although the City Council retained a role as company shareholder and a formal consultee on development proposals.

These intermunicipal struggles were more acute because of the changes in national government funds for development investment. The 'agreement' system following the *VINEX* allocated general packages of funds to groups of municipalities who consented to co-operate in relation to spatial development programmes agreed in *structuurplannen* and *streekplannen.* But this was later changed, with national funds being allocated to major projects, depending on agreement among municipalities. This provided the momentum for the creation of the *Noordvleugel* platform, which was used to promote Amsterdam's interests in the *Vijfde Nota* and the *Nota Ruimte.* A further metropolitan arena had also developed, the *Regionale Samenwerking Amsterdam* (Regional Co-operation Amsterdam), which included mayors, the Province governors of *Noord-Holland* and *Flevoland.* Some private-sector actors were invited to participate in relevant task groups. These arenas were able to broker agreements more easily than in previous arrangements, partly because politics had become less ideologically polarised (Salet and Gualini 2003). The major debates centred on infrastructure arteries, particularly the completion of the A9/A6 link to Almere, and the routes and stations of the private-sector High Speed Train companies (over which the public sector had limited control).[46] But these metropolitan region arenas remained fluid and informal, with new 'platforms' overlapping existing ones and all underpinned by complex coalition formation processes (Salet and Gualini 2003). Developing the persuasive mobilisation concepts and powerful coordinative momentum that had been possible in the past was increasingly difficult in such a governance landscape.

The City Council was also drawn into more joint projects with private-sector actors. This was particularly the case with the *Zuidas* project, the 'global city' promoted at the *Zuid/WTC* station (Salet and Majoor 2005). This development node had emerged on the city's doorstep rather than being a major project promoted by the council itself. The council only slowly swung around behind the *Zuidas* idea, vigorously promoted by some urban designers and developers, notably architect Pi de Bruijn, who had earlier designed the Arena Boulevard development in the Bijlmermeer area. A special public–private partnership agency was created in 1995 (the *Zuidas* coalition), and agreement was reached that all funds generated to the city from the development should be re-invested in the project. But there were competing images of how *Zuidas* should develop, with the concept of a 'top' location centred around office development co-existing with that of a more multi-activity locale with traditional city centre qualities. The key to realising this latter idea was to generate significant public spaces in and around the buildings, linked to easy routes to other parts of the city. The project promoters and the City Council argued that a critical part of the project should be a platform across the substantial trench through which the A10 and rail routes currently pass. Although there was only limited public debate about the *Zuidas* development, there was opposition to expanding this separation further. This generated the proposal for a 'dok' platform (with all the transport routes below ground), rather than a 'dijk' (a raised platform on which all the transport routes would be located). But this proposal depended for its success on persuading the Ministry of Transport about safety issues, and levering national government resources for a 'major project'. In addition, it was not clear that the high-speed train operating companies would select Zuidas as a major station. It is very near to Schiphol where a much-used station already exists. So *Zuidas*, although recognised at the national level as an important project, was a risky venture for the City Council.[47]

The City Council was also in a new situation with respect to the Districts. As explained earlier, these had been created in response to the citizen protests of the 1970s. Initially, the Districts had few staff and were very dependent on the main council. But, during the 1990s, some smaller Districts merged and overall, they built up in strength, taking over detailed functions for land-use regulation and local environmental management from the City Council, including some project development work.[48] District councillors were elected and often linked to the City Council through party networks, while officials had links through their professional networks. But, nevertheless, they increasingly came to challenge the City Council's views and actions, while at the same time acting as a channel through which citizens' concerns could be passed up to the City Council. Districts developed various governance processes and cultures of their own, but were important in maintaining the closeness between citizens and formal government that had emerged from the struggles of the 1970s. City councillors saw the neighbourhoods as their heartlands, and so were prepared to listen to issues raised by the Districts. But despite the *plaberum* procedure, and the rich consultation networks through which councillors and city planners tested out citizen opinions and reactions, there were emerging disjunctions in linking citizens' concerns with those arising in the various regional arenas, and in the special public–private development agencies evolving around major projects. This was exacerbated by continuing interdepartmental tensions, although these were diminishing by the mid-2000s. The preparation of a city strategic spatial plan, so long the arena where multiple scales, multiple values and multiple sectional interests were brought into conjunction and 'integrated', seemed to have lost its coordinative function and integrative capacity.

These evolutions in the governance landscape of the City Council created an uncertain situation for the city's strategic spatial planners. They had been a strong group, of around 20, in a grouping of over 280 technical staff, working to an alderman with a large portfolio, covering a range of functions from land administration to housing, urban renewal and water management, as well as spatial planning.[49] By 2004, staff losses were feared, as part of the council's cost-saving measures, and the planners were re-thinking their role. The increasing emphasis on major projects and on capturing resources seemed to undermine further efforts to develop a strategic conception of an urban region. The strategic concepts debated in the Dutch spatial planning policy community, whether of 'layers', 'urbanity' or 'Delta Metropolis', were unstable and seemed to have only limited persuasive power (van Duinen 2004; Zonneveld 2005a). National and municipal regulations governing development projects were also being criticised as inhibiting market investment.

But the DRO strategic planners refused to let their strategic orientation fade away.

Instead, they increased their efforts to develop a strategic understanding of the dynamics of the wider urban region, working with other municipalities in the context of the *Noordvleugel* informal arena. They also sought to connect strategic spatial concepts to the framing of major projects.[50] A new advisory mechanism was devised to frame the development of major sites and buildings, a form of area development 'brief' or 'envelope' (*bouwenvelop*). It was not anticipated that another *structuurplan* would be prepared for the city. Future strategies would, they hoped, be for the metropolitan region. A major review of national planning legislation was underway by 2004 which indicated that the formal tool of the *structuurplan* would disappear, to be replaced by a 'strategic vision' of some kind (Needham 2005; Zonneveld 2005b). The Amsterdam planners were working in the *Noordvleugel* arena on the preparation of such a '*structuurvisie*', using concepts of development nodes and layers (see Figure 3.10). They were deliberately seeking to break away from a highly specified concept of the spatial organisation of the urban region, to present a more subtle and flexible understanding, focusing on connectivities and the way urban geography evolves through continual changes rather than being systematically designed and managed by planning effort. The DRO had, by 2005, been able to redefine a role for itself as a major player in developing a strategic understanding of the urban area and as a kind of 'strategic think tank' for the metropolitan region, with an expanded research function. To enhance their capacities in this regard, the strategic planners made considerable use of university-based research, particularly AME/AMIDST at the University of Amsterdam, commissioning studies on the nature of 'urbanity', on the dimensions of 'accessibility', and on how to imagine the city as a 'portfolio' of neighbourhoods, each with its own changing dynamics.[51] But the shift in role to a 'strategic knowledge service' also meant that the planners had to change the images others had of Amsterdam City Council as a rather dominating partner, or even an 'enemy'.

So, as the City Council's strategic planners celebrated 75 years since the creation of the Amsterdam planning office (Jolles *et al.* 2003),[52] its role was under challenge. Its tradition of providing a comprehensive overview of the city's spatial development in regularly up-dated formal plans which guided public investment was ebbing away. The council itself and its various departments were learning to live in a crowded institutional space (Salet and Gualini 2003). Although the DRO planners were putting much effort into building their relationships with other municipalities in the wider urban area, they were criticised for neglecting both their relations with the City Districts and citizens, and the development of national and international linkages (Salet and Gualini 2003). But they nevertheless had the practices of a strong, active council to draw upon, and citizens expected the council to play a key role in shaping the city's future and expressing its identity. City planners noted that Amsterdam citizens expected a lot of their council, and were themselves acutely aware of a governance culture of active, engaged, critical commentary on the nature and future of the city, and of the actions of the City Council.[53] Although established governance relations and conceptions of the city were being destabilised, and market conceptions of urban development priorities were challenging those of liveability and environmental sustainability, these conceptions were still forced to co-exist and relate to each other in a governance culture that expected strong and accessible municipal government to manage the continually evolving Amsterdam. The challenge for Amsterdam City Council, and for all those involved in the governance of the diffusing, expanding urban area, is how to use the capacities and cultural expectations built up in the past to help to shape an emergent urban reality that can never be fully grasped or comprehensively 'managed'. As DRO planners already understood a decade before:

> Town planners today work with an obscure future in mind: they are certainly no longer exalted spirits who impose their ideas on a city. Town planning today is more a function of the city than the planners. Now the planners serve the city (DRO 1994: 218).

CONCLUDING COMMENTS

For most of the twentieth century, the Amsterdam City Council played a major role in building and shaping the development of the urban area over which it presided. It directed the substantial public investment flowing into the physical environment and made the rules for guiding specific building projects. In this context, spatial strategy-making in Amsterdam has played a crucial role in integrating different objectives and activities for urban development, and in coordinating and legitimating development projects. The city's spatial plans have also been important in expressing the identity of the city. Through this activity, as in the Netherlands generally, the strategic and development work of city planning has not only shaped markets, through focusing attention on particular locales and opportunities. It has both created markets and shaped the emergence of market players, particularly in residential development (Needham *et al.* 2005). Infusing this activity has been an emphasis on creating liveable environments for daily life, understood in terms of the qualities of the immediate residential environment and wider accessibility to the services and facilities of the city. Liveability and quality remain key concerns today, reinforced by concerns about environmental sustainability. This socially-focused development orientation shaped the creation of a strong municipal government capacity in a lively, cosmopolitan and richly textured civil society, often referred to in the 2000s as 'anarchic' and energetic in their challenges to government interventions, but yet supportive of its presence. In this context, spatial strategies for the city have had major effects, in the building of the city, in the quality of the built environment produced and in the expression of the city's identity. They have also had a valuable function in the context of the city's interactions with national and provincial government over development principles and land-use regulations applying at higher levels, and in accessing finance.

But the policy agendas and practices built up in the past sit uneasily in the emerging governance context of the twenty-first century. The power of national spatial development principles and local spatial strategies to create built space are limited not only by financial constraints but by new ways of organising the distribution of public investment funding. This focuses attention on major projects, rather than on long-term investment programmes, and on big schemes rather than the constant, careful management of change in the built environment. Further, it is increasingly recognised that the qualities and meanings of 'places' and connectivities which matter in the expanding metropolitan area are not just the product of building projects, but of all kinds of shifts in social, economic and environmental forces. A city council's influence on the interplay of such forces is much more complex and subtle to understand than the task of building new pieces of city. Thus market forces, cultural movements and government interventions intermingle in much more complex ways than imagined by City Planner van Eesteren in the 1920s. The *Gemeente Amsterdam* is trying to give expression to an identity as a cosmopolitan multicultural place, but this leads to interventions in cultural activities and the generation of 'ambience' as much as development projects. What may be disappearing is the integrative capacity to link multiple levels of government with citizens' concerns for liveability, pressure groups' campaigns for environmental sustainability and the focusing of business interests to achieve public interest benefits. This is arising not just because the spatial planners are locked into old traditions, although many critics argue this. In fact, the DRO planners are working hard to shift old paradigms and adapt to a new, more flexible way to understand how urban relations evolve and what needs strategic attention. Difficulties over integrative capacity also arise from the diffusion of governance effort among multiple arenas and the difficulty of articulating a way of thinking about the qualities and relations of a continually evolving metropolitan area.

The City Council still has a substantial role in urban development, through its considerable financial and land resources and its leverage over national investment funds. The governance practices and culture built up in the twentieth century in the Amsterdam area, as in the Netherlands generally, still embody a substantial capacity to 'summon up' conceptions of the city, and to debate them in a vigorous 'public realm' through which conflicting values, priorities and understandings can be brought into focus and developed into implications for specific strategic interventions. Amsterdam's urban governance, as viewed through its urban development activities, may be in a transformative period, one of uncertainty about how to use inherited capacities to build new governance relations and develop new conceptions of the city. It is possible that strategic spatial planning will ebb away into a more limited role for the council and the wider region. Yet this seems unlikely. The tradition and its practices remain a powerful force, backed by a feeling in government generally and in the wider society that spatial strategies are needed and that policies should have a clear spatial expression (Zonneveld 2005 a, b). The distinctiveness of this inheritance becomes clear when set alongside the Italian and English experience.

NOTES

1 The local names of organisations and of documents are given in italics.

2 *Dienst Ruimtelijke Ordening*; usually translated as 'physical planning department', but better translated as 'department of spatial ordering'.

3 This planning history is strongly featured in the Amsterdam Historical Museum's permanent exhibition.

4 See Faludi and van der Valk 1994; Jolles *et al.* 2003; Sutcliffe 1981. For helpful accounts of Amsterdam's governance and planning history, see Jolles *et al.* 2003; Mak 2003; Ploeger 2004.

5 In the Netherlands, most 'public facilities', including social housing for rent, are run by special semi-public trusts. In Amsterdam, the City Council retained its role as a major housing landlord until the 1990s, when its housing stock was handed to a social housing association. The council also leases sites for residential, industrial and commercial development (Needham *et al.* 1993).

6 Van Eesteren was a leading light in the international modernist planning movement, the CIAM movement (Faludi and van der Valk 1994; Gold 1997; Jolles *et al.* 2003).

7 The expansion of industry in Amsterdam created a working-class base which, in 1918, elected a socialist government.

8 Amsterdam City Council has had a more left-leaning majority than national government throughout the post-war period. The term currently used in the Netherlands is corporatist, but implying a broad perspective than just an economic nexus, and dominated by public-sector interrelations (Faludi and van der Valk 1994; Woltjer 2000).

9 The Netherlands was occupied by German forces from May 1940 to May 1945.

10 The IJtunnel, envisaged in the 1935 *Plan*, was finally completed in 1968, and the Coentunnel in 1966.

11 In 2004, there were 12 provinces and 481 municipalities. Provinces and municipalities are autonomous entities, but expected to pursue policies in line with each other and national government. The boundaries of provinces rarely change, but that of municipalities are often revised. However, there has been no major boundary change of Amsterdam City Council's boundaries since the incorporation of the Bijlmermeer area in the 1960s.

12 See Priemus 2002; Priemus and Visser 1995; Terhorst and Van de Ven 1995.

13 In 1982, Environment (*Milieu*) was added to the portfolio and title.

14 The *Rijkswaterstaat* had been in charge of land drainage and water management for a century and a half, and was often referred to by commentators and officials as a 'state within a state' (de Jong 2002).

15 TU Delft for design, transport and water engineering, University of Wageningen for agricultural development and landscape planning, and the University of Amsterdam for spatial planning.

16 Within the spatial-planning field, there were also divisions between urban designers, with a 'Stadtebouw' tradition, and social scientists (primarily geographers), or 'planologists'. Eventually, the two professional groups were brought together in a single professional association (*Beroepsvereniging van Nederlandse Stedebouwkundingen en Planologen* (*BNSP*) – Dutch Professional Organisation of Urban Designers and Planners) (Faludi and Van der Valk 1994).

17 Faludi and van der Valk (1994) suggest that these ideas were particularly influential among the

new planning teams at province level, influenced in the 1970s by the British structure plan experience.

18 Social democratic party networks were important, as well as alliances of municipal mayors, in producing this co-alignment.

19 The *Nota Ruimtelijke Ordening* are policy reports. These may lead to Key Planning Decisions, which require parliamentary approval.

20 In the Amsterdam area, these growth centres were in Alkmaar, Hoorn and Purmerend to the north, and Almere and Lelystad on the new polders to the east. Purmerend was later substantially downscaled. A key strategic Policy Note on housing was also produced in 1972, the *Nota Volkhuisvesting*, which guided how housing subsidies were distributed spatially (Faludi and van der Valk 1994).

21 By this time, the airport had been transferred from City Council ownership to a consortium in which the national airline, KLM, national government and the City Council were major shareholders.

22 See Jolles *et al.* 2003; Ploeger 2004; Pruijt 2004.

23 A report for the Chamber of Commerce by Utrecht economist Jan Lambooy seems to have had an important influence here (Ploeger 2004).

24 One impetus for the *Plaberum* was the greater need for coordination which this division produced.

25 No private housing was built in the city between 1978 and the later 1980s (DRO 1994).

26 By 2001, this had risen to 55 per cent, but has declined since.

27 Private housing associations have long been involved in producing housing development in the Netherlands, working closely with local authorities, who delivered serviced land at low cost (Needham *et al.* 2003).

28 No private housing was built in the city between 1978 and the later 1980s (DRO 1994).

29 Business interests influenced the Montijn Commission (Dijkink 1995).

30 It involved the mayors of the municipalities and a small executive staff, but was based in Amsterdam. By 2003, the number of municipalities had fallen to 16 (Salet and Gualini 2003).

31 This was confirmed in legislation in 1993.

32 Almere was able to generate major returns on its land development and did not want to share the proceeds.

33 New tunnelling technology enabled construction with less surface disturbance.

34 There had also been developments in the practice of agreements, and politicians sought a stronger emphasis on the location of new housing development (de Roo 2003; Faludi and van der Valk 1994).

35 'A' locations were areas with good public transport access; 'B' locations provided a mixture of access by public transport and car; and 'C' locations were areas close to motorway exits. Firms were assessed according to the mobility profiles of their workforce and channelled to A and B locations where possible (de Roo 2003).

36 These agreements were reached primarily within the respective sectoral policy communities. In the Netherlands, those involved claimed that all the experts know each other at the top level, facilitating the agreement system.

37 Needham *et al.* 1993, page 185. In the 1980s, office rents in Amsterdam were the highest nationally, and rising the most strongly. Amsterdam South (*Zuidas*) emerged as the top location in terms of office rents by 1991.

38 See Hajer 2001; Hajer and Zonneveld 2000; WRR 1999.

39 In Amsterdam, a *structuurplan* has the status of a *streekplan*. It sets the boundaries of green areas, defines urban zones and their qualities and provides a basis for the *bestemmingsplans* produced by the Boroughs, as well as, in theory, the framework for new development and urban renewal projects.

40 By the later 1990s, this concern had become less pressing as improved property market conditions within Amsterdam brought financial returns from the city's redevelopment projects.

41 There continued to be close links between planners in the city, the province and nationally.

42 Dutch planners took an active role in developing the *European Spatial Development Perspective* (CSD 1999), and its successors (Zonneveld 2005a).

43 My source here is a display sheet produced by DRO in 2004.

44 www.iamsterdam.com (accessed 21 August 2005).

45 British geographer, Peter Taylor, who attempted a classification of world cities, was asked to assess the city's potential in this respect (Taylor 2004a).

46 Amsterdam City Council preferred a northern route for the Amsterdam–Almere link, but this was expensive and connections further south were favoured by other municipalities and by the Ministry of Transportation, Public Works and Water Management.

47 My thanks to Stan Majoor for sharing his understanding from his doctoral thesis on *Zuidas* with me. (Stan's thesis is due for completion in 2006.) The story is still evolving. By 2005, a new partnership between national and local government and private shareholders was being proposed to take the project forward.

48 After much debate over whether the city centre 'belonged' to its specific residents or the wider urban polity, the city centre also became a district in 2002.

49 www.iamsterdam.com (accessed 21 August 2005).

50 In this, they were influenced by ideas developed in Barcelona on the role of major projects as strategic interventions (Calabrese 2005), a concept developing in Italian planning debates in the 1980s (see Secchi 1986).

51 See, for example, Bertolini and le Clercq 2003; Bertolini and Salet 2003.

52 It was created in 1928.

53 This active monitoring included making legal challenges to council plans and decisions.

CHAPTER 4

THE STRUGGLE FOR STRATEGIC FLEXIBILITY IN URBAN PLANNING IN MILAN

> The story [of planning in Milan] is that of the difficult quest to find an effective method of planning this area which lies at the vibrant heart of the Italian economy (Balducci 2001a: 159).

> Strong in terms of its economy and its rich society, Milan suffered from the lack of a truly strategic leadership throughout the [1993–2002] period (Dente 2005 *et al.*: 45).

INTRODUCTION

In moving from Amsterdam to Milan, the context changes to a much larger country and a much bigger metropolitan area. Italy, like the Netherlands, is a unitary state and the role of the state has been substantial in all spheres of life. But there has been no tradition of consensus politics, of partnership between the major spheres of society or of delegation of much policy activity to technical experts working within government. In other words, Italy does not share the traditions of welfare state corporatism of much of North-West Europe. Instead, political networks and clientelist practices have played a strong role in shaping governance cultures and attitudes to local administrations. The account told in this chapter is of continual struggles to confront and contain older governance practices to enable coherent policy attention to the challenges of securing some degree of social justice and environmental quality, as a long-established city explodes into a sprawling, economically dynamic metropolis.

Milan is one of central Europe's great cities. Positioned in several different geopolitical domains over the centuries, it has remained the dominant economic centre of Northern Italy, a capital for a wealthy agricultural region and zone of intense economic productivity and cultural sensibility. The city has been at the heart of key developments in Italian economic and political life in the twentieth century and is acknowledged as the country's commercial capital. The area of the administrative city of Milan, the *Comune di Milano*, is a dense complex of commercial, cultural and residential activities, with a population in 2001 of over 1.25 million people. But this area is located in a dynamically sprawling urban area that extends beyond the administrative boundaries of the *Provincia di Milano* (2001 population: 3.71 million) and across the *Regione Lombardia* (9.03 million) into Switzerland (see Figure 4.1). Flowing out over the flat plains of the Po valley, the *Padania* area, there are few physical constraints to urban development except for the

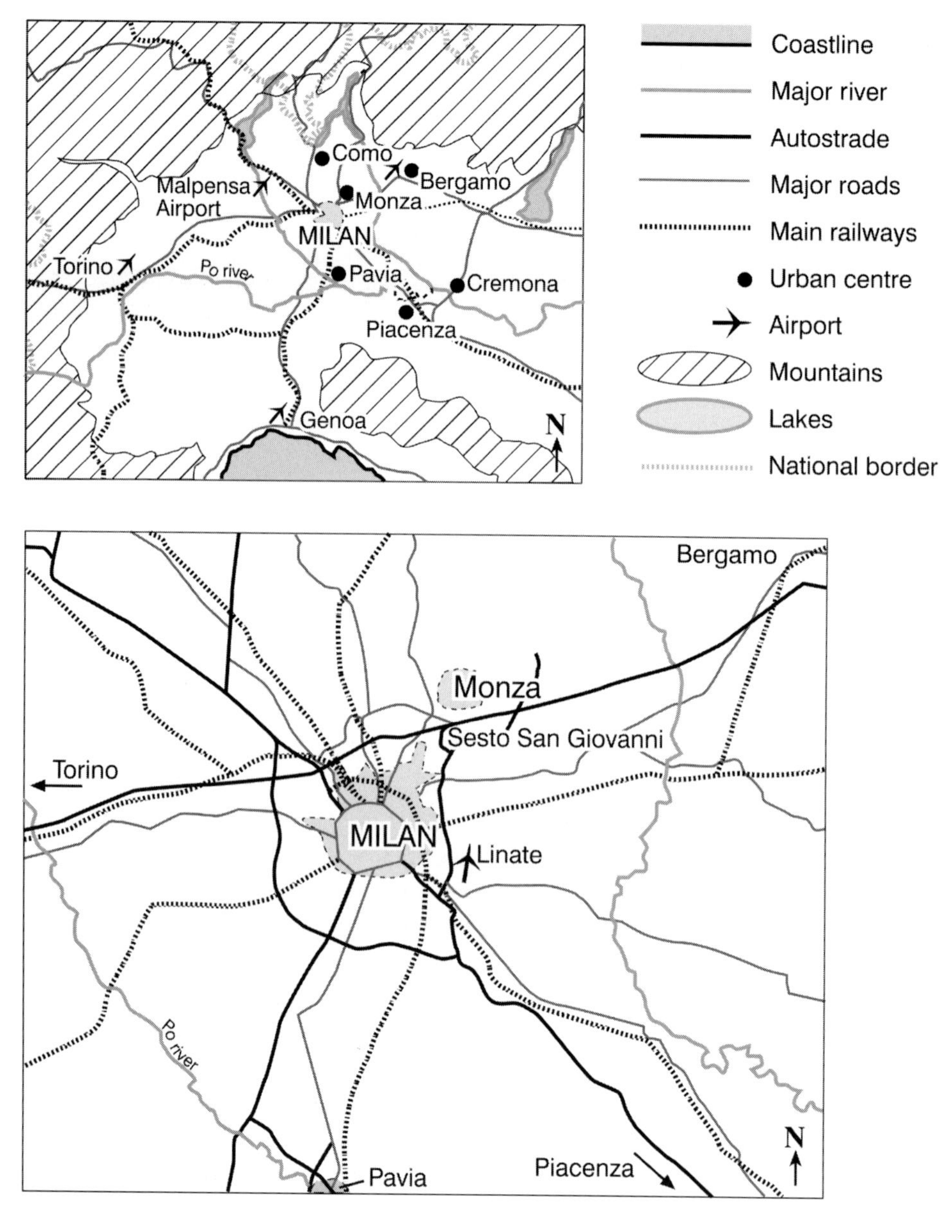

Figure 4.1 Milan's location

Alps to the north, the foothills and lakes, which provide attractive locales for Milan's elite, while the rich farmlands of the Po floodplain to the south are now protected from development because of their importance in food production.

During the twentieth century, the Milan region was at the heart of two 'economic miracles' (Foot 2001). The first was centred on industrial development, particularly heavy engineering. In a country whose economy was dominated by small family firms, Milan saw the build up of several major international manufacturing companies, including Alfa Romeo, Ansaldo, Breda, Falck, Innocenti, Marelli and Pirelli. These located on large sites that were then on the edge of a physically compact city core, with rooms for workers provided nearby (Foot 2001). The great days of these companies were in the 1950s and 1960s, the period of Italy's industrial boom and massive migration – from rural to urban areas and from southern Italy to the north. The population of Milan *Comune* increased in 20 years by nearly half-a-million and the Province by nearly one-and-a-half million. Housing this massive influx was a major challenge for Milan and the surrounding communes, in a political context where a strong working class communist/socialist politics challenged an equally strong Christian Democrat party, with close links to the Catholic Church. But, just as the communist/socialist political alliance gained the ascendancy in Milan in the 1970s, the industrial economic dynamic lost momentum in the face of international competition, as happened elsewhere in Europe. The big companies slimmed down, moved out or shut down, leaving a legacy of abandoned sites and workers threatened with unemployment. By the 1990s, the Milan metropolitan area had well over 12 million m^2 of abandoned industrial sites (Mugnano *et al.* 2005).

But as the industrial 'miracle' faded, so a new economic miracle emerged, drawing on Milan's old tradition of dynamic small enterprises, on its position as the country's commercial and financial hub, and on its rich cultural traditions, as expressed in a feeling for art and design. In many different ways, cultural and economic networks linked enterprises, many of them family-based, specialising in fashion, furniture and design products, to global markets. These networks of small firms and fragmented land and property ownerships shaped the political culture of the city's elites. Meanwhile, Milan's commercial and financial dynamic became increasingly involved in global networks, with international firms headquartering in Milan, building linkages with local firms (de Magalhães 2001). This economic and cultural climate supported a sense of innovative energy, of vitality and flexibility, celebrated by the socialist politicians of the 1980s in a flamboyant image of 'Milano da bere'.[1] As Foot's insightful account suggests, the spirit of a vigorous, cultured but strongly consumerist society drew into the limelight a much more neo-liberal political attitude (Foot 2001).

In Italy, politics is intertwined with business and civil society, with major economic opportunities strongly shaped by political dynamics. Urban development processes and urban planning are no exception (Vicari and Molotch 1990). Milan has been at the core of most of Italy's political developments of the twentieth century. It was the heart of fascism and of the resistance to fascism; the core of communist/socialist working-class mobilisation and also the base for the rise of the right-wing parties of the 1990s, *Lega Nord* and Silvio Berlusconi's *Forza Italia*. It was the place where Italy's old political cultures of 'clientelism' became systematised in the 1980s into a complex system of payments to political parties known as '*tangente*', earning the city the title of 'tangentopolis'. But yet it was also the place where this system was challenged, in the '*mani pulite*' campaign, which led in the early 1990s to the collapse of Milan's political classes and senior administrator level, and helped to underpin a strong movement in the 1990s towards new forms of technically efficient administrative practice.[2]

Public administration in Milan has been linked in very complex ways into these changing economic and political dynamics – in theory, modelled on a clear separation of powers between administrators and technicians, and between officials and politicians; in practice officials, politicians, experts, economic and social actors have been linked together through overlapping networks of family, political party, church, university and interest group. What has changed from one period to another is the manner of this intertwining, its distributive logics and the scale of the material and cultural benefits that have flowed from it. But even as the socialist politicians of the 1980s celebrated a dynamic new consumerist culture, so the role of the public administration was shifting and diminishing in the life of the urban region. In the 1970s and even the 1980s, municipal administrations could imagine that they were 'in charge' of the way their cities developed, structuring development opportunities and managing service delivery. Municipalities such as Milan owned large land and property resources within and beyond their boundaries and disbursed substantial resources (Vicari and Molotch 1990). Yet, increasingly, citizens and businesses despaired of the complexity and inefficiency of public administrations, and their inability to bring much-discussed projects to realisation. Public administration was something that business and citizen initiatives came up against, rather than being supported by. In contrast, and partly in response, all kinds of self-organising arrangements emerged in business and cultural arenas and within civil society (Cognetti and Cottoni 2004; Dente *et al.* 2005).

In contrast to the 1960s and 1970s, when urban planning and the city's plan were valued as a key arena for an ideological programme to shape how the city evolved, in the 1980s and 1990s there were few social movements focused around strategic urban development and management issues. The Milanese elite, with property and business interests in the centre of the city, were more interested in 'interior spaces', their apartments, offices, discussion arenas and exhibition halls. Few questioned the future of the urban core, despite considerable debate about urban futures in academia and in cultural magazines. Although aware by the 1990s of citizen concern about traffic problems, about health and safety, and the quality of public spaces in the city core, Milan's politicians and their supporters have, in recent years, had little interest in a strategic approach (Dente *et al.* 2005). Milan's experience of urban development strategy in the 1990s and 2000s has in this respect been very different to that of other major cities in Italy, which have been vigorously involved in producing strategic development plans.[3]

As in Amsterdam, politicians have looked to their planning departments and the tool of the city plan to articulate urban development strategies. But the political evolutions have created a very complex context for the practice of urban planning in Milan *Comune* itself. A key challenge for the planning function in Milan in the second part of the twentieth century has been to find ways of combining the regulation of development activity in ways that provide flexibility for all kinds of initiatives while, at the same time, paying attention to the wider public interest. The account in this chapter concludes with the most recent such attempt, in which the maturing of ideas about the limits of the prevailing comprehensive planning tool, the *Piano Regolatore Generale*, fed into the development of a range of new planning instruments in the *Regione Lombardia* and in the *Comune di Milano* at the end of the century. This coincided with a moment of political opportunity; the city's politicians were concerned that many much-discussed projects should actually get built, rather than being stalled in discussion and negotiation for many years. In a society that enjoys design ideas, there has never been a shortage of imaginative ideas about the future of Milan and the metropolitan area. But relating aesthetics and general principles to practical action is a different matter.

THE 'MIRACLE' YEARS OF THE 1950S AND 1960S

After the traumatic years of the Second World War and its immediate aftermath, Milan entered a period of growth and rising prosperity. Milan's major heavy industries expanded, with large waves of immigration from rural and southern Italy. As in Amsterdam, the development emphasis was on accommodating this growth by urban extension, to provide housing, urban services and public transport. The new estates extended the structure of urban neighbourhoods clustered around the city core into the municipalities surrounding Milan. The vigorous social and political life of the working class urban *quartiere*, much celebrated in memories of pre- and early post-war Milan, was often contrasted with the anomie of these new peripheral estates, though the reality was always more complex (Foot 2001). Meanwhile, the Milan elite and higher bourgeoisie[4] lived in the heart of the city, in *palazzi* and apartment blocks dating from the eighteenth and nineteenth centuries, with a few new additions over the years, enjoying easy access to cultural assets such as the *La Scala* opera house, the *Duomo* cathedral square and nearby high-quality retailing, good universities, small specialist services of all kinds and the company headquarters where many worked. 'The Milan core is a vibrant business and social "scene" morning and night' (Vicari and Molotch 1990: 614). A classic multifunctional urban area, Milan city core was then as it is now, well-served by public transport.

One result of the expansion of Milan during the years of the industrial boom was that the city owned a large stock of dwellings at controlled rents. This helped to sustain a left-wing political base in the city, continually challenging a right-wing politics sustained by the bourgeoisie and smaller enterprises. However, the municipal area was too small to accommodate the new estates, which meant that the *Comune di Milano* had to negotiate with surrounding municipalities to get access to building sites. What emerged was a politically driven scatter of peripheral estates, mostly of eight-storey apartment blocks, with working-class estates built in left-wing communes, and lower-density, higher-income estates in more right-wing ones. The land dealing and construction activity involved in these developments made small and large fortunes, with a good deal of speculative activity.[5] The resultant spatial structure continued to emphasise concentric development centred around the city core – the rings of pre-nineteenth-, nineteenth- and early-twentieth-century development, with ring roads where the city walls and an old system of concentric canals had once been, punctuated by large industrial plants, mainly on the north side, aligned beside the five railway tracks that converged on the edge of the city centre. This was the heartland of Milan's neighbourhood life. Beyond this, peripheral estates sprouted up in a much more compact form than in Amsterdam, and with much less attention to the articulation between residential areas, open space and traffic routes.

Urban planning, called in Italian '*urbanistica*', in this period focused mainly on designs for building peripheral housing estates (Foot 2001). Planning powers in the national law of 1942[6] required that urban development projects were located within the framework of a *Piano Regolatore Generale* (*PRG*). This combined a strategic focus on the spatial organisation of the city with the allocation of development rights to specific sites through detailed zoning, with schedules of standards and norms for each use zone/category. In 1953, the *Comune di Milano* approved its first *PRG*, prepared by a team led by architect-consultant Bottoni (Figure 4.2). The plan expressed many of the

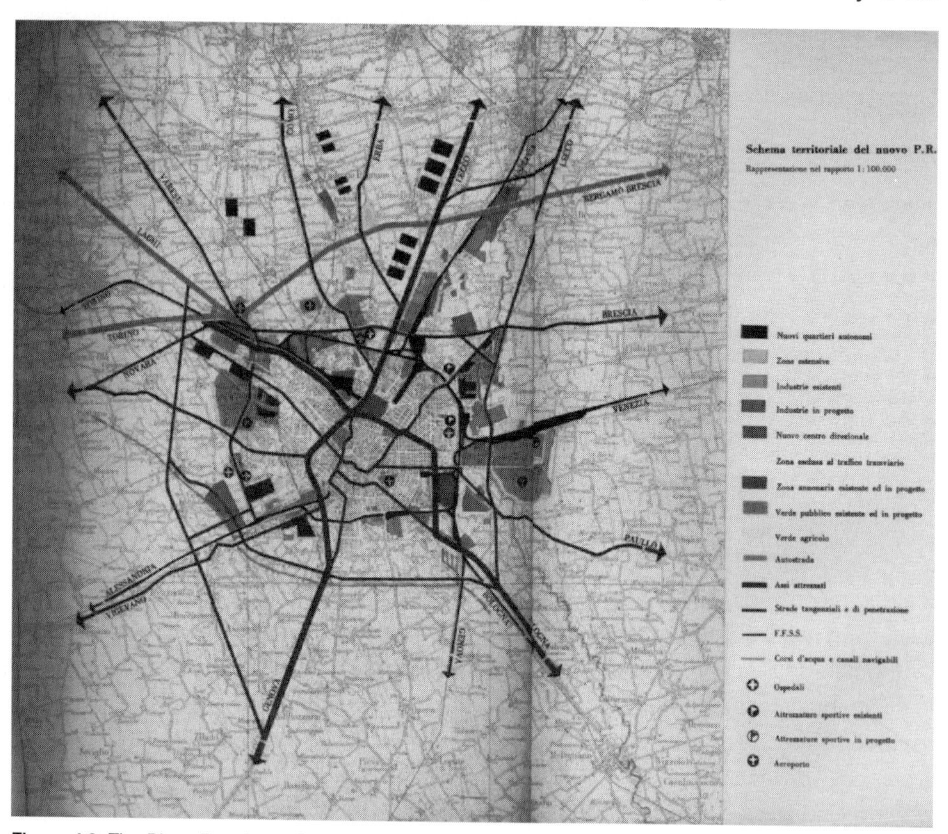

Figure 4.2 The *Piano Regolatore Generale*, Milan 1953

Source: Piccinato 1956: between pages 54/55, with permission of Inuedicione, Rome

planning ideas evolving in Italy in the immediate post-war period and was considered a leading planning exemplar in Italy at the time (Piccinato 1956). It focused primarily on urban extension, while emphasising the linkages between the different elements of the city and the location of the city area within a wider regional context. It also focused on the quality of living places as expressed in the built environment. This *PRG* served as a 'manifesto' for the council's political executive (the *Giunta*) (Gabellini 1988), focusing the *Comune*'s investment in infrastructure as well as urban extension. As in Amsterdam, it was assumed by politicians and planners that the *Comune* led development activity in the urban region and that the plan directed and coordinated the *Comune*'s development work. But, in practice, the plan had little leverage on the property bonanza developing outside the *Comune* boundaries, and the practice of making 'variations' to the plan's zoning provisions, '*varianti*', became common.

Because the *Comune* was building housing estates in the areas around the city, Milan politicians and planners had an interest in building a larger metropolitan area organisation. At that time, the next administrative level, the Province, seemed too large, and the region did not exist except as an enabling possibility in national law. Throughout the 1950s and 1960s, efforts were made to build a supra-municipal arena among the surrounding municipalities. At one time, over 70 municipalities were involved in this arena, but political disagreements continually upset negotiations. In parallel, a planning exercise was made to produce a *Piano Intercomunale Milanese* (*PIM*), a version of which was approved by national government in 1959 for an area comprising 35 municipalities. By this time, the municipalities in the *PIM* group were predominantly left-wing and sought control over the rampant land speculation in the Milan area. However, this plan was never fully approved by the municipalities involved and a formal arena for inter-municipal co-operation did not emerge. Nevertheless, the inter-communal effort continued to be supported by some municipalities as a voluntary research and advisory consortium of municipalities under the direction of mayors (Gualini 2003). Known as the *Centro Studi PIM*, this agency became an important actor in Milanese planning and survives to this day, though in a less-significant form.[7] It was the *PIM* that was the first to advocate a major investment in the city centre designed to expand the central node of the city through an underground rapid transit route connecting two of the main rail stations through the city centre. The *PIM* group recognised the conflicts of interests among groups and municipalities within the expanding metropolitan area, but sought to collaborate to reduce the tensions and develop a collective capacity to intervene in metropolitan urban development processes (Gabellini 1988).

In Milan itself during the 1960s, an increasingly ideological conflict emerged over the city's development trajectory. The left-wing view emphasised decentralisation and the creation of a multi-polar city centre.[8] The *PIM* idea for a public transport axis was developed into the *Progetto Passante* (the 'through line') (Vicari and Molotch 1990) linking major station terminals through the city centre on a north-west–south-east axis. The right-wing view emphasised the dominance of the traditional core around the

Duomo area and sought greater flexibility overall than that available in the 1953 *Piano Regolatore Generale*. An attempt was made to revise the *PRG* in the 1960s, to accommodate the vigorous market dynamics of a rapidly growing metropolis, but as a result of the conflicts, this was never approved (Gabellini 1988).[9]

Urban planners based in the Architecture School at the *Politecnico di Milano* played a key role in these planning developments. As design intellectuals, they were highly respected in Milanese and Italian culture. Several of the urban extension plans were prepared by them (Foot 2001) and they were centrally involved in the *PIM* work. The School's tradition was primarily that of city design, with a strong emphasis on urban form and architectural design. The design of dwellings, housing blocks, larger projects, neighbourhoods and cities were conceived as tasks to which similar skills and perspectives could be deployed. However, in contrast to the Netherlands and the UK, this planning culture was very weakly linked to any pragmatic consciousness of project management, public administration or of how land and development processes worked. The plans were primarily provided in consultancy mode for a client, who was then expected to deal with operational issues. This proved particularly problematic when the actual development process was governed by a complex practice of political negotiation, speculative pressure and attempts to subvert and ignore the regulations in the *Piano Regolatore Generale*. Clientelism, kickbacks to lubricate bureaucratic processes and the use of '*varianti*' to negotiate changes in plan zoning became normal practice, enabling speculative development to proceed with few constraints. As a result, in terms of actual development, only some development was in line with the plan. A great deal occurred outside the *PRG* framework or was negotiated through the *variante* procedure. Where a proposal in the plan did not accord with the specific interests of politicians, landowners or developers, little happened. As a result, and despite the hopes for comprehensive strategic guidance of the urban development process expressed in the 1953 *PRG*, there was little connection between urban development extensions and major infrastructure investments.

Thus a strong design-oriented and strategic approach to planning the urban area coincided with a very weak capacity to link framing concepts either to investment processes or to effective regulation of the location and form of development. The speculative profits made from urban development in this period fuelled left-wing criticism of rampant capitalism, while the lack of attention to development coordination created problems for industrialists, city residents and workers. This critique of the early post-war planning in Milan during a period of massive population expansion created a moment of opportunity for alternative ideas to surface (Balducci 2005a). In the 1970s, the political balance in the *Comune*, always in some form of party mix, shifted from the Catholic Christian Democrat party to the socialists and communists. This shift supported ideas about city planning and metropolitan development that had been evolving among left-wing groups in the 1960s, particularly in the arena of the *PIM* (Gabellini 1988).

PLANNING IN THE ASCENDANT IN THE 1970S

Milan in the 1970s was at the leading edge of innovation in strategic planning for urban regions. The later 1970s are remembered especially for the strong political support for a technically robust strategic approach to city development in a metropolitan context (Balducci 2001a; Gabellini 1988). The result was the 1980 *Piano Regolatore Generale* which legally remains in force into the mid-2000s. This was innovative, not merely as a new comprehensive plan for a major Italian city. It pioneered an approach that depended much more on the skills of social science analysis and much less on the morphological approach of urban designers. The plan was to be based on substantial analysis of social and economic conditions. The objective was to connect the analysis of socio-economic dynamics with the evolution of the city's physical structure, in a new, integrated way of understanding the city. The planning team sought to move beyond over-rigid zoning in order to provide more flexibility in the light of the complex dynamics of pressures for land-use change in a large metropolis. A key objective was to build a closer relation between strategic concepts and actual development. The team itself operated in a different way from that of a standard urban planning consultancy. Special teams were set up to develop knowledge and policy ideas on particular themes, involving a variety of stakeholders from across the city, including politicians. The resultant strategies and the plan became an important political platform for the rising Socialist politicians, especially in the second part of the decade (Balducci 2005a).

The origins of the new *PRG* lie partly in the tensions arising in the 1960s. The 1953 plan had not been implemented. Instead, there had been 'colossal speculation', social housing provision had been driven to the periphery of the city, the public spaces of the city had been neglected by the practice of continual '*variante*', and areas of the historic city had been 'massacred' (*UTERP* 1975, page 3). Academic commentators of the time saw the new plan as an attempt to confront a 'crisis' in the welfare state and in the economy (Ceccarelli and Vittadini 1978). With strong, and to some extent cross-party, political support, the work on the plan was given a high priority. The politicians saw it as making a major statement about the city, to guide the broad spectrum of *Comune* activity. A comprehensive plan was also a valuable support for a new strategy to promote more social housing for many workers and their families, given the overcrowded living conditions in the city.[10]

The new *PRG* evolved over an eight-year period. The council gave authority to form a special office to revise the plan in 1972. This was set up in 1974 as the *Ufficio Tecnico Esecutivo per la Revisione del PRG* (*UTERP*). During 1973, a great deal of effort was put into collecting all kinds of data about the city, through a special data-collection unit. Once the technical office was in place, staffed by a multidisciplinary and multi-party team of mostly young professionals, plan preparation proceeded energetically. The team itself had strong links to the *Politecnico di Milano* and other universities, although party links were more important. Close connections were made with the city's neighbourhood councils, created in the 1970s.[11] Linkages with labour unions and with party networks were also significant, to sustain support and to develop understanding (Gabellini 1988). The team maintained close links with the *Piano Intercomunale Milanese* (*PIM*) team, providing a channel through which the *PIM* ideas of the 1960s flowed into the *PRG* work. This helped to ensure a broad understanding in the new *PRG* of the relations between the area of the *Comune di Milano* and the wider metropolitan region.[12]

A draft *PRG* was available by 1975. The analysis accepted that the city population would fall and that industries would move out. The key was to improve the liveability of the areas of the city, to resist gentrification, to preserve sites for industry to help resist further tendencies for closure and out-migration, to improve transport and service provision and provide more green spaces (Gabellini 1988). By the late 1970s, debate about the city and its planning had become strongly polarised and politicised, in line with the increasing role of parties in the organisation not just of Italian politics but of Italian economic and social life (Foot 2001; Vicari and Molotch 1990). For both left and right, the centre of Milan was sacrosanct, with its streetscape and skylines (Vicari and Molotch 1990). National conservation legislation also limited development possibilities in the older parts of the city. For the Christian Democrats, a centre–right party linked to the Catholic Church and to many of the traditional Milanese elite, the whole Milan area pivoted around the city core ('*cuore*' means heart in Italian). They therefore argued for investments that increased the regional centrality of the centre and for land use regulations which allowed flexibility for the development of tertiary sector enterprises. Market forces should be given primacy in regulating urban growth and in generating and distributing benefits for the public good (Gabellini 1988). The Communists emphasised housing and service provision, neighbourhood quality of life, and increased work opportunities through the protection and expansion of industry. The Socialists, who increasingly dominated the *Comune*, largely supported the Communist position at this time. The *PRG* thus expressed a left-wing view of a working-class city of neighbourhoods with a historic core, accessible by good public transport.

Those involved with the making of the plan believed in the power of the local state (embodied in the *Comune*) to shape urban development. With strong political leadership and the involvement of several municipality departments, they anticipated that projects for the management of public spaces and the provision of services and facilities would be delivered as specified in the plan. *Comune* organisation, as in Italy generally, was formally similar to that in Amsterdam. There was a City Council and an executive team (*Giunta*) consisting of a Mayor (*Sindaco*) and political heads of functions (*Assessore*). *Comune* functions were organised into Departments. Department heads and many senior positions were held by administrators with legal training, with professionals expert in particular fields in a subordinate position. Technical capacity was frequently supplemented by the use of consultants, many based in universities. The new *PRG* was seen by its advocates as a politically oriented but largely technical mechanism to integrate the disparate groupings that existed within this formal structure, replacing the party and social networks that had previously been the integrative device, cutting across the professional communities.

The planning team recognised that some modification of planning instruments was necessary to provide flexibility in the plan's regulatory influence over land-use change. They understood that a *PRG* should be a comprehensive zoning instrument. But given the difficulties of predicting exactly what public and private development proposals would arise and where, they sought to provide much more flexibility in the range of zones and in the norms and standards as applied to zones, particularly in the more central areas, where the morphology of already-built-up areas was an important guide to the nature and shape of new developments. This approach was seen to be particularly innovative in the Italian context (*UTERP* 1975). The result was a detailed zoning map for the whole of the Milan municipal area (Figure 4.3), in contrast to the emphasis in the 1953 plan on urban extension. It was assumed that the *Comune* would lead development activity, according to the logic of citizens' needs. Property market activity would fill in some of the development primarily in existing areas and within the constraints of the existing urban form.

This flexibility went with a very precise specification of norms and standards for each zone. Through developments in national legislation, permits to develop were not only restricted to projects that conformed to the land-use and cubic-space specifications in each zone. They also, under a national law of 1977, had to pay necessary urbanisation charges related to the provision of public services (schools, health centres, open spaces) (Ave 1996). The draft Milan *PRG* prefigured the national changes, and specified requirements. These powers were further strengthened in 1978 by a national law that introduced *Piani di Recupero* (renewal plans), which allowed municipalities to expropriate land in areas where land and buildings needed renewal and to re-allocate the funds from such projects to support further urban renewal projects (Ave 1996). As a result, the new *PRG* seemed to be backed by very substantial powers for the public management of the urban development process.[13] This was reinforced by a national and municipal political orientation that gave a strong emphasis to urban planning (Gabellini 1988).

Compared to the 1953 *Piano Regolatore Generale*, the new *PRG*, available in final form in 1978 and formally approved in 1980, was very ambitious. It focused on the whole urban area, not just areas of urban extension. It demanded a strategic, coordinative capacity within the *Comune* and a technical capacity to manage development projects and regulate development according to politically-agreed, policy-oriented technical norms and standards. But such an urban governance capacity had little tradition in the Italian context. Despite the political support initially given to the new *PRG*, the politicians hesitated over approving it. Many commercial interests and some politicians raised problems with the plan. Some planning academics also criticised the ambitions of the enterprise. Ceccarelli and Vittadini, writing in 1978, saw the plan as an ideological dream,

Figure 4.3 The *Piano Regolatore Generale*, Milan 1978/1980

Source: Ceccarelli and Vittadini 1978: between pages 80/81, with permission of Inuedicione, Rome

liable to be undermined by the '*rito ambrosiano*', the Milanese practice of flexible, incremental adjustment to influential lobbies:

> The Milanese urban society, sclerotic, aging and conflictual, will always tend to defend its privileges with respect to the metropolitan area, but these privileges will be continually challenged and put in crisis by commuters, transient residents and underpaid workers, essential to the city's functioning. In the coming years, governing in a Milan of this nature will not be an easy undertaking, but for many 'living here' will become ever more difficult (Ceccarelli and Vittadini 1978: 87, author's translation).

Socialist Mayor, Carlo Tognoli and his *Assessore* for *Urbanistica*, Paolo Pillitteri,[14] initially maintained support for the plan. The *PRG* was eventually approved in 1980, and almost immediately set to one side by the politicians.[15] Despite perhaps the strongest concentration of political power and technical competence as compared with other major Italian cities, developing a policy-driven mode of governance of urban development processes involved a transformation of the political and governance culture. In the 1980s, the Socialist domination of political life in the city took governance processes in a different direction, in which the role of party networks in creating economic and real estate opportunities became even stronger than in the 1960s. Yet the 1980 plan remained the legally relevant regulatory document into the mid-2000s. In the 1980s, the use of the *variante* procedure once again became the norm, and the plan's role as a strategic guide for the *Comune* as a whole evaporated. This collapse of support for the plan was not just a matter of politics. The period of the production of the plan was the era when deindustrialisation took hold, and a new momentum in the tertiary sector was experienced. Despite the strong commitment to quality of life in neighbourhoods, the plan could do little to resist the run down and then closure of the big industries on which the industrial conception of Milan rested. In a major break with the strategy of the 1970s, a new strategy and practice was promoted that enthusiastically embraced the new economy, celebrating it in planning terms in a 'turn' from 'plans' to 'projects'.

THE POLITICS OF PROJECTS IN THE 1980S

Faced with continuing industrial decline, but a parallel expansion of Milan's long-standing commercial and financial economic dynamics, the Socialist political party which came to dominate national and Milanese politics in the 1980s developed a celebratory and entrepreneurial attitude to the promotion of the city. Promoting growth and the expansion of the tertiary sector became the primary strategic focus (Gabellini 1988). The city core became once again the critical focus of attention. Milan was presented not so much as a city of neighbourhoods but as a great European city and centre of advanced tertiary activity (Bolocan Goldstein 2002; Boriani *et al.* 1986). A metropolitan perspective was retained, but within the context of re-inforcing the centrality of the Milan core. The policy of dispersal of activities across neighbourhoods was dropped in favour of central city projects and, as de-industrialisation progressed, major projects in former industrial areas. The 1980 *PRG* was viewed as an obstacle to the maintenance of the competitive position of the city. In this viewpoint, the new political strategy had the support of real estate and building industry interests which had been strongly opposed to the *PRG* (Balducci 2005a). The new emphasis in urban development was therefore on even greater flexibility in the approach to norms and standards and a proactive emphasis on the promotion of major new projects that would develop the assets of the city and the region (Gualini 2003). These projects were linked especially to the opportunities available on the obsolete industrial sites and around the main rail stations, which had substantial reserves of unused land.

Intellectual support for this political 'turn' was provided by leading planning academics at the *Politecnico di Milano*. Two sometimes conflicting themes, grounded in the Milanese experience, reverberated in planning debates in the 1980s. One criticised the nature of the Italian *Piano Regolatore Generale* (Mazza 2004a, b). The other promoted the strategic role of major 'projects' in shaping transformations in urban morphology and dynamics (Secchi 1986).[16] Thus the 'turn' to a project emphasis was not just a convenient response to a new political and economic project. It could also be grounded in a considered position on how to understand, revive and renew large and complex urban agglomerations.

In practice, however, the 'turn to projects' allowed planning attention to shift to the aesthetics of building projects, and away from their impacts on urban dynamics. Little technical attention was given to the real-estate dimensions of projects, the assumption being that the public sector would define the project opportunities open to private investment. In practice, the political rejection of the premises of the 1980 *PRG* and the support for greater flexibility in land-use regulation led to a kind of ad hoc deregulation, pursued in the form of '*varianti*'. As the real-estate market picked up in the later 1980s, a flow of smaller sites became available for development (Gualini 2003). The major public-sector focus was on the larger sites, either in public ownership or owned by the major industrial companies, on which there was a great deal of discussion but very little actual development.

Although the concepts of the 1980 *PRG* were set aside and its comprehensive approach neglected, two important urban development strategies were produced in the 1980s that had significant material outcomes in the 1990s. The first was the *Documento Direttore del Progetto Passante*. This revived the concept of a rail link between the Garibaldi station, the city centre at Piazza delle Repubblica and on to the Porta Vittoria and Rogoredo stations. The aim was to create a north-west/south-east spine across the city centre, linking development opportunities at both ends and creating a multi-polar city core. This project was supported by an economic feasibility assessment, although this was largely neglected as the project garnered support. Instead, its presentation in attractive images promoted in the media helped to justify the project. The second strategy was *Il Documento Direttore delle Aree Industriali Dismesse*, finalised in 1988. This identified further areas for major redevelopment projects, shifting land uses from industry to a mix of commercial and service activities, open space and apartments.[17] Both strategies marked a break with the earlier comprehensive planning approach, producing instead an agenda of project sites. Many architects, planners and Milan's design elites became involved in proposals and debates about appropriate design ideas for projects on these sites. This in turn generated popular protest in affected areas, where neighbourhood councils were often still strongly committed to resisting gentrification, preferring the industrialisation strategy embedded in the 1980 plan. Some academics also criticised the lack of any urban and regional development logic which could justify the particular mix of development activities proposed (Tosi 1985; Vicari and Molotch 1990), although others read the implications of the projects as generating a more polycentric urban form (Secchi 1988).

For the rail authorities,[18] the owners of former industrial sites and the *Comune* as landowner itself, the project agenda not only represented a way of presenting the city in a modern, European context, but an understanding of the city in terms of opportunities to realise real-estate returns. The turn to projects offered a potential market logic, in contrast to the 'basic needs and quality of life' logic of the 1980 *PRG*. New partnership possibilities began to develop between public- and private-sector actors, in a way unfamiliar until then in Italian urban development (Bolocan Goldstein 2002). Yet, although real-estate interests supported the new orientation, they were not actively involved in project development as such, recognising that the market potential of the sites could only be realised by complex political negotiation (Vicari and Molotch 1990). For the first part of the 1980s, under Mayor Tognoli, respected for his strategic leadership and grasp of urban dynamics, and very well-connected to Bettino Craxi, who was rising to the position of national prime minister, such negotiation seemed likely to lead to major changes in Milan's urban development and real-estate opportunities. The extraordinary paradox of the energetic promotion of major projects in Milan is that hardly any of the ideas emerged from the architects' journals into actual concrete development projects, although many smaller projects did proceed, if slowly, through the *variante* procedure. As Gualini argues, 'at the beginning of the 1990s, Milan's score in the pursuit of its strategic goals appeared to be dramatically low' (2003: 275).

The reasons are complex and much discussed (Gualini 2003). One was the challenge of agreeing actual building designs and transport routes, with project proponents in disagreement, for political, design and real-estate reasons. The consequence was that, if and when a project scheme was finally approved, it often contained conflicting elements. For example, the *Progetto Passante,* intended to strengthen the north-west/south-east axis of the city core, co-existed uncomfortably with the proposal for the third metro line, which aimed to connect the central station, via the Piazza della

Repubblica and on to the Rogeredo station, and hence had a similar objective. National funds were made available for the first part of the *Progetto Passante*, only, in conjunction with the metro project and the transport logic for the truncated project became increasingly weak. Another reason was the difficulty of assembling sites where ownerships were fragmented. In addition, many of the proposals were vigorously contested by Milan's lively neighbourhood groups, linked to the leftist parties increasingly sidelined by the ascendant Socialist Party.

A further reason for the lack of project realisation was the sheer complexity of the process of assembling regulatory approvals and investment agreements through the various departments and permit procedures within the *Comune* (Ave 1996). The project agenda was a politically driven one, with little attention given to the municipal organisation and wider governance processes through which it was to be realised. In particular, the different sectoral departments of the *Comune* and the region, through whom much of the investment funding for the projects had to flow, did not give the project agenda much emphasis in their own investment priorities (Balducci 1988). The inability of key actors within the public sector to reach agreements led to major failures in some high-profile project initiatives. This inability in turn shifted the power relations between the public and private actors. Projects only proceeded if powerful private actors had a strong motivation and held sites in a single ownership, as in the Pirelli (Bicocca) case (Gualini 2003).

By 1985, the political majority in Milan had moved from a Socialist majority to a multi-party council, making it difficult to negotiate strategic agreements through these contestations (Balducci 2005a). Public administration in Milan was thus unable to generate the political direction for either the realisation of major projects or the revision of a comprehensive plan. Behind the scenes, there was a further dimension to the governance landscape. Just as in the 1960s, the project agenda was politically driven, and accompanied by a political system for negotiating 'kickbacks' or '*tangente*' payments by developers and real-estate operators in exchange for building contracts and development opportunities. This kind of 'clientelist' payment, given to businesses through links with party networks, was widespread in Italy until the 1990s. What characterised the practice in Milan was its systematic nature, with calculated payment amounts, distributed proportionally among the parties in relation to electoral support (Foot 2001). In this way, it became routinised and co-opted all the parties. As Vicari and Molotch (1990) show, far from being a pro-development regime driven by real-estate interests, as so often found in the United States (Logan and Molotch 1987), the Milan pro-development regime was driven primarily by the Socialist Party, in hidden alliance with all the other parties and other important public-sector agencies, since they all stood to benefit from the kickbacks.

During the 1980s, debates about planning and development in Milan concentrated increasingly on the city itself. Although the *PIM* agency continued to provide valuable research and data, regional attempts to promote collaborations (*comprensori*) among communes for the delivery of particular services were having little success (Gualini 2003). Meanwhile, the wider metropolitan area continued to expand across the region, while Milan's population steadily fell. The urbanisation of the wider region was facilitated by transport investments. These included major national highways (the *tangenziali*), but also many road schemes initiated by municipalities themselves, which, when connected, created an expanding road network (Balducci 2005a).[19] There was little co-alignment between government levels, and little coordination between municipalities and between different sectors of administration. Where integration occurred, whether vertically or horizontally, it was pulled together through party networks.

Then, suddenly, the 'party system' collapsed:

> Massive and deep-rooted systems of political and economic corruption were unmasked by the dramatic 'clean hands' (*mani pulite*) investigations in the city, which began with the arrest of a mid-level Socialist official in February 1992 and the disappearance of the Socialists from the political scene that they had controlled for so long' (Foot 2001: 157).

This collapse reverberated across Italian political and government life, leading in Milan to the removal not only of a whole class of politicians from all the main parties, but also a clear-out of some of the long-standing technical staff (Balducci 2005a). In terms of governance capacity, the break with the past was much greater than in 1980. However, much of the urban development agenda of the 1980s lived on into the 1990s. The emphasis on the tertiary sector and on major projects remained. The new themes to emerge in the 1990s were a search for technical competence, an emphasis on actually realising projects, and a search for a more flexible and effective approach to managing urban development processes than either comprehensive city-wide zoning or politically driven project promotion.

BUILDING NEW GOVERNANCE CAPACITIES: ALTERNATIVE MODELS IN THE 1990S

The political crisis in Milan spread across the country and removed not only a generation of politicians, but, at least in the short term, the networks that had linked levels of government and actors in government, the economy and civil society. In this context, in Italy generally, the 1990s was a period of innovation and experimentation in building new modes of governance and new approaches to urban and regional development policy. Officials in all levels of government, with legal–administrative and professional backgrounds, used the moment of opportunity to initiate more technical, policy-driven approaches to the management of government (Dente *et al.* 2005). As elsewhere in Europe, there was also a new emphasis on collaborative partnerships and 'round-tables' ('*tavoli*') with all kinds of actors from social and business life. The emphasis on technical competence and partnership formation expressed a search for new sources of legitimacy, as well as new policy ideas, for the politicians who were elected after the political crisis of the early 1990s. These governance developments were reinforced by the European Commission, particularly with respect to negotiations over the allocation of the Structural Funds (Cremaschi 2002; Gualini 2004b). In parallel, the European discourse of economic competition between cities encouraged initiatives in strategic planning at the city level.

A key innovation in Italian government structures and functions was the increased role for regions. Italy remains a unitary state, and regions have had legislative and resource-management powers since 1972, disbursing national government funds (Gario 1995). In the early 1990s, the range of powers and competences was strengthened, including the power to legislate in the field of *urbanistica* (urban planning) and to enter into programme agreements for planning and coordinating projects and policies. Regions already had a role in disbursing the funds made available by national government for service-delivery programmes. The Provinces and other ad hoc groupings of municipalities ('*comprensori*') were expected to take on roles in the coordination of programmes. The traditional landscape of municipalities, which vary in size from big cities with populations of around one million and many small communities with populations of a few thousands, was left intact, but with encouragement for collaboration among them. National legislation in 1990 also provided for the creation of metropolitan areas (see Table 4.1). For municipalities, the main innovation was the introduction of elected mayors in 1993 (Magnier 2004). These reforms suggested that the development of more horizontal, issue-oriented and technically-informed networks would fill the gap created by the collapse of the party networks and the processes of multi-level political fixing of the party system.

The reform momentum was picked up vigorously in the urban and regional planning field, as regions and municipalities sought to promote better conditions in both urban and rural areas. The requirements of access to the European Structural Funds and the way these were developed at national government level proved particularly important in promoting new policy-driven and technically competent governance practices (Gualini 2004b). A new generation of graduates from the planning programmes that expanded in Schools of Architecture and Engineering in Italy took up posts in municipalities, regions and special agencies. In the rich region of Lombardy and Milan itself, however, European funding was of little significance. More important was the national programme for the renewal of obsolete industrial areas, the *PRU* (*Programmi di Riqualificazione Urbana*) (Bolocan Goldstein 2002), and the new regional powers to legislate as regards the instruments of urban development planning. Of particular importance in Milan was the introduction of a new national planning instrument, the *PII* (*Programmi Integrati di Intervento*), with the regions being given the power to specify how this power could be used. Neither the *Comune* of Milan nor the Province were interested in the formation of a metropolitan area, as municipalities did not want to lose powers over re-zoning the many

Table 4.1 Formal levels of government: mid-1990s

Levels	*Relevant to Milan*	*Powers relevant to urban development and planning*
National		Provides enabling legislation for urban planning and for municipal government organisation Provides funds for special programmes Allocates resources to regions
Regional	Lombardia	Makes transfers to municipalities from nationally provided funds for service delivery undertaken by municipalities Powers to pass legislation defining planning procedures and instruments Approves *piani regolatori*
Province	Provincia di Milano	Allocates regional budgets for some services, especially for road building and technical education Encourages inter-municipal coordination and collaboration in service provision and other initiatives
Comune	Milano	Prepares planning instruments Approves proposals for private development projects Invests in infrastructure and some development projects Provides services
Sub-areas	Nine 'decentralisation zones'	Provide local services Express views on development proposals

industrial sites in the area, with their considerable development potential. Within the area of the *Comune*, the neighbourhood councils introduced in the 1970s were re-structured into nine '*Zone di Decentramento Comunale*', which had service delivery functions but did little to strengthen the connection between citizens and their local government.

While other Italian cities embarked vigorously on strategic spatial-planning initiatives,[20] Milan in the mid-1990s was more introverted, emphasising technical competence and administrative procedure in a rather traditional way (Dente *et al.* 2005). Departmentalism became even stronger than before, without the party coordination mechanisms. Milan was increasingly portrayed as deficient in urban qualities and in public-administration capacity (Gualini 2003). The new political forces that surfaced after the collapse of the major parties were populist and increasingly business-oriented. From 1993 to 1997, the elected mayor (Marco Formentini) and the political majority were from the new *Lega Nord* party, which promoted a regional idea rather than a specific policy agenda. In the planning field, politicians focused on progressing many stalled projects into development (Balducci 2001a). Mayor Formentini had originally anticipated the preparation of a new *Piano Regolatore Generale*, an initiative that had also been on the centre–left agenda in the late 1980s. But there was little momentum behind such a project, which was seen as a very complex enterprise (Balducci 2004), and real-estate pressures were relaxed in the property slump of the early 1990s. Instead, the Planning Department during the Formentini administration[21] sought to provide some strategic logic to the promotion of an agenda of projects. An important driving force for these initiatives was the ambition to shape emerging regional legislation and practice as regards the rules governing the *PII* and the disbursements of funds for the *PRU*.

This led to two significant initiatives. The first, the study *Nove Parchi per Milano* (Mazza 2004c), focused on major development areas beyond the city centre and challenged monocentric conceptions of the urban area (Oliva 2002). It also involved a new mechanism for negotiating public benefits from development:

> the valorisation of a key urban resource [the derelict areas within the urban fabric] is turned into an ... appealing urban design vision ... meant to upgrade living standards through the supply of green recreational areas, to be realised by allowing higher densities on some parts of sites in return for the provision of parks on other parts (Gualini 2003: 276).

This pioneered the idea that public-interest benefits should not just be calculated as payments related to standard site-based requirements, but could be negotiated as specific contributions in kind to creating public realm assets. This was a new approach for Italian planning, but required a strong private investment interest and an effective capacity for public-sector management and coordination (Bolocan Goldstein 2002). The *Nove Parchi* study was undertaken by a team of academics, providing both a strategy for project development and design ideas for specific projects. It was used in the Planning Department as informal guidance in negotiations over particular sites. However, the projects it generated added yet more sites to the existing project agenda, and there was still no clearly articulated strategic logic within which the development expectations and requirements of the different projects could be located. This situation was exacerbated when Milan's proposals for the *PRU* programme were drawn up, with only limited overlap between the *PRU* projects, the *Nove Parchi* projects and the inherited agenda of projects (Bolocan Goldstein 2002) (Figure 4.4).

In general, however, the *Lega* period was largely one of pragmatic actions, characterised by few new initiatives and little contact with other social groups within the city.[22] After the rich, if complex, networks of the old party system, this period seemed in retrospect one of political and administrative isolation, from the rest of the city as well as the wider Milanese urban area (Gualini 2003; Newman and Thornley 1996). Despite the

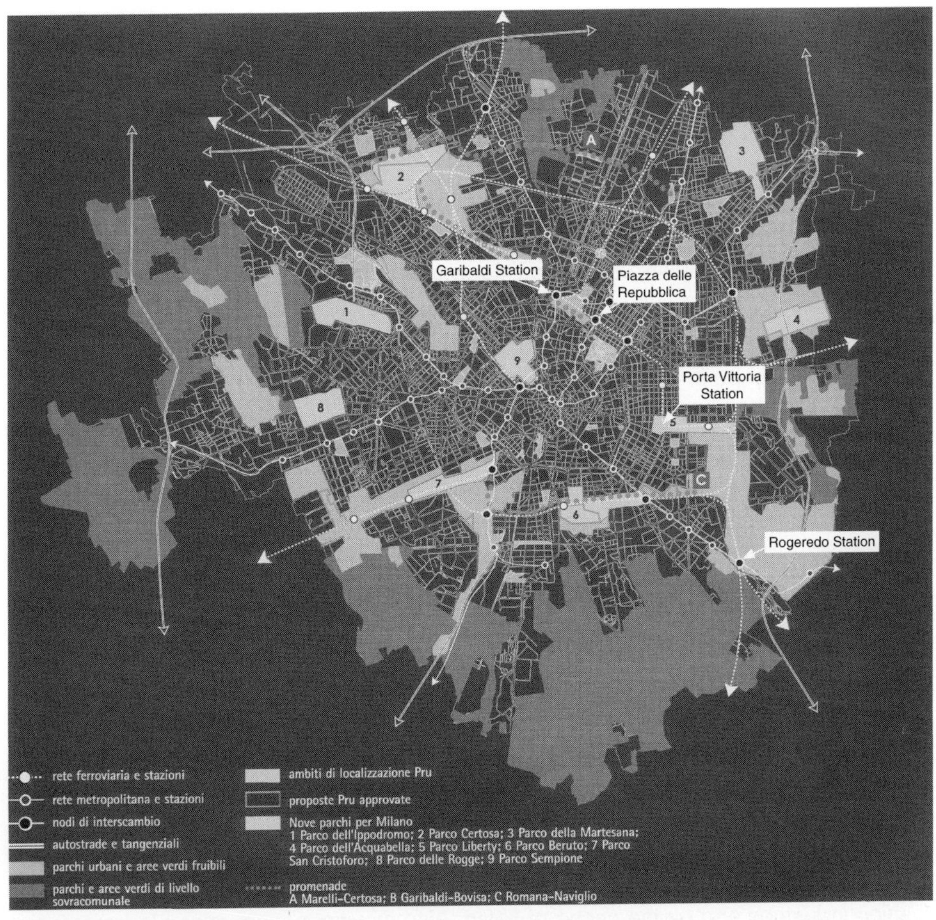

Figure 4.4 Project sites by the mid-1990s

Source: Bolocan Goldstein 2002: 101, with permission of Inuedicione, Rome

Note

The sites of *PRU* projects and those proposed in the *Nove Parchi di Milano* are indicated in lighter shading. The *Nove Parchi* projects are numbered

strong emphasis in national and regional initiatives on coordination and collaboration, the reality of Milan *Comune* remained strongly sectoral, with each department operating largely independently under its *Assessore* and little coordinative power in the municipal *Giunta*.[23]

By the elections of 1997, there was a movement towards a more strategic and interactive view of the city, its development and its articulation with other parts of the region. This was promoted particularly clearly by the manifesto of the centre–left coalition.[24] Its manifesto emphasised the strategic relation between Milan and the region, the interaction with the neighbourhoods, and introduced a new priority of building collaborative, participative processes for strategy formation and development management. This participative agenda and its development in the planning field derived from a body of ideas being developed in the *Politecnico di Milano*, linked to international debate in the academic planning field. inspired in part by work by Judith Innes and others at the University of California, Berkeley. However, centre–left forces were narrowly defeated by Silvio Berlusconi's *Forza Italia*, leading to a businessman Mayor for the city, Gabriele Albertini, a *Forza Italia* City Council and *Giunta* with business backgrounds in most of the city's main Departments. It was left to the municipalities around Milan to develop the collaborative agenda at a strategic level, through inter-municipal initiatives (Pasqui 2002), with some help from the Province.

Meanwhile, the Milan urban area continued its relentless growth and expansion, sprawling across the Lombardy region and beyond, with services and infrastructure being added incrementally and unevenly. Urban analysts increasingly referred to this sprawling landscape in terms such as as '*la citta frammentata*', '*la citta diffusa*', '*la citta infinita*' or '*un immagine caotica*', with proposals for alternative urban morphological ideas.[25] It was not just that the distances between the Milan core and the rest of the urban area were becoming ever greater. These images attempted to capture a reality with multiple layers of socio-spatial networks, with a diversity of nodal patterns. The urban agglomeration was not merely 'polycentric' but should be imagined in terms of networks (Tosi 1990).[26] However, Milan's commercial and property-owning elite was little concerned with such issues. The property market revived towards the end of the 1990s and boomed again in the early 2000s.[27] Property owners in the city core were not particularly development-oriented and were content to see their assets appreciate over the long term (de Magalhães 2001). Nor were the elite much interested in the rest of the city. They lived a life in city-centre apartments and larger homes on the coast to the south or in the mountains to the north. Families in search of bigger dwellings, better services and a cleaner, safer environment moved out of the city. Coming into the city centre, in contrast, were many young people from all over Europe, attracted by the design and fashion industries and by the caché of Milan. Also moving in were migrants from poorer parts of the European Union and from the Balkans and North Africa. This latter movement was producing some degree of socio-spatial segregation in what had been socially mixed neighbourhoods around the high-value city centre.[28]

Yet these various socio-spatial shifts in the city and the region had little impact on perceptions of the city among *Comune* politicians. In the minds of politicians, elite groups and many citizens, the city did not need an explicit expression. It existed as a sort of taken-for-granted force, an ambience so powerful that the threats posed by the diffusion and fragmentation forces were barely noticed, except in the impact on city-centre daily life, especially pollution, congestion, safety and the quality of public spaces. As one of those involved in the planning innovations in the *Comune* commented, for the elite Milan was the square mile around the Piazza del Duomo. The rest of the area, including the inner and outer neighbourhoods and the wider urban area, was just 'territory'.

In contrast, the public administration found itself in a new situation. The collapse of the old political parties and their networks broke not only all kinds of linkages between government and the wider society; it also destroyed any remaining respect for, and expectations of, public administration. One consequence has been the expansion of all kinds of self-governing initiatives in civil society, many of these drawing on neighbourhood mobilisation experiences of the 1970s and 1980s (Dente *et al.* 2005). Some new movements, more issue-oriented, were also appearing. Business groups, which previously relied on party networks, and then the new, more right-wing politicians, to promote their interests, also began to get more assertive in promoting their concerns about the development of the urban region economy.[29] These developments in the public realm created potential opportunities for more collaborative governance practices (Bolocan Goldstein 2002) and had some impact on *Comune* service-delivery practices (Dente *et al.* 2005).

The new mayor, Albertini, in power from 1997 to 2006, pursued a largely pragmatic path, but nevertheless sought to link the public administration to the wider society, and to demonstrate efficiency and effectiveness by getting projects completed. Albertini viewed the city as if it was an enlarged '*condominio*' (apartment building), which needed efficient management (Dente *et al.* 2005). Under his leadership, there has been little political interest in a strategic and coordinated view of the city and its development. A departmental re-organisation was undertaken in an attempt to improve performance, focused on an output-oriented emphasis, with coordination to be achieved through regular meetings of the team of *Assessori*. In the planning department, this encouraged innovations in the management of development projects. The main emphasis as regards the planning function in the late 1990s was once again on providing a more flexible approach to the regulation of development, to facilitate the realisation of projects both large and small, but in a way that was technically competent and transparent. There was also a political concern to ensure that Milan was in the forefront of the development of the new regional legislation for planning. This was aided by links between politicians and officials at regional and municipal levels and through church networks that replaced the old Christian Democrat party networks. To pursue this agenda, the *Assessore* for Planning, Maurizio Lupi, was advised by senior officials in his Department to seek technical advice from internationally renowned planning theorist Luigi Mazza of the *Politecnico di Milano*, who for the past 15 years had been writing about technical ways of introducing greater flexibility and discretion into Italian planning law and practice. The result was a strategic attempt to innovate new, strategically situated, project negotiation practices.

THE SEARCH FOR TECHNICALLY DRIVEN STRATEGIC FLEXIBILITY

MOTIVATIONS

The focus of the innovative approach developed in the late 1990s/early 2000s in Milan was on combining a highly selective approach to spatial strategy with the introduction of new instruments into the practices of the Milan *Comune*. For the planners of the *Comune*, the innovations represented a major change in concepts and practices:

> The Comune di Milano was . . . [required] to reconsider its *modus operandi* and its entire organisational set-up: this process of drastic reorganisation of the local government 'machine' being an indispensable prerequisite to be able to define and implement a new model of technical, administrative and economic assessment (Collarini *et al.* 2002: 129).

Planning commentators saw the initiatives as part of a continuous innovative tradition in planning approaches and instruments in Milan (Palermo 2002). However, the *Comune*, as 20 years before, was hardly ready for the 'drastic reorganisation' implied in the new planning tools. The *Giunta* was dominated by businessmen, who focused primarily on conditions in the core area of the city. There was little political interest in the wider metropolitan area, or even in coordination between the different departments of the *Comune*. The prevailing attitude was incremental rather than strategic (Dente *et al.* 2005). In this context, the challenge for planning innovation was to set in motion technical instruments and a momentum which would, in time, encourage and provide support for a 'turn' to a strategic approach and a metropolitan perspective. It can be seen as a kind of 'strategic planning by stealth' in a very difficult institutional context.

Although cautious about a strategic initiative in the field of urban development, Mayor Albertini was keen to re-establish links between the major economic actors in the city, neglected by the *Lega Nord* administration. During 1997, a consultation exercise about city issues and priorities was undertaken with these actors, though this had little impact on subsequent policy.[30] As with his predecessors, the Mayor's primary focus was on realising projects, which was seen to need greater flexibility in land-use regulation processes. In parallel, initiatives were underway to develop regional legislation to underpin the new approach to coordinating action on development sites through the *Programmi Integrati di Intervento* (*PII*). A key instrument was to be a strategic framework document, to guide the specific instruments that gave development rights to land and property owners. Following intensive debate over the previous decade among academics and within the national professional association[31] about the need to divide the functions of the traditional *Piano Regolatore Generale* between a strategic guiding framework and specific zoning instruments, the idea was emerging in the *Comune* for a requirement that the *PII* should be set in the context of a *Documento di Indirizzo* (a directing policy statement). This was taken up by the *Regione* in changes to the legislation, and renamed a *Documento di Inquadramento*. The Mayor and *Giunta* approved the preparation of such a policy statement by the Strategic Planning section in the Planning Department in 1998 (see Table 4.2).

As the initiative developed, it led not only to the production of a strategic framing document, the *Documento di Inquadramento*, but to new ideas about flexible zoning and the allocation of development rights, new practices for evaluating development project proposals and the introduction of a new coordination instrument, the *Piano dei Servizi*. All these instruments were designed to shape a practice for negotiating the interactions

Table 4.2 Chronology of the production of Milan's strategic planning instruments

Date	*Event/activity*	*Formal decision*
1997	Election of Mayor Albertini *Stati Generali* consultation with key interest groups and associations	
1998	Giunta approves initiation of work on a *Documento di Indirizzo* for Milan *Assessore* Lupi invites Luigi Mazza to act as leading adviser to the Working Group to prepare the *Documento di Indirizzo* November: Working Group starts work	
1999	May: First draft available (*Ricostruire la Grande Milano*)	Regional Law 9/1999 approved
2000	January: Presentation of the draft to the *Giunta* February: Various social and economic organisations in the city asked to comment May: Presentation to the *Comune* Council June: Council approves the *Documento di Inquadramento* June: Seminar held by the Italian Society of Urbanists (*SIU*) on the *Documento* July: Seminar held by *INU* for Milan Architects Society October: Seminar held by *INU* on the Milan experience in relation to the development of the region's planning law October: Seminar held with the *Associazione Interessi Metropolitani* (*AIM*)	Milan *Comune* Council approves the *Documento*, using its autonomous power to do so provided by Law 9/1999
2001	*Comune* organises a meeting on Milan's development	Regional Law 1/2001 introduces the instrument of the *Piano dei Servizi*
2002	Mazza introduces the idea of a simplified approach to zoning in the city to politicians	*Comune* Council approves simplified zoning approach
2003	Work on the *Piano dei Servizi* initiated Mazza presents ideas for a coordinated approach to the *Piano dei Servizi* to the Mayor and Giunta	Regional law proposed, introducing *Piano di Governo del Territorio*

continued

Table 4.2 continued

Date	*Event/activity*	*Formal decision*
2004	Mazza resigns from advising the *Comune* on strategy June: Draft of the *Piano dei Servizio* completed	
2005	Work continues on the *Piano dei Servizi* Preparations for the *Piano di Governo del Territorio* underway	Regional Law 12/2005 approved, enabling preparation of *documento di piano, piano dei servizi and piano delle regole*

Source: See Pomilio (2001, 2003), updated by author

between public and private actors over development projects that responded to market initiatives, while at the same time influencing where these initiatives arose and extracting significant public benefits. The objective was to replace the political 'fixing' of the past with the technical assessment of project impacts, driven by clear policy principles.

Professor Luigi Mazza, of the *Politecnico di Milano*, acted as consultant adviser to this work, to provide intellectual orientation to the development of the approach. He was widely respected in Italy and internationally as a leading scholar of planning systems, with extensive practical consultancy experience with municipalities, including an advisory role in the preparation of the *Nove Parchi per Milano* study. He had written extensively on the necessity to separate the strategic and zoning functions of plans, on the need for flexibility in devising planning strategies, on the interactions between planning strategies and regulation tools, and on the shaping of land and property market opportunities.[32] Mazza was also known for his political independence from the political networks of previous periods, although having a clear commitment to certain planning principles. This led him to emphasise the role of planning instruments and practices in shaping land and property markets and in regulating development to ensure public benefits. There were tensions in this position between Mazza and *Assessore* Lupi, for whom the new planning instruments were envisaged as mechanisms to make the practice of producing '*varianti*' to zoning plans more speedy and more transparently legitimate.

FRAMING A STRATEGIC UNDERSTANDING

Working with the head of the strategic planning section, planner Giovanni Oggioni, and under the general direction of the administrative departmental head, Emilio Cazzani, Mazza and a small team of officials and secondees from the *Politecnico* prepared a draft *Documento* (Strategic Framework) very quickly. Mazza was hired in 1998 and the first draft of the *Documento* was produced within six months.[33] In contrast to the strategic-planning initiatives in other cities in Italy, the emphasis was not on producing a new comprehensive strategy for the city but on providing the groundwork upon which such a conception could emerge (Comune di Milano 2000; Mazza 2001). As one planning commentator noted:

> it occupies, albeit in an imperfect way, that theoretical and technical space which, in other contexts and in different languages, has been identified as a strategic plan and a structural plan' (Gabellini 2002: 132).

Mazza emphasised the importance of developing a strategic understanding of urban dynamics. Rather than attempting a comprehensive plan, any strategy should focus on the emergent urban development tendencies which were shaping the spatial patterning of the urban area and how these could be influenced strategically by public investment initiatives and regulatory interventions. Given that much of the city was already built, the focus of strategic effort should be on the areas and sites where change was expected. The zoning function could be approached by assuming that existing use rights would remain. This meant that small-scale projects in line with existing uses could proceed without the need for any kind of '*variante*' procedure.[34] Mazza was well-aware of the political orientation of the *Comune* and the limitations this placed on any major strategic planning initiative. Instead, he saw the opportunity of producing a *Documento di Inquadramento* as a narrow window for technical innovation through which to build a practice that could grow into greater significance if and when the political opportunity arose for a stronger role for spatial strategy in the orientation and organisation of Milan's public administration (Palermo 2002).

In contrast to the collaborative initiatives being developed by Mazza's colleagues in the municipalities to the north and south of Milan (Pasqui 2002), the work in preparing the *Documento* was a technical planning exercise. The main effort of the Planning Department[35] centred on major development projects on sites owned by the *Comune*, urban design in conservation areas, responding to private development initiatives (through the *Programmi Integrati di Iintervento* process) and the provision of building and related permits. The team preparing the *Documento* worked alongside these major functions, and sought to maintain close links with their *Assessore*, and, less directly, the mayor. This working practice reflected a traditional relationship between technical staff and politicians.[36]

The resultant *Documento di Inquadramento: Ricostruire la Grande Milano* (Comune di Milano 2000) is in two parts. The first part was in the form of an essay written by Luigi Mazza (2004c). He presented the *Documento* as having two purposes: providing a new, more flexible yet clear procedure for planning practice and a frame of reference for the *Comune*'s urban policy. As a strategic frame of reference, Mazza emphasised the importance of linking strategic concepts to specific project proposals in an interactive rather than a linear way. A key purpose of the strategic frame of reference was to provide a policy-driven strategic context for decisions about the *PII*, that is, for significant development proposals that do not directly conform to extant zonings.

A critical relation for Mazza was to connect the assessment of a project to a strategic understanding of emerging urban dynamics. Arguing against the idea that projects should conform to a previously agreed strategy, he proposed an understanding of strategy as a concept that could be reviewed and reconstructed around every new development project:

> [a project is a project if] proceeding from its specificity, it is able to reconstruct around itself a comprehensive vision and assess this vision against the other comprehensive visions that the city was able to produce and that the project itself has the power to suggest (Mazza 2004c: 47, translated by original author).

Critical to the framework was a strategic understanding of the evolving urban dynamics of the city of Milan in its wider regional context, the 'city as it is evolving'. The second part of the *Documento di Inquadramento* emphasises that its strategic orientation is aimed at increasing Milan's qualities as a national and international economic and service node, combined with a 'traditional ability' to integrate the activities of those visiting and working in the city. This is called a 'relational strategy', intended to position the city on a growth trajectory in relation to other European cities, a clear mobilisation of the discourse of urban 'competitiveness' (Comune di Milano 2000: 63). However, Mazza emphasised the relentless growth of the wider urban area (*La Grande Milano*) and the extent to which the *Comune* area was losing momentum relative to the areas outside. In this geography, the city was losing dynamism to the metropolitan area. This led to an argument that more development opportunities needed to be created within the city. However, rather than continuing with a model of the monocentric city, an alternative spatial idea was necessary, with sufficient reality to provide a stable concept to which land and property development actors could relate. Drawing on ideas already developing in the major studies of the 1980s, in the *Nove Parchi di Milano* study and in the *PRU* agenda of projects of the 1990s, Mazza proposed that development projects should be encouraged to cluster along a key emerging transport axis within the metropolitan area. The axis flows from west to east, from the new airport at Malpensa via the city centre to Rogoredo and the existing Linate airport, stretching out to the east towards the airport of Bergamo. This is complemented by a projection to the north-east, linking through the Bicocca site to further obsolete industrial areas overlapping with the *Comune* of Sesto San Giovanni to the north. The resultant structuring image is of an inverted T, a '*t-rovesciato*' (Figure 4.5).

The intention was that public and private investment should be concentrated along this axis (Balducci 2001a). In this way, a metropolitan area perspective was inserted into the map of project proposals and development sites inherited from the 1980s. The idea of this axis, referred to as a '*dorsale*' (backbone), with a '*cuore*' (heart) in the city centre, is to link developments around the airports in the periphery of the region to the city centre, a development of the earlier idea of linking the railway stations, and to encourage

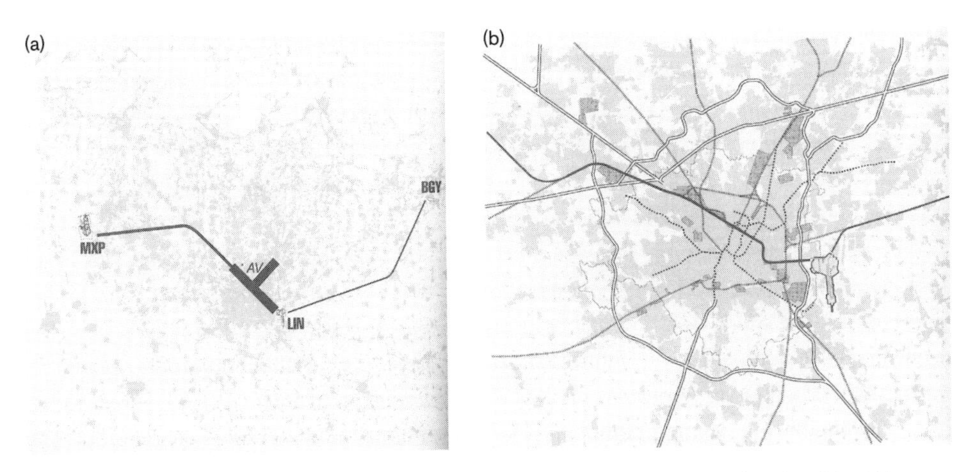

Figure 4.5 The *t-rovesciato* axis and the major development sites: (a) The scheme for the urban '*dorsale*'; (b) the new urban spatial model

Source: Comune di Milano 2000, Figures 6 and 8, with permission from Milan City Council

development opportunities along the axis to come forward. Combined with this strategic shaping of the urban area, the *Documento* emphasised the importance of providing a 'better urban and environmental quality' within the city. This combines a concern with providing more green areas, better services and accessibility across the city, with a positive approach to emergent initiatives from all kinds of sources, and suggestions for developing alliances with other stakeholders across the urban area. Mazza's introductory essay ends with comment on the organisational implications of a strategic approach, with suggestions for more attention to regulatory norms and standards, to administrative reorganisation, and perhaps the creation of an arena for debate about the city, its future and the significance of proposed projects. The second part of the *Documento* enlarges on the issues outlined in the introductory essay, and draws in many of the existing policies and ideas already in circulation within the planning department.

Although grounded in earlier planning ideas about the key structuring elements of the city and the region, the '*t-rovesciato*' image came from Mazza, rather than from any discussion among key actors in Milan. For the Working Group, the objective of the strategic image was to influence politicians and officials in the *Comune* rather than private investors, and to make them more aware of the relations between development projects within the city and the wider region. Following the strongly emphasised view that the *Documento* was not to be seen as a strategy but as a step towards a strategy, many issues were left under-specified. This was particularly so with the investment proposals of other *Comune* departments. In a situation where inter-departmental coordination was generally lacking, the *Documento* aimed to raise challenges and questions. For the technical staff of the planning office dealing with projects, the *Documento* identified '*linee de forza*', emergent lines of force that were already shaping the city. While they began to develop the potential meaning of the strategic idea, *Comune* planners explored notions of multi-nodal corridors and radial lines of force, harking back to older images of a core city radiating out across the region, with multiple urban nodes.

Given the emphasis on maintaining political support for the approach, discussions with the *Assessore* and the *Giunta* were critical, with only limited time given to other stakeholders before the *Documento* was approved by the *Comune* Council in June 2000. Several seminars on the *Documento* were then organised for planners,[37] for regional officials and with the *Associazione Interessi Metropolitani* (*AIM*). Mazza also wrote about the exercise in the planning press (Mazza 2001). As a result, by 2001, the work on the *Documento* was attracting considerable critical attention among planners, with a special issue of the professional journal, *Urbanistica* (Bonfanti 2002).

For the Italian planning community, the *Documento* was a completely different kind of product to the usual plans and schemes. It was a policy text, filled with careful argumentation. It had no illustrations apart from minimalist sketches of the strategic ideas. Some saw the approach as taking 'flexibility' to extremes, representing a market-led approach to deregulation of public control over land-use change. Others revived the old arguments between a market-controlling and a market-driven approach to planning which had divided planners in the past (Salzano 2002). Some wondered if the technical, policy-oriented emphasis would be strong enough to squeeze out the old clientelistic practices. Another line of criticism focused on the process, arguing that it had been far too narrowly-based and therefore could not build persuasive force across government departments or with private actors.[38] Of particular concern was the neglect of the neighbourhood dimension of Milanese life, once so important in the 1970s; although by the 1990s, the links between citizens and their City Council seemed to have become increasingly remote. Other critics claimed that too much attention was still being given to the central area and too little to developing new nodal centres and decentralisation of functions and relations with the local context, echoing the arguments of the 1960s and 1970s (Mugnano *et al.* 2005).

While the academics debated the principles and ideology underlying the *Documento*, planning staff in the *Comune* were absorbed in developing the different practice culture that it meant for them. Instead of checking projects for conformity with plan zones and norms, they now had to assess them in terms of their performance in relation to evolving urban dynamics and rather general policy principles. Not surprisingly, they felt the need for some kind of more precise specification to guide their work in project development. As time went on, they were also constrained by the limited interdepartmental awareness of, or support for, the arguments in the *Documento*. The position of the planners weakened when *Assessore* Lupi became a member of the national parliament.[39] By 2004, the primary leverage of the *Documento* was in the negotiation of projects in the context of *Programmi Integrati di Intervento*.

NEGOTIATING THE PUBLIC INTEREST IN DEVELOPMENT PROJECTS

In Italy, until new national and regional planning laws were developed in the 1990s, a land or property-owner or developer who wished to undertake a building project had a right to develop subject only to a building permit, if a project was in line with the uses and norms indicated in a prevailing *Piano Regolatore Generale*, which generally took the form of a detailed zoning map. The 1980 *PRG* for Milan was perceived as remarkable in its day because it introduced considerable flexibility into the specification of zones and norms (Palermo 2002). Nevertheless, most larger projects tended to deviate at least in some respects from *PRG* specifications, and in Milan, with the rejection of the basic policy thrust of the 1980 *PRG*, deviation was the norm. Larger projects therefore proceeded through the '*variante*' procedure. But this could be time-consuming, involving assessment within the planning office, consultation with other municipal departments, consultation with affected land and property interests, and with the public in the area of the proposed project. The new planning legislation of the Lombardy region introduced by Law 9/1999 not only enacted the power to prepare a *Documento di Inquadramento.* The key role for the *Documento* was to provide a framework within which a simpler procedure for approving '*varianti*', the *Programmi Integrati di Intervento* projects, could be followed.[40] The *PII* process applied to projects initiated by private actors, rather than the major projects pursued by the public sector. With reduced public-sector funding, the law anticipated greater reliance on such privately initiated projects, as elsewhere in Europe. The key innovations of this new procedure as developed in Milan were a parallel rather than a sequential process for consultation and technical assessment, a policy-driven and negotiated approach to deciding how public interests and private objectives could be combined in a project, and greater reliance on technical assessment of the merits of a project.

The new procedure emphasised intense consultation with the main stakeholders at the early stages of a development proposal. A key innovation for the planning officers was the use of informal '*tavoli*' or round-tables, drawing in representatives from service agencies, developers and property owners, and citizen groups if relevant. These middle-level collaborative arenas have become increasingly valued by those involved, leading to creative problem-solving and considerable learning about the challenges of development coordination. The role of the Planning Department staff is to identify who needs to be involved and to set up the consultation processes, in parallel with undertaking or commissioning external technical assessments, including environmental impact assessments. Planning staff also acted as guardians for both strategic policy and for the negotiation of public-interest benefits (*beneficio*). In default of a formal strategy, the planning staff used the *Documento* to give strategic orientation in these consultation processes. The negotiated project, including the package of agreed public benefits, and the technical assessments, are submitted to a special panel, which reviews the material and makes a technical report. This panel, the *Nucleo di Valutazione* (Evaluation Panel) was composed, in the early 2000s, of a mix of technical experts, legal administrative officers, and three independent consultants, including Luigi Mazza and Lanfranco Senn, an economics professor from Bocconi University, who provided guidance on assessing the balance between public and private benefits from a development proposal. Overall, panel members were selected to emphasise technical expertise. The panel makes an assessment based on policy and technical issues, and its minutes are available to the public.

The overall package is then submitted to the *Comune* Council, for approval as a '*variante*'. Generally, the advice of the Evaluation Panel is followed. Once over this hurdle, a project proposal still has to proceed through other relevant approvals, including acquiring a building permit and permits relating to conservation requirements, before finally receiving full Council approval. Only then can the developers proceed to construction. Despite the potential for delay in these subsequent processes, *Comune* planning staff believed that there had been a significant speeding up of the development approval procedure, and a significant negotiation of public benefits.[41] Although most of the many development projects proceeding to completion in the real-estate boom of the early 2000s were on the earlier *PRU* sites, by 2005 the new procedures were beginning to produce development on the ground.

Meanwhile, further flexibility for smaller development projects was provided through innovations in the general approach to zoning, approved by the *Comune* in 2002 on Mazza's recommendation. Zoning gives land and property owners rights to develop. The new zoning approach established three broad zones based on existing uses: the historic core, governed primarily by conservation legislation, the rest of the built up area, and the areas that could be considered as undeveloped (*vuota*) (including sports fields, airports, etc.). This replaced over 40 zones specified in the 1980 *Piano Regolatore Generale*. In this new approach, within the built area, everyone has the same development rights, as expressed in a standard plot ratio. If a developer wishes to build at higher densities, and if the policy framework suggests this is appropriate, it is necessary to accumulate rights from those who would like to develop but for whom the policy framework indicates that this is not appropriate. Through developing a market in development rights,[42] over time, the land value map of the city would change as development concentrated along the development axis proposed in the *Documento*. A key element of this simplified zoning approach is that developers could only accumulate plot ratio rights from other owners if the latter sold their property to the *Comune*. This new approach clearly had major implications for property values, and for the *Comune*'s approach to acquiring and managing sites and properties.[43]

The *Comune* planning staff involved in these processes were generally very positive about the innovations. They emphasised the substantial increases in understanding among those involved, both of the impacts of development projects and of each other's situations. Nevertheless, these new processes and powers presented complex challenges. On the one hand, there were questions about the nature of development rights and the balance of compensation from those whose rights are reduced in the shift from the old to the new zoning approach. More immediate, however, was the problem of determining what public benefits to negotiate for and how to manage those acquired. Some knowledge was needed of existing provision and deficiencies, which in turn involved coordination within the *Comune*, to determine provision, needs and where the financial payments and land transfers should go. To maintain trust in the technical emphasis of the *PII* process, the planning staff needed to have some reassurance that, once transferred, the contributions would be used for the purposes assigned in the development negotiations. The *Piano dei Servizi*, an instrument introduced in the regional law of 1/2001, and argued for by Milan planners, was the mechanism used to pursue these issues.

The idea of a *Piano dei Servizi* (a Service Plan) was to provide a transparent and legitimate statement of the demands to be made by municipalities in negotiating contributions to the public benefit (*beneficio*) from developers. This sought to shift away from the rigidity of the earlier specification to developer contributions related to particular land uses/plot ratios, avoid the potential for corrupt payments (*tangente*), and provide greater clarity and certainty to developers. It was thus, in theory, a key tool for integrating investment resources with regulatory powers. In Milan, Mazza argued that the *Piano dei Servizi* should be seen as a key corporate planning document for the whole *Comune*, coordinating the investment plans of each service department.[44] But, by the early 2000s, the different departments of the *Comune* were operating in an even more sectoralised way than before, with their separate agendas, their own service plans and their own networks of experts and contractors. A coordinated *Comune* investment plan was not likely to succeed in this context. Rather than being used as a corporate coordination mechanism, the *Piano dei Servizi* was prepared within the Planning Department by a team of young, well-qualified but temporary staff, led by planner Giovanni Oggioni.

This team, strongly influenced by the *Documento di Inquadramento*, understood the *Piano dei Servizi* as more than just a shopping list of *Comune* projects for which developers' contributions were sought. They set out to understand the spatial pattern of existing service provision, in order to identify areas of service and transport deficiency. They conducted analyses of accessibility to different services, using simple distance-decay measures, focusing on services provided by the *Comune*. Information on the provision of services was not readily available, which led team members to contact other departments through horizontal, middle-level contacts. They then found that most of those involved in service delivery did not maintain spatially-referenced information about their services and investment plans.[45] Team members also began to perceive the very different ways of thinking that prevailed in other departments, as they came across a landscape of largely autonomous organisational cultures. Nevertheless, as many planners have found in situations where there are strong resistances to central coordination within a municipality, through these discussions, a kind of informal middle-level networking between departments began to develop, with the potential for the planners to provide a helpful information service to other departments. This expanded out, as team members sought out academic teams who had studied social groups and service provision in the city,[46] and also began to make links to the 'decentralisation zones'. The first ideas for a *Piano dei Servizi* were available by June 2004, with further development completed in December 2004. Increasingly, team members came to see the *Piano dei Servizi* as a knowledge system, to encourage service providers to revise their own programmes in the light of deficiencies and opportunities revealed by locational analysis and by the practice of negotiating public-interest benefits through planning processes. By late 2005, some departments and 'decentralisation zones' were contributing to, and making use of, this 'knowledge system', which was also proving useful in the project negotiation process.

But while this work helped to draw attention to what should be negotiated as public benefits in *PII* development projects, there were further problems about how these benefits should be used and managed. The *Comune* was already having trouble managing its existing facilities, with pressures to reduce public spending. The new *PII* procedure was generating more assets.[47] By 2004, some of those involved also feared that the spectre of clientelism might be re-appearing, with land assets acquired through the negotiation processes being used to serve the purposes of political patronage rather than the provision of the benefits for which the contributions were negotiated. If this happened on any scale, developers' trust in the legitimacy of the negotiation process could easily evaporate.

INNOVATION IN GOVERNANCE PRACTICES

Rather than a major effort to 'summon up' a new, persuasive image of the city of Milan for the twenty-first century, as many other Italian cities attempted, the initiatives described here were focused on transforming administrative practices. They were technically focused and centred inside the fine-grain of government processes. Their objective was to attack the policy–action gap that the previous plan-focused and project-focused city development strategies had encountered. The idea was to bridge the gap by a combination of appropriate legal instruments, technically informed judgement and interactive network-building with relevant stakeholders. In effect, the promoters of the initiative were seeking to build up the kind of technically focused policy community around the planning function which was so well-developed in Amsterdam. Underpinning the initiatives in governance process was a sophisticated recognition by the planning academics of the complexity of urban dynamics and the multiplicity of the driving dynamics shaping urban futures. The role of the state was seen as important but much more limited than that imagined by the Milan planners of the past. The key to a strategic role in urban development was therefore to develop a deep and robust understanding of urban dynamics, to focus attention on critical structuring elements and to use the operational tools available in a coordinated and consistent way. The heart of the planning function thus lay neither in the preparation of a comprehensive strategic plan, nor in project design or master planning, but in the making of a strong relation between strategic ideas and operational tools, both with respect to public investment in development and the exercise of land-use regulation. Technical judgement rather than a comprehensive plan was thus to be the critical mechanism connecting policy to the allocation of development rights where significant changes to the urban fabric were involved.

In seeking to change practices inside the 'bureaucracy' of a municipality such as Milan, continual efforts were needed to co-align the emerging regional legislation with local practices, with strong political and technical links between the two government levels. The planners also needed to keep key *Comune* politicians on-side with the technical arguments. By the mid-2000s, the new practices had had material outcomes, in terms of a flow of projects through the new processes and into active construction. They had produced significant learning about new practices among those closely involved, some of which had helped to shape regional and national legislation. However, while politicians valued the technically focused administrative improvements, there was little sign of a momentum for a more strategic and coordinated approach to urban development within the Milan *Comune* as a whole.

Nevertheless, within, the planning department, practices continued to evolve, anticipating proposals for further regional laws. These consolidated the separation of a strategic framework (the *Documento di Inquadramento*, now called *Documento di Piano*), from the formal specification of development rights and constraints, the latter to be contained in a new *Piano delle Regole* (Plan of Regulations). Along with the *Piano dei Servizi*, these three documents could then provide the basis for a new type of overall plan, the *Piano di Governo del Territorio*, which would finally remove the old *PRG* and the practice of continual '*varianti*'[48] (Figure 4.6). Increasingly, those working on these documents saw them as connecting to different stakeholder constituencies. The

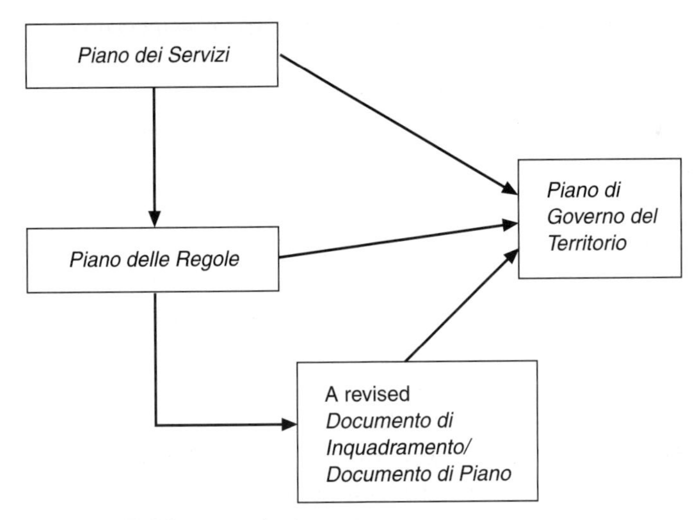

Figure 4.6 Building a new planning strategy

Documento di Inquadramento was especially important for political direction, though future revisions might be prepared in a more interactive way with economic and civil-society actors. The *Piano dei Servizi* was envisaged as a knowledge resource and an arena for internal coordination with service departments in the *Comune*, and with the 'decentralisation zones'. The preparation of a *Piano delle Regole* was expected to involve real-estate interests. By late 2004, the major effort was to progress work on the *Piano dei Servizi* and to initiate a revision of the *Documento di Inquadramento*. This revision had become urgent due to the uplift in the property market in the early 2000s. As a result, most of the development opportunities indicated in the development axis (the *t-rovesciato*) had been taken up, and market interest had shifted to the southern rim of the city, where there was also strong resistance to development focused around the conservation of landscape and natural resources.

But the wider context remained a difficult one for any coherent approach to spatial strategy-making in the city. There were thick boundaries everywhere – between departments in the *Comune,* between municipalities, between citizens and their City Council and between many economic interests and the *Comune*. Experts were grouped around governance functions primarily in terms of a sectoral focus, and even then professional networks were not as strong as other social and economic ties and the re-forming political networks. In such a governance context, many groups in the metropolitan area pursued their activities in all kinds of self-organising ways and largely without reference to the arenas of formal government, about which they had few expectations. There were lively debates in all kinds of arenas about urban conditions. Observers commented that Milan had a strong economy and civil society and a weak government capacity (Dente *et al.* 2005). The Province attempted some initiatives in intermunicipal strategy formation, but support for these was uneven (Balducci 2005b). By 2004, the Milan Chamber of Commerce was also exploring the need for a more strategic approach to the city's development, strongly influenced by concepts of globalisation and economic competitiveness (Bassetti 2005). But the political leadership in the *Comune* of Milan had little interest in these initiatives. Drawing on a long tradition of the autonomous city state, they saw the *Comune* as a powerful and largely autonomous actor, operating in a political, economic and cultural milieu in which the *Comune* stood for the city, a city which was itself strong, dynamic and in charge of its own destiny. They felt no sense of threat, challenge, crisis or any political opportunity in an explicit effort in spatial strategy-making. The capacity for different groups across the metropolitan area to take initiatives to promote their own projects seemed to confirm the leadership in their view that a major strategic initiative to re-imagine the city and its direction was unnecessary. As one commentator noted:

> The city is undergoing a major transformation, but this is the result not so much of a precise vision of the priorities to emphasise in its development, but rather of diffuse, micro changes, of many smallscale projects without any relation to each other. Paradoxically, the very lack of vision leaves space for a plurality of actors, for actual

> experiments in grassroots project development, in producing urban improvements, in liveability, in sociability, which in the discourse of urban planning remain at the level of declarations of principle (Mugnano *et al.* 2005: 191, author's translation).

The innovations to regulatory practice in Milan at the turn of the century are thus likely to be a useful asset, if and when a wider momentum builds up in Milan for a more coordinated and strategic approach to urban region development. But it remained very uncertain how far they would provide the nutrients to encourage the evolution of such an approach. Instead, the innovations are perhaps best understood as a 'practice-in-waiting' for a window of strategic opportunity. Even without a well-developed strategy, they provide a mechanism for keeping social and environmental issues 'in play' in negotiations with developers. However, making them work as intended, in a technically competent and uncorrupt way, is demanding and organisationally complex. It requires the development of new skills among planning and administrative staff in the *Comune* and a change in culture towards policy-focused performance assessment rather than norm-conformity. It also means that developers and infrastructure providers have to think and act differently as they work out how to secure their interests. It implies that public administration across the *Comune* departments is able to manage the public assets created in effective and policy-related ways, and with reasonable efficiency.[49] If this is not achieved, real-estate interests will lose confidence in the capacity of the public administration to deliver in the spirit of the innovations, and turn instead to easy projects, many of which are outside the *Comune* area. Or they will look for informal ways of getting what they need to build a project, especially as the driving forces promoting technically-focused innovations in public administration weakened in the 2000s.

CONCLUDING COMMENTS

Milan is a rich and lively city, filled with economic opportunity and diverse and dynamic cultural networks. Its prestige as a design and fashion ambience attracts people and firms across the world, who enjoy its fluidity and opportunity. All kinds of relational networks create and link diverse groups of interest and activity, some centred in economic relations, some in cultural fields, some linked to old families and organisations, others to new enthusiasms. This generates an innovative energy in civil society, as well as among the various business communities (Vicari Haddock 2005; Vitale 2006). What is lacking is a well-developed 'public realm' within which the opportunities and challenges generated by all this inventive motion can be 'called to mind' and debated in ways which can mobilise collective action to promote synergies and limit the downsides of dynamism. Despite the lively debates in all kinds of arenas, the connections between these debates, with each other and with the formal government, are not well-developed. This was not so in the past, and seems to be in part a consequence of the collapse of the party system as the primary mechanism for integrating governance capacity, and with it a deep decline in confidence in the capacity of formal government. But it also arises from an introverted complacency among elite stakeholders about the position of the city, in its region, in Europe and internationally. This combines with a political activism around particular issues, rather than a broad political platform about conditions in the city. The 'greyness' of Milan's ambience referred to earlier is primarily a reflection of a weak 'public realm'.

In such a context, a governance initiative promoting an urban development strategy based on a capacity to 'see the city' in some explicit way encounters little support in the society. Nor does a business-oriented political leadership in a general European and Italian climate where neo-liberal policy ideas are very strong provide much encouragement for any attempt to generate a collective focus around some explicit articulation (or 'vision') of the city to which the multiple networks around the city might connect. For many, the city's identity exists as a deep, culturally embedded presence. It does not need to be 'summoned up' anew and refreshed. The key strategic intervention, for Mayor Albertini and his team, has primarily been to open up opportunity for the energy within the city's economic and cultural life to flourish, and in particular to make the complex government bureaucracy work with greater technical efficiency. It has been a step too far for politicians and most officials to consider that these ambitions might be promoted by a more integrated approach to government organisation, and a stronger focus around place qualities rather than separate service functions. Instead, issues about place qualities and public benefits are being raised in the painstaking technical and organisational work of creating a 'knowledge system' through the *Piano dei Servizi*.

There is no lack of discussion about the city, its challenges and qualities. These debates have generated a rich intellectual grounding for a relational approach to urban dynamics and planning processes, although many new ideas are readily recast in the terms of the old images of a monocentric city versus a city of neighbourhoods. But these debates only weakly connect to each other, and do not reverberate around the diverse arenas and networks that exist within the city, or link well into formal government arenas. They are just a small part of the city's cultural energy. As a consequence, many aspects of urban life, though known systematically by some and experienced by all, are invisible to many and do not inform collective action agendas. Connections between neighbourhood life and wider city issues seem in particular to have become very weak. Yet city residents have concerns about traffic congestion, about parking provision, about the quality of the street environment, about safety and security, about developing socio-spatial segregation, about the increase of pollution, about water supply, about out-migration from the city and in-migration to it. Several departments in the *Comune* have responded incrementally to some of these issues (Dente *et al.* 2005). But there is no mobilisation machinery that taps into these concerns to articulate a coherent demand for more attention to an integrated approach to the 'liveability' of different places in the city.

The Milan case thus provides an example of governance *incapacity* for collective action to develop around a coherent view of 'the city' or urban 'region' as a basis for guiding interventions to open opportunities and limit the adverse effects of a dynamic economic and cultural urbanity. The formal machinery of local government in the city operates as a relatively isolated nexus in an innovative society. In this respect, it is not unlike some other old European cities that now find themselves in a sprawling metropolitan region (Motte 2005; Salet *et al.* 2003). But how far does this governance incapacity matter and to whom? Some argue that the innovative capacities in the economy and civil society more than compensate for the weakness of formal government. The design and fashion economy needs little from the spatial organisation of the city in any case. The confusion of public-sector administration creates plenty of space for alternative action and adjusts more readily to new demands where stronger government systems might create rigidities and barriers. The 'Milanese way', encapsulated in the notion of the '*rito ambrosiana*', is also difficult for outsiders to penetrate and, as a result, helps to resist the advance of 'globalising' and internationalising forces.[50]

But others argue that the lack of a strategic governance capacity to 'see the city' in an interconnected way generates the continuing neglect of many issues and ignores future major problems in the making. Multiple innovations and bottom-up initiatives compete and clash with each other. Some social groups, especially the elderly and recent immigrants, have a difficult life in the city. Major development projects compete with each other, and undermine each other whenever the property market sags. Problems of housing affordability drive people out of the city, as does the increasing congestion and pollution. The taken-for-granted, and largely monocentric, city imagined by a complacent elite is slowly disappearing, as it did in Amsterdam. These arguments encourage a new effort to re-imagine the city in a strategic way, as a way of mobilising attention to problems with place qualities and liveability appearing across the urban area. Some call for a major collaboration exercise in generating a 'strategic vision' for the city in its region as a way of creating a 'public realm' of debate to connect together the diverse networks that co-exist in and around the city (Balducci 2005a).[51] The long-term value of the technical innovations described in the last stage of this story will depend very much on which of the above arguments prevails in the evolving governance culture of the city and metropolitan area.

NOTES

1 This celebrates a consumerist culture, and refers to the practice of socialising in bars and restaurants (Foot 2001).

2 For accounts of this experience, see Balducci 2001a, b; Dente *et al.* 2005; Foot 2001; Gualini 2003; Vicari and Molotch 1990.

3 See Dente 2005; Fedeli and Gastaldi 2004; Martinelli 2005; Nigro and Bianchi 2003; Pugliese and Spaziente 2003.

4 Mostly long-standing land- and property-owning families.

5 It was here that Silvio Berlusconi began building his fortune and his business empire, in the development projects *Milano Due* and *Milano Tre*.

6 This law was seen as innovative at the time (Ave 1996).

7 The consortium was established in 1961 (Gualini 2003). From 1961–1971, the *Assessore* for *Urbanistica* in Milan was the president of *PIM*, encouraging a flow of planning ideas between the city and the sub-regional area (Balducci 2005a).

8 This later became a 'polycentric' concept (Secchi 1988).

9 Gabellini (1988) notes a struggle between an administrative/technical view of Milan *Comune* officials and a more political approach by the then *Assessore* for planning in the *Comune*, Hazon, who was also President of the *PIM*.

10 This housing strategy, approved in 1975, was named after its political promoter as the *Piano Velluto* (Balducci 2005a).

11 A national law of 1975 enabled neighbourhood councils to be established, and 20 were set up in Milan at this time. These had formal rights to be consulted on land use and housing issues. Over time, these councils became dominated by party networks and citizens lost interest. The councils were merged into nine large '*zone di decentramento comunale*' in the 1990s (Vicari and Molotch 1990).

12 The *PIM* promoted its own plan, but this was never approved.

13 However, the *Comune di Milano* lacked the funds to buy the land and owners refused to sell.

14 Both were closely linked to the rising Socialist politician, Bettino Craxi.

15 In contrast, many of the ideas of the Milan *PRG* were realised in the plan for Bologna produced in the 1980s, which did have significant effects on that city's subsequent development.

16 Both were criticising the preoccupation with elucidating the rules of urban morphology as the basis for constructing urban zoning plans and for reading city dynamics.

17 See Balducci 2001a, 2005a; Gabellini 1988; Gualini 2003; Oliva 2002; Pasqui 2002; Vicari and Molotch 1990 for these initiatives.

18 *Ferrovie dello Stato*, now *Trenitalia*.

19 Major national and regional transport projects had a more chequered history. A major northern transversal route, the *Pedemontana*, remains as a project (Novarina 2003), while the development of the new airport of Malpensa, initiated as a project idea in the 1970s, was limited by the lack of major connecting routes and by the resistance of surrounding municipalities to allowing development adjacent to the airport.

20 See Fedeli and Gastaldi (2004); Martinelli (2005); Nigro and Bianchi (2003); Pugliese and Spaziente (2003).

21 Called at this time the *Ufficio di Pianificazione e Progettazione Urbana*, in the broader *Settore Urbanistica*.

22 The *Lega*'s economic links were primarily with small firms and some professionals, not major economic actors.

23 See Balducci 2005a; Bolocan Goldstein 2002; Gualini 2003.

24 Which proposed planning academic Alessandro Balducci of the *Politecnico di Milano* as *Assessore* for planning. Balducci had spent some time in Berkeley, connecting to Innes' work on consensus-building practices in strategic urban planning (Innes 1992).

25 See Boeri 1993; Bonomi 1996; Indovina amd Matassani 1990; Macchi Cassia *et al.* 2004; Pasqui 2002.
26 By this time, a relational understanding of urban dynamics was well-established among planning academics in the *Politecnico di Milano*, drawing on contributions in Italian social sciences which had a significant influence in the development of a relational geography internationally.
27 One reason was the shift from equities to property as an investment medium at this time.
28 Traditionally, social segregation in the city had been limited, but in the 1990s segregation tendencies began to develop as poorer migrants moved into a social housing stock occupied in particular by elderly people (see Zajczyk *et al.* 2004).
29 Key organisations here are the *Associazione dei Industriale Lombarda* (*ASSOLOMBARDA*), the *Associazione Imprenditori Edili* (*ASSIMPREDIL*), the *Camera di Comercio di Milano* and the public transport agency, *ATM*.
30 A process referred to as *Stati Generali* was set up, intended to draw in different elements of society, recalling revolutionary French models (Balducci 2001a).
31 INU – Instituto Nacionale de Urbanistica.
32 Mazza's various papers have been collected into four volumes: Mazza 1997, 2004a, b, c.
33 See Balducci 2005a; Gabellini 2002; Pomilio 2003.
34 His model was the British 'discretionary' approach to land-use regulation (Curti 2002).
35 The *Direzione Centrale Urbanistica*, changed in 2003 to the *Direzione Centrale Pianificazione Urbana e Attuazione, P.R.*
36 This was one of the issues that attracted critical attention from the planning community (Balducci 2001a; Gabellini 2002). Mazza, however, defended the approach as the only one possible in the Milan political context at the time.
37 Through the two organisations representing planners, *INU* and the Italian Society of Urbanists (*SIU*).
38 See Balducci 2001a; Curti 2002; Gabellini 2002.
39 He was replaced by *Assessore* Vergha, who had been one of the team working on the 1980 *PRG* in the 1970s.
40 Legislation for these was introduced at national level in 1992. This was converted into regional law in 1999.
41 Data is kept on the flow of projects through the procedure and the public benefits negotiated. This shows that, starting from 2001 until 31 August 2005, 127 projects involving over 6.7million m^2 in area had entered the assessment process as *PII* projects. Of these, 22 had been rejected or withdrawn as not appropriate, 33 were in the initial phases of assessment (one million m^2), successful negotiations had been completed on 53 (2.5 million m^2), with a further 21 (2.69 million m^2) still under investigation (data provided by the *Comune di Milano*). Significant 'public benefits' have also been negotiated. Nearly half of the land area involved was dedicated to public uses (including street space, etc.), and a total of nearly €400,000 had been promised as developers' contributions for service provision, fees and negotiated contributions. This last category made up 28 per cent of the total.
42 That is, a form of transferred development rights.
43 These concepts had not been formally incorporated in a *Piano delle Regole* by late 2005.
44 Mazza resigned his consultancy with the *Comune* in early 2004 because these ideas were not taken up, but continued as a member of the Evaluation Panel.

45 An exception was the *Comune*'s *Uffici Tempi*, which had its own information on services, through a web-based information system on the time/place of both public- and private-sector services, sponsored by the public transport company, *ATM*.
46 Including the team of Guido Martinotti at the *Universita di Milano – Bicocca*.
47 In addition, the new zoning approach involved municipal purchase of all sites that were to be transferred from one ownership to another as part of the proposed market in development rights.
48 These proposals are contained in a regional law 12/2005, but Milan's proposed approach to the allocation of development rights had not been approved by late 2005. It was hoped to include it in the emerging *Piano di Governo del Territorio*.
49 Dente *et al.* (2005) point to significant improvements in the administrative capacity of the *Comune*.
50 See Balducci 2004; Cognetti and Cottino 2004; Novarina 2003.
51 For these arguments generally, see Balducci 2001b; Curti 2002; Foot 2001; Gabellini 2002; Oliva 2002; Vicari Haddock 2005.

CHAPTER 5

TRANSFORMATION IN THE 'CAMBRIDGE SUB-REGION'

> One cannot make a good expanding plan for Cambridge (Holford and Wright 1950: viii).

> Previous policies have sought to protect the historic character of Cambridge by dispersing housing to villages and towns beyond the Cambridge Green Belt. However, efforts to limit employment growth within and close to Cambridge and to encourage spin-out to other centres have only partially succeeded.... The planning framework which nurtured the emergence of the Sub-Region as the home of the 'Cambridge Phenomenon' is no longer sustainable (CCC 2003: 98–99).

INTRODUCTION

The previous two cases were of large cities with long histories as major urban centres in Europe. The story of the emergence of the 'Cambridge Sub-Region' shifts the focus to places beyond such centres. Yet Cambridge too has a long history as a significant city in the European imagination and especially in the consciousness of the British elite. Over the past half-century, the area has been drawn into the nexus of an expanding London metropolis, with the centre of London only 50 miles (80 km) away. More than this, however, Cambridge has become a major growth node in London's outer metropolitan area, in which, by the turn of the century, English government capacity at all levels was being tested in a struggle to achieve a 'balanced' and 'sustainable' approach to managing growth.

Like much of southern Britain, the Cambridge area has a geography of medium and small-sized administrative centres and market towns and villages, in a landscape of undulating green fields, meadows and woodland copses celebrated in English literature, painting and poetry. But it is history, not geography, that has created Cambridge as a special place within the culture and politics of the nation's elites. Until the end of the twentieth century, Cambridge was as much a university as a market town, a training ground for the political and administrative class, for many in the business world, and for the higher echelons of the educational establishment. Over the centuries, the university has claimed large areas around the old city core for its own, and remains a major landowner and developer in the area, as well as a substantial employer and generator of activity. For centuries, the university acted as guardian of a contemplative, 'ivory tower' tradition of the role of an academic institution, an existence apart from the bustle and noise of commercial and industrial society. Then in the late 1960s, sections of the university turned 'entrepreneurial' (Allen *et al.* 1998). The result was the emergence of a dynamic cluster of high-tech and biotechnology companies, which, by 1985, was named by a group of consultants as 'The Cambridge Phenomenon' (SQW 1985). Meanwhile, with improvements in infrastructure and massive increases in car ownership and use, general metropolitan growth pressures around London threatened the treasured landscape (see Figure 5.1). The story of this case centres on the struggle to manage the space demands of this transformation while retaining the traditional landscape imagery. The arenas of the British planning system and its power to regulate the amount and location of this development are central to this account. The case illustrates the strengths of the system (Brindley *et al.* 1989), the power of national government to determine the parameters of local discourses and practices and the increasing difficulty, in the British governance context, of coordinating regulatory power with resources for development investment.

It is also a story of the power of spatial strategy to shape attention and maintain a degree of local control over the scale and form of urban development in the face of external pressures. The first guiding strategy, drawn up in mid-century by one of the well-known planners of the period, William Holford, sought to 'cap' the growth of the city and disperse growth pressures elsewhere in the region. The 'ghost of Holford'[1] still shadows the imaginations of key actors now shaping the area's future. The story of the Cambridge Sub-Region is both an exemplar and a test of a national commitment in the early 2000s to a 'sustainable' approach to managing growth pressures through a strategically oriented spatial planning (ODPM 2003). The ambition is to find ways to accommodate growth pressures, whilst limiting environmental resource use, and providing accommodation for those on middle- and low-incomes as well as the increasingly affluent. But there are many challenges to overcome, affecting all levels of government, if this ambition of 'sustainable development' is to be achieved. As with the previous accounts, the story starts in mid-century, and concludes with a major emphasis on the period between 1995–2005 when the new, growth-oriented strategy for the Sub-Region took shape and began to be translated into key development projects.

PRESERVING THE IVORY TOWER: THE DEFENCE OF CAMBRIDGE

In the mid twentieth-century, the Cambridge area was a key test for the powers of the new national 1947 Town and Country Planning Act, and the ideas of the planning movement that informed it. By the late 1920s, growth pressures were building up in the area, threatening the conception of the city as a quiet university town in a relatively remote region of the country (East Anglia). In addition, the growth of motor traffic was causing difficult problems of congestion in the city's medieval road system, which was also the crossing point of two significant regional routes, the A10 from London to the north Norfolk coast and the A14 from east-coast ports to the south Midlands.[2] Ribbon

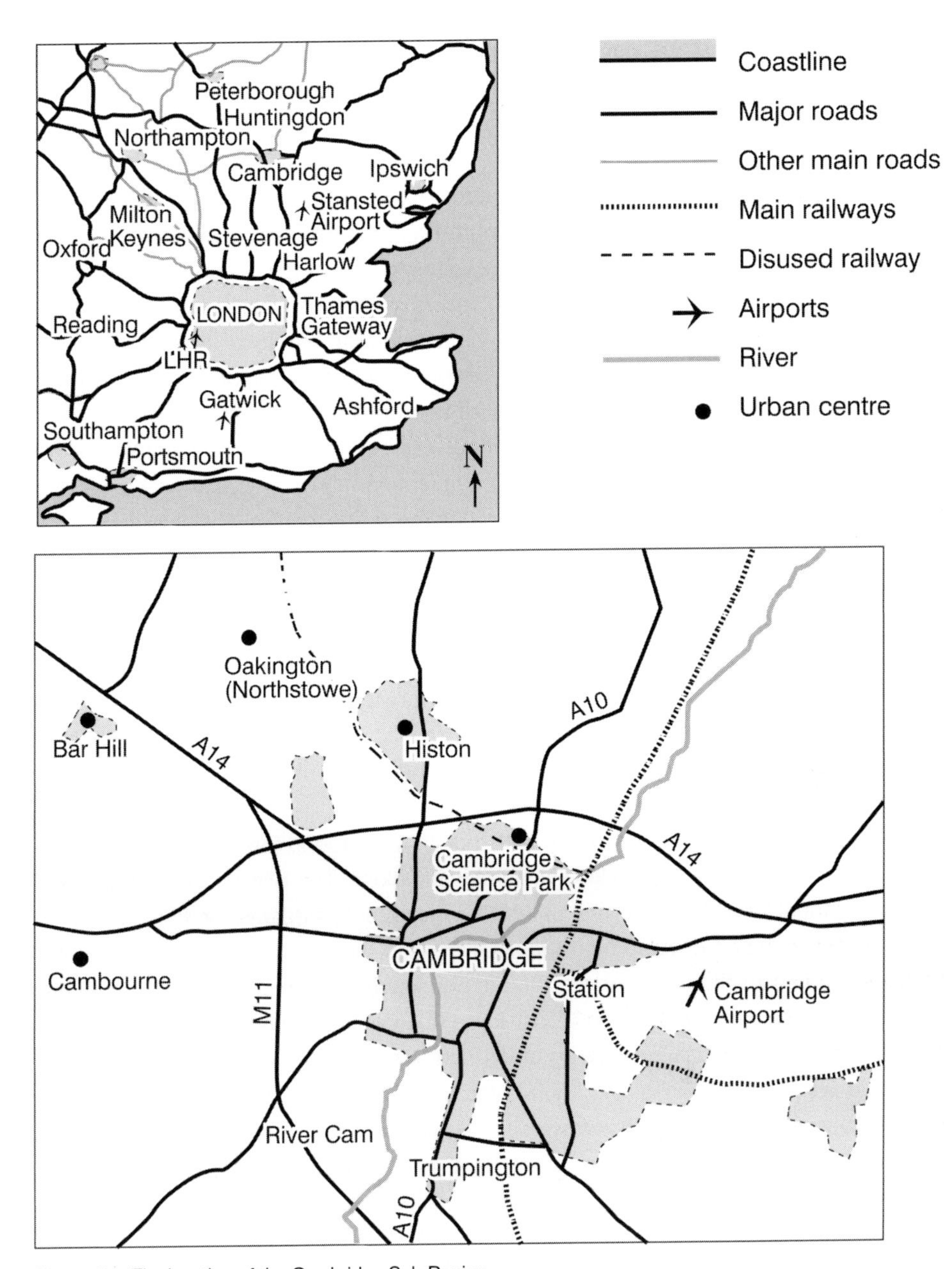

Figure 5.1 The location of the Cambridge Sub-Region

development to the south and west, enabled by the extension of sewerage and the sale of housing plots by the university and colleges, helped to mobilise local concern about the landscape impact, which in turn drew in key figures in the national town planning movement of the period, as well as national politicians and civil servants. This interest was in part a reflection of the significance of Cambridge in the geographical imagination of politicians, civil servants and professionals. But it was also a result of the active and well-connected campaign of the Cambridge Preservation Society (CPS), founded in 1928 (Cooper 2000).

The advocates of a coherent planning approach to preserving the traditional Cambridge environment were active in establishing a Regional Planning Committee in 1928, with a remit to 'make a general inquiry into the present state of the county . . . with the . . . purpose of preserving its native character and providing for development' (Davidge 1934, foreword). This resulted in 1934 in the *Cambridge Regional Planning Report*, produced by planner William Davidge, a former president of the national Town Planning Institute. The report expressed a well-established imagination: an area of villages with market towns beyond, all centred on Cambridge, a university centre and a market town.[3] Although this was clearly a well-established conception, a key proposal was for a ring road around the city, to resolve the traffic problems in the centre. But this involved a bridge across the river Cam and Grantchester meadows, (romantically associated with the English poets, Lord Byron and Rupert Brooke), and was hotly contested. In addition, the university and colleges objected to any interference with their rights to develop their own lands. In this early period, four planning issues emerged that still have resonance in the twenty-first century: the attempt to combine preserving the city's character with accommodating development; the transport dilemmas; the intense contestation over proposals; and the creation of informal networks and arenas through which to promote planning ideas and strategies.

During the Second World War, further growth occurred in the city as industry was moved out of the London area to less-vulnerable locations. The Ministry of Defence also created a number of airfields, several of which subsequently became the focus of development attention. Such development greatly increased the fears of those trying to safeguard the particular ambience of the city. It also caused problems for the university, which had always relied on a cheap labour force to service the various colleges in which staff and students had common dining areas and 'rooms'. The new industries provided better-paid work opportunities. More jobs also created demand for more housing, and more facilities. Meanwhile, the road system still directed both local and regional traffic through the centre of the city, where it intermingled with the movement of students and staff between colleges and the flow of shoppers coming in from the surrounding villages and market towns.

The British 'town and country' planning system before the war had been evolving as a mechanism to regulate the 'sprawl' of development surrounding urban areas. There was a strong strain of anti-urbanism within the movement. However, the ability to regulate development was limited by difficulties over compensation to land and property owners for loss of development rights (Ward 1994). During the war, as in the Netherlands, the issue of the spatial pattern of development became linked to ideas about post-war reconstruction of war-damaged areas and of poor housing in the major urban areas. This spatial planning effort became located as a significant part of the creation of the welfare state, with its ambition to deliver better living and working conditions for all. The national strategy was to shift industrial development from the congested southern parts of the country to the northern parts hit by the depression in the 1930s, and to regulate development around all settlements to prevent sprawl. Densities in the overcrowded cities were to be reduced, with accommodation to be provided outside cities in free-standing new towns. Rights to develop land were nationalised, with a 'once-and-for-all' compensation settlement. Development rights had therefore to be obtained from the state, via the planning system. This was a revolutionary move, only possible in the particular post-war conditions of a collapsed property market (Cullingworth 1975). Concerned to prevent development sprawling around cities, the 1947 Town and Country Planning Act extended 'development control' to both urban and rural areas, and gave the major powers over the regulation of development to the county level of government.

In the Cambridge area in 1947, the County of Cambridgeshire covered the area of Cambridge City[4] and a band of about 100 villages around it, organised into rural districts. Beyond this were other boroughs and rural districts in a circle of market towns, some of which later became part of an enlarged county. In the 1940s, there were not only major disputes about the proposal for a Cambridge bypass but also concerns about whether the primary planning authority should be the city or the county. The university was also resisting the imposition of planning controls. A Joint Planning Advisory Committee was established for the county area, including county and city councillors, as well as university representatives, the university having formal seats on both county and city councils. This Committee evolved into the County Planning Department, with planner Leith Waide as head.[5] Waide had good links with the national planning movement and with those developing the national legislation. Through these, the national Ministry of Housing and Local Government (MHLG) agreed to fund a study to provide a framework for the preparation of a county development plan.[6] The Committee commissioned Professor William Holford to undertake the study, jointly with Myles Wright, drawing on staff resources from the newly-staffed County Planning Department. Holford and Wright had both worked for the national Ministry (Cherry and Penny 1986; Waide 1955).[7]

Holford's biographers state that he was reluctant to take on the Cambridge commission: 'In Cambridge, powerful interests dominated the area to be planned. [It] reeked of history and tradition, and . . . possessed micro-political systems of distinctive character and utmost complexity' (Cherry and Penny 1986: 141). What is striking about the plan is the way it speaks directly to these 'micro-political systems', by focusing on the key areas of dispute and providing carefully constructed arguments to support the proposed strategic framework. Nearly half the report discusses the pros and cons of different road proposals. The remainder considers the general development strategy, the situation of the university and the colleges, and development in the city centre, as well as what should happen next (Holford and Wright 1950).

The Holford Plan, as it subsequently became known, largely adopts the 'preservationist' viewpoint, so actively promoted by the Cambridge Preservation Society. Its central concern is to preserve the special identity of Cambridge in its rural setting:

> Incomparably beautiful in many things, miserably defective in others, Cambridge is still one of the most pleasant places on earth in which to live. Moreover it is now perhaps the only true 'University town' in England. The question is whether it can control its own destiny in the face of a multitude of unplanned events that will tend to change it. When these changes come, and even before they take place, can they be arranged to maintain and enhance the essential character and virtues of the town? (Holford and Wright 1950: vii).

The key proposal was to limit the growth of Cambridge to 100,000 (or even 125,000, Holford and Wright: viii), allowing for some growth beyond the then-estimated population of 86,000, in order to sustain needed services and retail provision. In effect, this meant deliberately restricting housing development. However, increased car ownership and use were accepted as inevitable. Therefore, measures were needed to deflect through-traffic from passing through the centre itself.[8] The proposed Outline Development Plan (Figure 5.2) aims to control:

> the physical spread of Cambridge and nearby villages, with the aim of maintaining their present general character while allowing for necessary changes and some general growth. Sites for housing and other new buildings have been chosen to encourage reasonably compact development, to keep the sequence of open spaces along the river and to prevent neighbouring villages becoming merged with the town (Holford and Wright 1950: viii).

The argumentation of the plan establishes the case for limiting the city's growth and explains the basis for proposed developments and improvements. The emphasis is always on improving conditions and providing a good-quality environment for the 'ordinary citizen', who enjoyed living in Cambridge as it then was. To achieve this, Holford and Wright argued that any industrial development not related to particular Cambridge needs and initiatives should be deflected to other parts of the country. This required persuading national government to amend the rules applying to the distribution of industry.[9] However, and significantly as it later turned out, university development was considered an exception to this restriction. The university and its colleges were major landowners in and around Cambridge, particularly to the south and west. They had a number of expansion projects in mind and were deeply embedded in city, county and national governance arenas. Some key university figures were also active in the Cambridge Preservation Society (Cooper 2000). In the Holford and Wright plan, university developments were allowed to escape the emphasis on compactness, being allocated a 'reserve

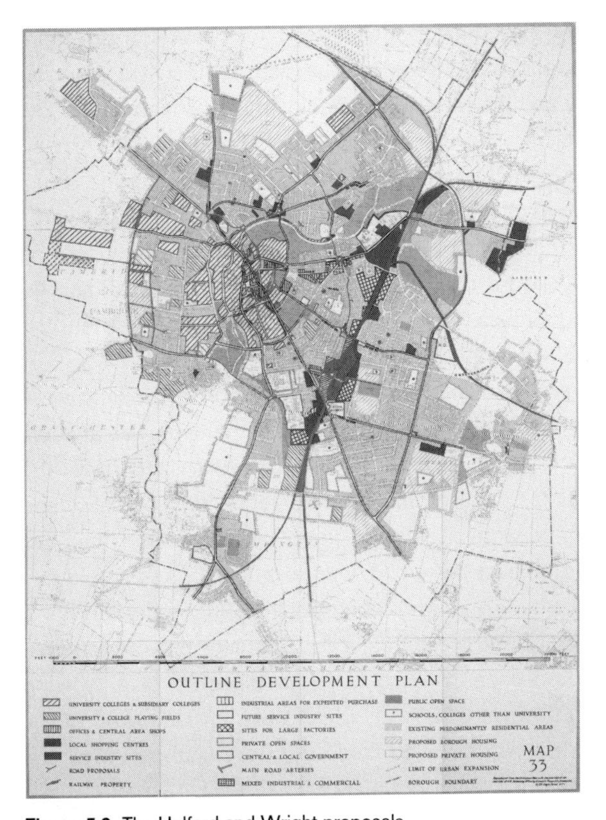

Figure 5.2 The Holford and Wright proposals

Source: Holford and Wright 1950, end pocket, with permission of Cambridgeshire County Council

of development land' in the west. Nevertheless, such development was expected to be contained within a setting of small villages, fields, woods and meadows, with 'green wedges' penetrating into the heart of the city. This concept became a prototype for what became the Cambridge Green Belt.

Such a proposal to limit growth was not accepted without controversy, both from those against any further development at all, and from those believing the town should develop further. These controversies were played out through letters to the national press and in formal objections to the plan. With the backing of national government for the strategy, both with respect to the ideas for limiting the growth of the city, and for restricting industrial development, the Holford plan became the basis for the first County Development Plan for Cambridge, approved in 1954 (Brindley *et al.* 1989). This was despite:

> several weeks of a long public inquiry, a considerable area of *The Times* and other leading newspapers and finally . . . a day in the High Court (Waide 1955: 83).

The Cambridge Preservation Society largely supported the Holford strategy. But there were tensions between the county and the city as to who had control over planning strategy, and between the university and the local authorities over restrictions on the use of their lands. The City Council also believed that the city should be allowed to expand further (Cooper 2000; Waide 1955). The strategy was, however, in line with the broad thinking within the planning movement at the time, which at this period had a powerful influence on general opinion and on the national Ministry. The approach also suited the balance of university interests. The Holford plan in effect squeezed out competition for both labour and development opportunities, while allowing the university and the colleges substantial freedom for manoeuvre on their own lands.

The 1954 County Development Plan anticipated an expansion of population in Cambridge itself from 86,340 in 1948 to 100,000 in 1971, and in the county as a whole from 164,700 in 1948 to 187,400 in 1971 (Cooper 2000). To explain and justify the plan's approach, the County Planning Officer commissioned respected journalist and planning commentator, Derek Senior, to write a 'guide' to the plan, to 'expound' the plan, so that interested people locally and all over the world 'who love Cambridge' could see the plan 'in the round', free of all the technical and legal requirements necessary for a formal plan statement (Senior 1956: 1). He presented a development plan as a framework which would steer growth processes. 'A plan is not a blueprint or a working drawing, but a statement of policy' (Senior, 1956: 2), which gave an indication of the approach that would be adopted by national and local planning authorities to the exercise of their powers with respect to the Cambridge area:

> It follows that a development plan is essentially a compromise – between what we have and what we should like; between conflicting claims on the same land, labour and materials; between incompatible ideas and between differing scales of value. The test of a good plan can never be whether it completely solves one problem, fully meets one need, or wholly satisfies one claim, regardless of other problems, needs and claims. The test must always be whether a different compromise would yield a total result for the same expenditure of time and resources (Senior 1956: 2).

This is an early statement of the conception of a development plan as a statement of policy, rather than as a masterplan blueprint or a specification of development rights, reflecting the distinctive approach to planning system design developed in the UK since 1947 (Davies *et al.* 1989).

The Holford and Wright plan largely shaped the pattern of urban development in the following 40 years. In 1957, its ideas about open spaces and green wedges were converted into the principle of the Cambridge Green Belt, relatively tightly defined around its inner boundaries, but extending 3 to 5 miles (4 to 8 km) around the city; very wide, compared to other urban green belts in England (Elson 1986). Meanwhile, national policy pursued a strategy of deflecting industrial activity from the 'congested' London and the South East to the northern industrial areas. New Towns were initiated across the South East, beyond the metropolitan green belt (the nearest to Cambridge being Harlow and Stevenage), and Town Expansion Schemes were negotiated with the London County Council. Huntingdon and Haverhill, among the market towns around Cambridge, set up such schemes.[10]

In the period of economic prosperity in the 1950s and 1960s, growth pressures once again built up (CCC 1961). The county had prepared a 'Town Map' or detailed development scheme, for Cambridge itself,[11] following the approval of the County Development Plan. But the university and other interests continually sought more room for development. While making minor adjustments, the county maintained the position that growth pressures should be deflected to the market towns beyond the green belt and the ring of villages, or even further afield to areas in the north and east of East Anglia which were suffering from economic problems. Some concession to the need to provide for growth within the Cambridge area was made in the proposal for a new settlement beyond the green belt to the north-west, at Bar Hill, approved in 1964. This started a practice that has since continued of creating more 'villages' around Cambridge rather than expanding existing villages too much or letting Cambridge's expansion swallow up the distinct village identities.[12] The growth being accommodated in these developments was in part 'council housing' (subsidised housing built for rent by local authorities), but mainly consisted of estates of private housing, with services provided by the public sector.

Later, with substantial growth pressures across the whole of the London metropolitan region, proposals were made for further large new towns. Two of these affected Cambridge. One was the substantial transformation of the City of Peterborough to the north, beyond Huntingdon, and Milton Keynes to the west, though much less accessible to Cambridge due to the difficulty of east–west road and rail travel. These seemed to provide a strong deflection of growth pressures away from Cambridge. The only major road proposals achieved in the 1960s within the Cambridge area itself were a northern bypass (now the A14), a few road improvements and a bridge over the Cam to create an inner ring road. However, during the 1960s, some major transport proposals began to take shape. The first of these was the construction of the M11, from London to Cambridge, creating what was in effect a wide bypass to the city to the west, onto the A14 Huntingdon road. The second was the electrification of the rail system, allowing a one-hour journey to London. The third was the complex decision process about a third London airport, initiated in the 1960s, resulting finally in the decision that this should be sited at Stansted, 25 miles (40 km) to the south of Cambridge.[13] The Cambridge area was being drawn into the orbit of the rapidly expanding London metropolis.

By the mid-1960s, some local actors recognised the scale and significance of these growth pressures. Cambridge City Council continually challenged the county strategy (CCPO 1977). City Architect and Planning Officer, Gordon Logie, argued that:

> The situation revealed is a startling one.... Nostalgia for the past is very strong in Cambridge and many will argue against change of any kind. If so, they will be profoundly misguided. The best of Cambridge as we know it today has been built on changes far more sweeping than any proposed in this report; the worst of Cambridge is the product of inertia and lack of positive thinking (Logie 1966, Introduction).

Logie postulated several possible future scenarios, but favoured accepting substantial growth generated both by 'science-based' and other research-linked industries, and the expansion of commuting from London. His strategy broke away from Holford's 'compact city' approach, to propose 'tongues of development' along the main radial routes. By the late 1960s, a working group including the local authorities, the university and the national ministry were examining the future size of Cambridge (CCityC 1968: 24, para. 158).

Another ally in the struggle to break growth constraints was the East Anglia Economic Policy Council, set up under the auspices of the national Department of Economic Affairs, created by the 1964 Labour Government (Cullingworth 1972). The members of the council included business, university and property interests, as well as local authority councillors from the counties of Cambridgeshire, Norfolk and Suffolk. Their first report (EAEPC 1968) highlighted growing infrastructure deficiencies, with growth pressures being experienced across the region. A particular emphasis was given to the need for better east–west routes, including the northern bypass for Cambridge. To grasp the spatial organisation of the region, the study subdivided the area into 'city regions', then being promoted in the discussions on local government re-organisation.[14] The study emphasised the scale of growth pressures affecting the 'Cambridge Sub-division'. Along with Logie, the study proposed that development should be in close proximity to the city, rather than dispersed to the outer villages and market towns.[15]

Yet despite the momentum to reconsider the growth limits on Cambridge, the Holford strategy was upheld in a national decision in 1968 to refuse an Industrial Development Certificate for IBM, a major company developing computing technologies, to set up its European headquarters in Cambridge. This seemed so misguided to those in the university who recognised the potential for 'science-based' development in new technologies that it gave momentum to ideas developing for a university-initiated science park development (While *et al.* 2004). The university set up an inquiry into the value of promoting science-based industry in Cambridge. The resultant report was completed in 1969, and in the same year, Trinity College proposed the Cambridge Science Park on land the college owned near the A10/A14 to the north of the city (Garnsey and Lawton Smith 1998; SQW 1985). This was the start of the economic growth dynamic which has since grown into a 'globally-significant' 'cluster' of new industrial activity, centred on high-tech innovation.

This early stage of the Cambridge Sub-Region story illustrates well many of the characteristics of British planning at this time, as developed in a situation of constant and well-informed public and elite attention. It shows the regulatory power of a spatial strategy to shape physical development opportunities if all levels of government give it support. It illustrates the significance of national government in local development policy but also the power of well-placed local actors to influence national policy. It shows the tensions in government policy between the spatial strategy backed by strong powers of land-use regulation, and the spatial consequences of major infrastructure investments. One shapes the geography, the other changes it. In the Cambridge case, a clear spatial strategy was developed, which was continually challenged and reviewed, with discussion structured by attempts to conceptualise the urban region, its particular qualities (essences) and its dynamics. In these debates, key stakeholders (the university, the large farming landowners) and lobby groups (the spokespeople of the Town Planning movement and the Cambridge Preservation Society) were co-involved in structuring the arguments, along with representatives of the city, the county and national government. These debates flowed out into media stories and letters to the newspapers, and were reinforced by an articulate citizenry. The continual contestation helped to raise development standards, since those advocating development had to demonstrate its positive qualities. In these contestations, the overall strategy of the limit on the city's growth was continually both reinforced and challenged by demands for space for economic activities and for more housing. Yet for the next 25 years, the strategy was maintained.

Growth Management Through Regulatory Planning

During the 1950s and 1960s in Britain, urban development processes were managed within a framework of development plans which allocated land for new development, specified areas for comprehensive development (largely bomb-damaged areas, town centres and areas of poor housing), defined green belt boundaries and indicated major transport improvements. At the national level, the policy framework emphasised dispersal of industry, and later offices, away from the congested South East to the older industrial areas, and to the new and expanded towns. It was assumed until the 1960s that the primary driver of development was the public sector, through these projects and its social housing programme. However, in areas with strong economies, the private sector undertook most of the development (Hall *et al.* 1973). The development plan, and the concept of settlements contained by green belts, provided a limited supply of sites. This helped the expanding private house-building industry by maintaining the high value of development sites. A close nexus slowly evolved between the development industry and the planning system. Development sites were steadily released through reviews of development plan allocations and green belt boundaries (Ball 1983; Healey 1998b). The strategy of 'urban containment' provided certainty to the industry, reducing its risks, while the regular adjustments provided 'flexibility'. The house-building industry continually demanded more flexibility and argued over which (and whose) sites to release for development, in a rhetoric that rarely acknowledged how residential land and development markets were themselves structured by the practice of a 'drip-feed' release of sites.

By the mid-1960s, the growth forecasts of the early post-war period were proving far too low as the national economy prospered, leading to major growth pressures in the South East, and to strong pressures for further land release. The perspective of planning activity, which had become focused around managing redevelopment projects and regulating development in line with rather conservative development plans, was under pressure to enlarge, to take on a broader awareness of social and economic dynamics as these played out at the regional scale (Wannop 1995). In this context, a new planning act[16] introduced two levels of development plan, the structure plan and the local plan. The first was very similar in concept to the Dutch *structuurplan* but, unlike the *bestemmingsplan*, the local plan remained advisory, as rights to develop were only given with the grant of planning permission. In parallel, regional and sub-regional studies were promoted to produce strategies for accommodating the growth pressures foreseen (Cowling and Steeley 1973; Wannop 1995). The Cambridge area, and East Anglia generally, was not in focus in this search for ways of accommodating growth. The Strategic Plan for the South East 1970 (SEJPT 1970) sought to shift some of the growth dynamic from the west of London to the east, and foresaw the M11 stretching into Cambridgeshire, but the major growth areas were to be in the Southampton area, around Milton Keynes, in the Reading area (and near London Heathrow airport), in South Essex (along the Thames Estuary) and in the Gatwick/Crawley area (around London Gatwick airport). These ideas shaped the development allocations in structure and local plans in the South East of England for the next 20 years.

However, the urgency to accommodate growth slipped away in the 1970s. The property boom in the early 1970s was suddenly cut short by the rise in oil prices produced by the OPEC oil crisis. There followed a period of economic recession, in which the industrial foundation of the British economy was to be radically eroded, to be followed in the 1980s by the expansion of producer services and the financial sector. In this context, the strategy of forcing businesses out of London and southern England was challenged by local authorities in these areas suffering from industrial closures, and by firms claiming that their locational choices were no longer contained within the national economy but were being made in a European or global context. IBM, refused permission to set up its headquarters in Cambridge, was just such a company.

In parallel, national government in the mid-1960s had initiated a process of reorganising local government tasks and boundaries, in the search for an administrative structure that matched more appropriately the functional relationships of localities. This led to the reorganisation of local government in 1974. In this reorganisation, a two-tier structure of formal local government was created, with advisory regional councils remaining in existence. In planning matters, counties were responsible for strategic planning (the structure plan) and for transport, while districts were given responsibility for the preparation of local plans. Planning staff were expanded at both levels. Counties were also responsible for education and social welfare services, while districts were responsible for social housing provision and local environmental management. In Cambridgeshire, the City of Cambridge became a district, completely surrounded by the previous rural districts consolidated into a new district of South Cambridgeshire (see Figure 5.3). The representation of the university on the County and City Councils was abolished. Cambridgeshire County Council thus emerged as a very strong player in growth management in the area, although challenged by the district councils, whose planning powers were also strengthened. Cambridgeshire was grouped with the counties of Norfolk and Suffolk in the administrative region of East Anglia.

Economic activity in the area, and in East Anglia as a whole, was much less affected than the rest of the country by the 1970s recession, as it had never had a strong base in the traditional heavy-manufacturing industries. Growth pressures were felt across East Anglia, modest in scale relative to some other parts of southern Britain, but substantial in the more rural context of the region.[17] This was accommodated, as Holford and the 1954 County Development Plan had intended, in the nearby market towns, in other parts of the region (Peterborough, Norwich, Ipswich), and within villages around the major centres. The 'cap' on population growth in Cambridge was successfully maintained, in part because household sizes were falling. Most new housing was provided through the private-sector residential development industry. This operated by acquiring land, obtaining permission and then constructing 'housing estates', producing

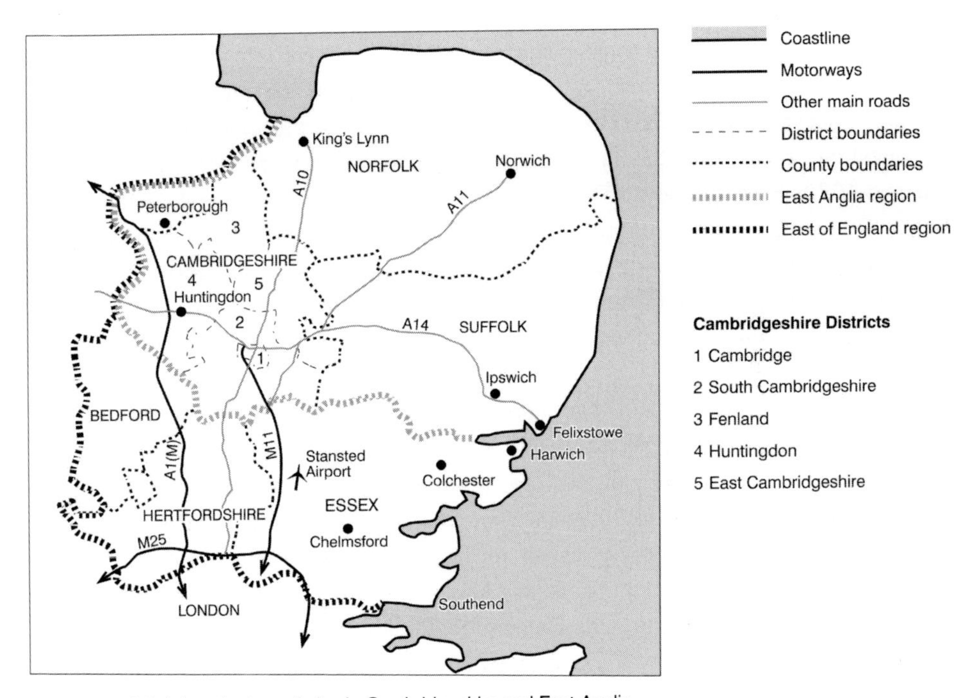

Figure 5.3 Administrative boundaries in Cambridgeshire and East Anglia

standardised terraces, semi-detached and detached dwellings in batches. The planning authorities during the 1960s had to learn fast how to regulate this production process, not only to ensure that it was restricted to land allocated in plans, but to negotiate for the provision and maintenance of public spaces and payments for infrastructure provision.[18] Achieving quality in residential development thus became increasingly market-driven, in a market highly structured by very limited supply and dominated by an oligopoly of producers.[19] Throughout southern Britain, the poor design of residential development and the lack of adequate services fuelled what would anyway have been opposition to large-scale new development (Rydin 1986; Short *et al.* 1986).

Meanwhile, in the Cambridge area, the expansion of science-based industry was underway. In 1971, responding to the university's policy on promoting science-based industry, the restrictive policy on industrial location was modified at national government level with respect to Cambridge, 'to include greater provision for science-based industries largely in the interests of the University' (CCPO 1977, page 48). This formalised the policy of selective restraint on employment-generating activity. Only businesses that directly serviced local activities (including university expansion), or which were linked to science-based initiatives, could escape the restrictions of the strategy for dispersing employment away from Cambridge. In the shadow of this apparently restrictive policy, employment in the area steadily grew.

Local government reorganisation placed Cambridgeshire County in a strong position as regards the strategic orientation of the regulatory power of the planning system. However, throughout the 1970s, resources for public investment became more limited and uncertain due to national fiscal crises. The focus of national 'urban policy' in England centred on the areas of social and economic difficulty in the major urban centres. Environmental arguments were also beginning to emphasise the importance of conserving resources. Investing in the infrastructure for growing areas was less of a priority. By the mid-1970s, the strategy of 'urban regeneration through peripheral restraint' had become strongly established, supported not only by those arguing for help to urban areas with problems of poverty and obsolescence, but by all those seeking to protect the countryside from development (Healey *et al.* 1988). This general ambience coloured the County Planning Department's careful approach to the development of its first structure plan.

The work on the structure plan had been preceded by a more developed study undertaken for the East Anglia Economic Planning Council in 1974. The council had commissioned a special study of the Cambridge area, as one of a series of city region studies, looking forward 20 to 30 years. The resultant report, Strategic Choice for East Anglia (DoE 1974), focused on issues relating to the economy, quality of life and distributional issues, and sought to identify how much change there would be and who would be most affected. As regards Cambridge, the study concluded that growth pressures around Cambridge would increase, but that development should be encouraged elsewhere. It thus reasserted the restraint policy. There had also been two consultancy studies in the Cambridge area, one on transport and the other which focused on retail provision.[20]

The County Planning Department's structure plan team, with around 20 staff, undertook a number of studies that served as technical input to the development of alternative strategies. This followed a common practice in England at the time (Drake *et al.* 1975). The planners sought to develop strategies based on analyses of social, economic and environmental conditions. In recognition of the special sensitivity of the Cambridge area, a Sub-Area Study was also produced in 1977, as a basis for consultation on options for the scale and location of growth. This recognised the special character of Cambridge:

> [this] lies not only in its wealth of historic buildings, but more particularly in the relationship between three distinct elements, the historic town centre, the surrounding ring of colleges, and the associated public and private open spaces adjacent to the river Cam (CCPO 1977: 1).

This study emphasised the role of the town as a sub-regional centre, a centre for small businesses unrelated to the university and a tourist destination. It reflected a politics that sought to escape from the 'University town' concept of the city. From the 1940s to the 1970s, political control of Cambridge City Council shifted between Conservative and Labour, but from the late 1970s, the emerging Liberal Democrat party became an important third group, and by the 1990s, a Liberal–Labour majority dominated the council. Labour councillors emphasised provision for poorer citizens while Liberal Democrats took up the 'green' agenda. Both sought a richer and more inclusive recognition of the city's qualities and dynamics.

While opening up various options, the 1977 report provided a careful argument for maintaining the restrictions on growth in the Cambridge area. However, within the Cambridge area, the strategy of deflecting growth to the villages beyond the green belt was called into question. Two problems with this strategy were appearing. First, it increased commuting, both from the villages into Cambridge, and across the South Cambridgeshire area, as there had been a scatter of industries located in villages because of limited space within Cambridge itself. Second, many of the villages lacked services, particularly schools and health centres. The Cambridge Sub-Area Study, and the subsequent structure plan, therefore argued for more housing development to be located in and around Cambridge itself, as well as in larger villages where increases could support better services. There was also a strong emphasis on road improvements in the area and on expanding public transport and cycle provision. In this context, the study outlined alternative strategies for locating development. This focused on an agenda of locations, several of which were former airfields owned by the Ministry of Defence. This agenda was regularly re-visited over the next 30 years, as pressures to accommodate further development built up (see Figure 5.4).

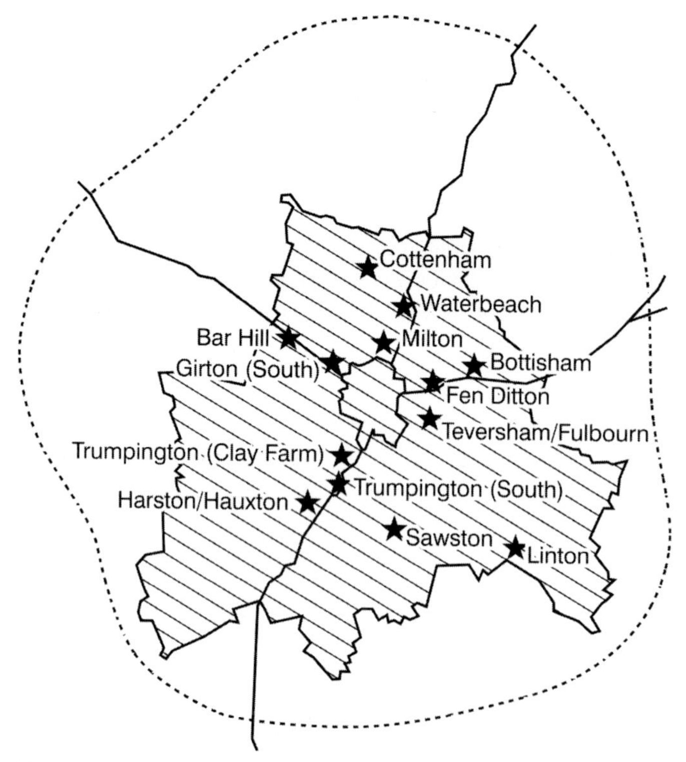

Figure 5.4 Development locations identified in 1977

Note

The dotted line denotes the wider region and the shaded area the 'sub-region' in which possible growth locations (the stars) are identified

The authors of the study were clearly uncertain about growth prospects. In the late 1970s, businesses and the development industry were relatively cautious in estimates of future growth potential, but the county planners could see that growth momentum might pick up in the future. They were aware of substantial growth potential in electrical engineering and instrument-making and in other activities linked to university scientific activity. They also foresaw that the function of Cambridge as a major regional centre was likely to generate more employment (CCPO 1977). They therefore argued that the structure plan should consider short-term allocations for development, in ways which would not compromise future development needs. A critical concern was to reduce commuting, as a way to achieve both social objectives (better work/life conditions) and economic/environmental ones (the reduction of congestion). The key to this was locating housing nearer employment centres, which meant within or around Cambridge itself. This in turn meant some modification to the inner boundaries of the green belt. The growth dynamic of the sub-area was to be restrained by limits on non-essential employment-generating developments. The public consultation required by national legislation for structure plans was framed around the discussion of specific growth locations.

The county planners approached the consultation process very systematically, doing more than the legal requirement. In this way, the resultant structure plan was formally legitimated, having followed correct procedure, been approved by two legitimate political bodies (the county and the national government), after careful technical assessment and extensive public discussion.[21] Both the university and the City Council sought less restriction on employment expansion in the city, and some of the villages wanted more employment opportunities locally. But overall, the predominant message was that the growth rate should be lower than that previously provided for (CCC 1979). The 'drip-feed' land release of development sites thus proceeded in an open and transparent way, providing landowners and housebuilders with a relatively stable policy context in which to acquire an interest in potential development sites.

The Structure Plan was approved in 1980 (Figure 5.5). It carried through much of the strategic direction established previously, limiting employment growth by selective restraint policies within the Cambridge area, dispersing other growth to the north and east. It consolidated a classification of settlements into four types related to their capacity to accommodate housing development, a 'planning vocabulary' used ever since. The plan was structured into statements on policy topics (settlement, employment, housing,

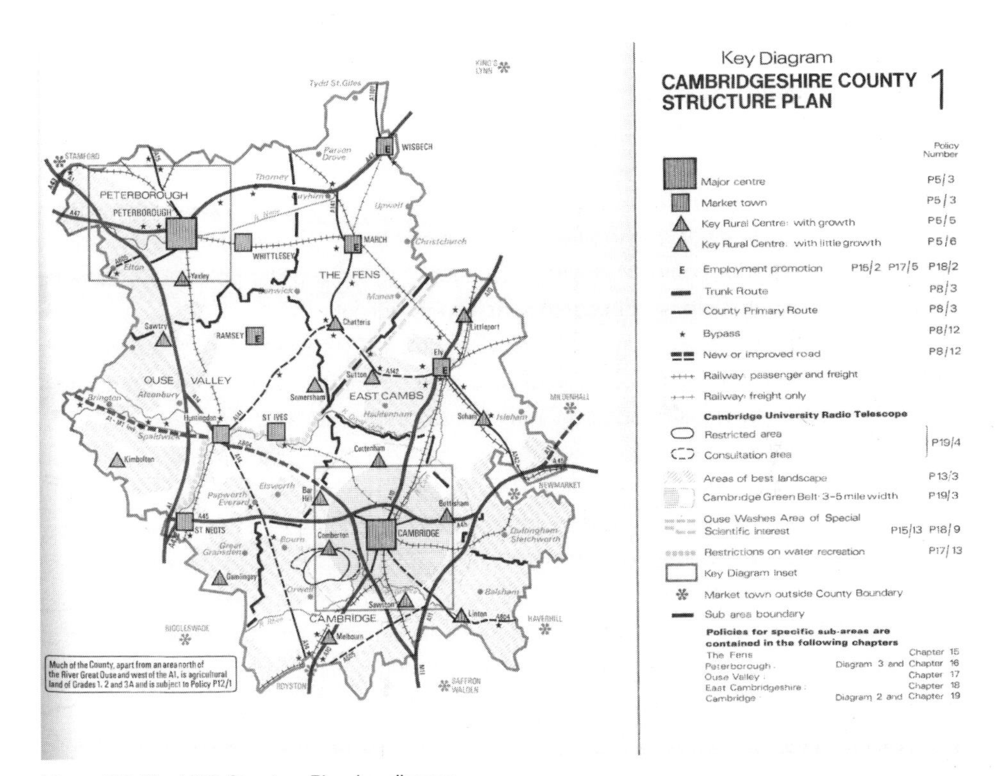

Figure 5.5 The 1980 Structure Plan: key diagram

Source: Cambridgeshire County Council 1980, with permission

shopping, transport, etc.) and statements on 'sub-areas'. For the Cambridge Sub-Area, the locations where provision was to be made for housing development were listed, with the numbers of houses anticipated in each area for the period 1981–1991. Provision was to be made for 8,400 new dwellings in this period in the Cambridge area itself (CCC 1980). The green belt was to be maintained, though the proposal for a substantial extension to the south was rejected. The selective restraint of employment-generating development was also retained.

In this way, a clear strategic frame was re-articulated, embodying the evolving county strategy over the past 25 years. Settlement growth was to be contained within clear boundaries by planning regulation, with strategic planning practice focusing on struggles between planners, conservationists and developers about how much land to allocate, when and where. The practice not only served the interests of the house-building industry, but it also helped to support the investment value of housing and hence owner-occupiers.

Structure plans were intended to provide a strategic framework for the preparation of more specific local plans. The Cambridgeshire planners attempted to be as specific as possible, but were aware that over-specification would be resisted by the districts and by the national ministry. It was left to the districts to prepare local plans for the main development sites. The county planners then turned to the preparation of a Green Belt Local Plan.[22] However, this plan was continually held up by pressures from Cambridge City and the university to release sites on the inner edge of the green belt, and overtaken by reviews of the 1980 Structure Plan. The Green Belt Local Plan was approved by the County Council in 1987, but national government intervened to prevent adoption until the 1986–1989 structure plan review process had been completed. By this time, a neo-liberal national government had created a more market-oriented environment for growth management and a development boom was underway. Aligning the policies of districts, county and national government became increasingly difficult.

The Cambridgeshire county planners in the 1970s built on a tradition that had evolved locally, but was also infused by a wider planning approach that had been developing in the post-war period. A new social, economic and environmental consciousness among the planning profession at large was particularly strong among the strategic county planning teams created after the 1974 reorganisation. Rather than articulating visions of place, these teams sought to analyse the dynamics of local economies, of the relations between people, jobs and travel, and the balance between development and environment. Yet the economic analysis built primarily on trends in sectors of employment. There was little understanding of the dynamic evolution of different kinds of companies and their time–space relations. The only discrimination was between 'local' and 'non-local' businesses. Nor did the planners foresee that the electrical engineering and scientific research activity they noticed was developing into a powerful economic dynamic. They also believed they could control the commuting and other development pressures arriving along the extending M11 and through the growth of Stansted Airport.

This was not just a technical failure of prediction. It also reflected the mood of the late 1970s, dominated by economic difficulties.

It was this mood that helped to elect a national Conservative government in 1979. The Structure Plan was approved at the point of a major 'turn' in British politics, towards a pro-growth, pro-market philosophy. Development initiative, in this philosophy, was to be released from over-regulation and every area was to have a '5-year supply' of housing land, based on market estimates of demand (DoE 1980). This destabilised the 'drip-feed' practice of making sites available, signalling that many more development opportunities might be negotiated through the planning system. The rhetoric of entrepreneurial initiative as a driver of economic prosperity also reinforced the context in which a new identity was emerging for Cambridge as the locus of a major economic 'cluster' of high-technology industries.

BREAKING THROUGH: SUSTAINABLE DEVELOPMENT AND A 'HIGH-TECH' CLUSTER

'THE CAMBRIDGE PHENOMENON'

During the next two decades, county and district politicians and planning officers struggled to provide a coherent and consistent strategic framework for growth management, in a situation where strategies were continually challenged by developers seeking to break through restraint policies, and where the power to resolve conflicts lay with a national government itself uncertain about its planning policies. This struggle was experienced throughout prosperous southern England, where a deregulating government encountered its own heartland supporters who were increasingly assertive in protecting their local landscapes from further development. This struggle was particularly acute in the Cambridge area where the growth dynamic generated by science-based high-tech industry and the steady incorporation of the area into the London commuting range took off in the 1980s. In this situation, the county and district growth-management strategies had to be continually adapted to meet changing and unstable national policy positions, a process that became increasingly frenetic in the 1990s (see Table 5.1). Within the British planning system, the formal power of national government lies in the capacity to review plans and specific development decisions. Inquiry processes undertaken under the auspices of national government are required for development plans and wherever a developer 'appeals' a planning decision on a specific development. In these processes, consistency with national policy has always been a major concern. These provisions allow a national government with centralising tendencies to exert a powerful influence on local development strategies.

The Thatcher administrations of the 1980s was just such a government (Gamble 1988; Thornley 1991). Suspicious of both local government and the planning system, early national policy statements in the planning area emphasised the relaxation of

Table 5.1 Chronology of planning strategies, 1974–1995

Levels of government	*1973*	*1975*	*1980*	*1985*	*1990*	*1995*
National			Circular 9/80	Circular 14/85 PPGs introduced	This Common Inheritance	PPG3 (Housing) PPG 6 (Retail)
Regional		EA Reg. Plg Council Rpt		SCEALA Reg. Policy Statement	RPG6(1)	
County	Retail study	Cambs Sub-Area Study	SP approved Green Belt LPlan started	SP review and EiP	SP review and EiP approved Transportation study	
District: city			Small area local plans	District-wide LPlan ----------	----------	----------
District: SCDC				District-wide local plan ----------	----------	----------
Other major events and decisions	Cambridge Science Park opens			Cambridge Phenomenon study Stansted Airport expansion agreed	Major schemes 'called in'	Cambourne new village agreed Duxford superstore refused

'bureaucratic' planning regulations and demanded that greater priority be given to 'market' assessments of when and where development should take place.[23] With tough policies restricting public expenditure on public services generally, it was also expected that private developers would pay for the service and infrastructure needs that they generated. In theory, this provided a helpful context for those promoting substantial further growth in and around Cambridge. The economic arguments were looking increasingly strong. By 1985, 400 high-tech firms employed 16,000 people, about half in eight major companies (SQW 1985). The university made good use of its status as an 'exception' to growth-restraint policies. Following the opening of the Cambridge Science Park in 1973, a further science park, in Melbourn, was opened in 1982.

The scale of this development marked a shift in the economic culture of the sub-region. A dynamic cluster of new technology industries and business services, with a culture of scientific research innovation and entrepreneurial initiative to exploit the results of research, was emerging in what had been imagined as a university town and a sub-regional centre. A 'paternalist' university culture was transformed into an entrepreneurial ambience at the leading edge of new technology industries (Allen *et al.* 1998). The Cambridge example became an iconic myth of the economic success of Thatcherite neo-liberalism (Crang and Martin 1991). By the late 1990s, the Cambridge 'cluster' was being recognised not just as of national importance, but comparable (if on a smaller scale) with a select group of such innovative 'clusters' internationally.[24] The notion of a 'cluster' implied not just a group of companies in the same industry, but an ambience of interaction and exchange of ideas and contacts that fostered innovative development, both scientifically and economically (Crouch *et al.* 2001). But Cambridge could use its university tradition to provide a 'prestige' address for companies (Morrison 1998). A Cambridge location meant accessing this address as well as tapping into a distinctive ambience. This in turn could be used by companies not connected to the university to argue for an exception to the county's selective dispersal policies.

The contradictions in the county-planning strategy between allowing selective growth related to university science-based industries in Cambridge and the dispersal strategy were pointed out in an influential report by consultants Segal Quince Wicksteed in 1985. This identified and named what was happening in the Cambridge economy ('the Cambridge Phenomenon') and identified the pressures that selective growth was creating, particularly in the housing market and in traffic congestion.[25] The 'Cambridge Phenomenon' study presented high-technology industry as 'clean and green', as opposed to the traditional image of industry. This idea was materialised in the 'science park' development concept. The Cambridge Science Park became a model that spawned a wave of science and business park developments across southern England in the later 1980s (Massey *et al.* 1992).[26] In the Cambridge area, two science parks and six general-purpose business parks were opened in the period 1985–1990, with a further six parks opening in the late 1990s (SQW 2000; While *et al.* 2004).

This 'naming' of a new economic dynamic in Cambridge served to position its development impetus as a significant asset in the new 'Thatcherite' neo-liberal politics (Crang and Martin 1991). But economic growth in Cambridge was not just the result of the impetus of this 'cluster'. It was also a consequence of the emerging geography of southern England, in which the city was an increasingly important regional centre for economic, social and administrative services, a major tourist destination, and ever closer to the London metropolis. The 'drip-feed' release of development sites in the sub-region inevitably came under severe pressure in such conditions.

Through major infrastructure investments, Cambridge was being increasingly absorbed into the London metropolitan area. By 1990, Cambridge had bypasses to the west (the M11), the east (the A11) and the north (A14). Electrification was in progress, linking the city to two central London stations in less than an hour. The prolonged inquiry into the third London airport concluded with a decision in favour of the expansion of Stansted in 1985, although it took a further decade and the growth of the low-cost airlines before the scale of activity there took a quantum leap. All of these generated demand from both households and firms to locate in the area. In the flexible and booming development climate of the late 1980s, the area also attracted the interest of land and property developers. These quickly responded to any sign of a weakening of national and/or county policy restraining development on greenfield sites. Projects focused not just on science and business parks, but on major housing schemes and large out-of-town retail complexes. These development pressures led to a spate of major inquiries, about road proposals, about out-of-town centres and about 'new settlements'. It was in the arenas of these inquiry processes, as much as in the revisions to prevailing structure and local plans, that planning policy in the Cambridge area was progressively adjusted.

CO-ALIGNMENT IN A CHANGING POLICY FRAMEWORK

Strategic planning in the Cambridge area had proceeded since the 1940s with a degree of sophistication and continuity that realised the best hopes of the designers of the 1947 Town and Country Planning System. The planning work was well-informed, carefully argued, and combined flexible recognition of local circumstances and debates about the scale and nature of growth with a robust capacity for using regulatory powers effectively (Brindley *et al.* 1989). This capacity came under severe pressure in the 1980s. The rolling forward of development land allocations within a strong regulatory framework was undermined not just by the promotion of a market-led view of development at national level. Resources for public investment were also cut back, infrastructure agencies were privatised, and local authority staff cut. The Regional Economic Planning Councils were abolished. The team available to prepare structure plans in the county was reduced from around 20 in the late 1970s to around five in the late 1980s.[27] New regional advisory bodies were formed, but staffing was limited. A major consequence of staffing cutbacks was that, instead of in-house technical work, increasing use was made of consultants, a practice that escalated in the 1990s and 2000s. But the major problem for all levels of the planning system was uncertainty over strategic policy.

At national level, government expressed its 'policies' in statements[28] on particular issues, such as green belts, housing allocations, retail provision and developers' 'obligations'. These statements had overriding standing at appeals against the refusal of planning permission and were a mechanism through which national government could steer local planning policy and developers' expectations. In the mid-1980s, the emphasis was on allocating land for growth. In Circular 9/80 (DoE 1980), as previously noted, planning authorities had been required to allocate a '5-year supply of housing land' in their plans. In Circular 14/85 (DoE 1985), they were expected to grant planning permission for development unless to do so would 'cause demonstrable harm to interests of acknowledged importance'.[29] This put a premium on the quality of argumentation and on consistency with national policy, both in planning documents and in relation to specific development decisions. In 1986, at a time when development pressures and neo-liberal ideas about de-regulation were at their height, Cambridge County Planning Department began preparing a revision to the County Structure Plan. But by the time the revised plan was approved, in 1989, national government had realised that market-led development in boom conditions caused major problems of infrastructure overload and led to strong public resistance with electoral consequences. At the end of the decade, the development boom collapsed, leaving developers arguing for greater certainty and stability in government policy, to protect them from their own poor judgements.

By the late 1980s, concerns for environmental quality also began to creep into the government's agenda. In 1990, the then Minister for the Environment, Chris Patten, secured agreement for a cross-departmental report, *This Common Inheritance* (SoS 1990). This argued that the concept of 'sustainable development' should pervade all government policies. In the fields of transport and land-use planning, this brought notions of demand management and growth control back into favour. For many councillors and planners in the Cambridge area, the 'sustainable development' philosophy signalled a revival of their role in growth management, and a significant counterweight to the 1980s emphasis on market-led development strategies. Many councillors and planning officers recognised the importance of supporting the expansion of university-related, science-based industries, but they were also concerned about weakening the selective restraint strategy. Such weakening allowed companies with no ties to the area or links to science-based industry to add to the growth pressures in the area. The 'sustainable development' philosophy provided an argument for a return to a clear selective restraint strategy. This philosophy was also attractive to more traditional Conservatives, to business interests aware of the value of a quality environment for their companies and workforce, and to the rising electoral power of Liberal Democrat and Labour councillors at county and city level. But it was not clear until the mid-1990s how stable the national government commitment to sustainable development actually was.

Between 1985 and 1996, the county revised its structure plan twice, a Regional Report was produced, there were several public inquiries with major policy implications, a Green Belt Local Plan was finally approved, and Cambridge City and South

Cambridgeshire districts produced their first district-wide Local Plans. As noted above, the first County Structure Plan review sought to allocate more capacity to accommodate 'high-tech' growth in the Cambridge Sub-Area but retained the overall emphasis on dispersing growth as far as possible to the north and east of the county, where economic difficulties remained, while resisting commuting pressures from Hertfordshire, Essex and London. At the same time, the rural setting of Cambridge was to be maintained. But concern about the increase in commuting from the villages, market towns and further afield into Cambridge led the county planners to propose the allocation of development sites for housing and business development near the city, concentrated in locations either adjacent to the city boundary or in 'village-scale' new settlements.

The result was a small increase in the numbers of dwellings to be accommodated in the 'plan-period', mainly in the South Cambridgeshire area (see Figure 5.6). The County Structure Plan (CCC 1989) proposed that two new settlements should be accommodated, one along the A10 to the north and the other along the A14 either to the east or west, both beyond the green belt. The Structure Plan also indicated that one major out-of-town retail superstore might be accommodated, to relieve the retail pressure on Cambridge. Inevitably, these allocations aroused intense controversy, from those with alternative sites as well as those seeking to resist further development. The county planners would have preferred to indicate with some precision where these projects should be located, but districts argued that it was their job to locate sites precisely in their local plans. Following a practice emerging nationally, they argued that the County Structure Plan should primarily focus on the criteria for selecting locations. Shortly after the plan was completed, the county also undertook a transportation study. This

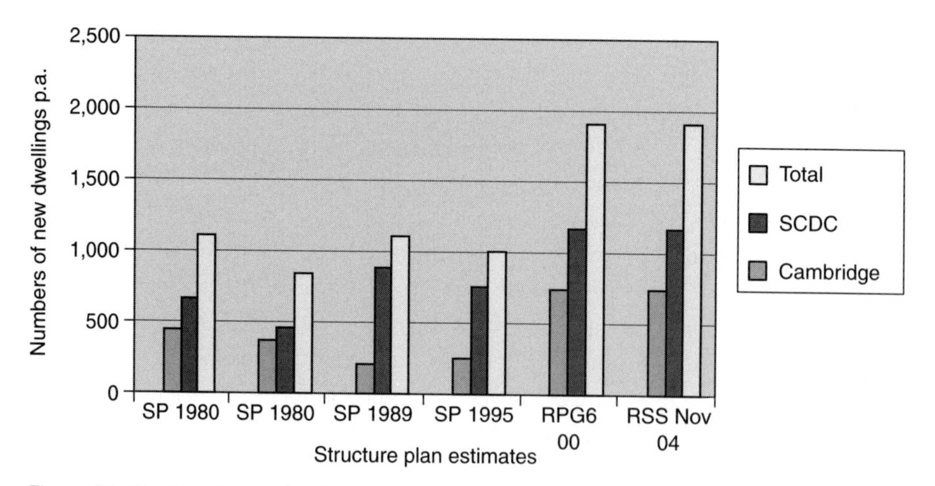

Figure 5.6 Allocations for new housing in planning strategies, 1980–2005

Source: The estimates for 1976–1981 and 1981–1991 are from the 1980 Structure Plan; those for 1986–1991 from the 1989 Structure Plan, those for 1991–1996 are from the 1995 Structure Plan, those for 1999–2016 are from RPG6 2000, and for 2001–2021 from the November 2004 draft RSS and were still the same by the time of the 2005 EiP

proposed further road schemes, but also developed proposals to accommodate a shift from road to public transport modes around Cambridge, with proposals for a light-rail route from Oakington near the A14, along a disused rail route via Histon to the mainline and Cambridge station, and south via the new Addenbrooke's hospital to Trumpington. This study also proposed a system of 'park-and-ride' sites around the city and recommended road pricing measures, all of which attracted controversy.

This strategy was then rolled forward into the production of Regional Policy Guidance. By the 1990s, the production of such guidance, in the form of national planning-policy statements, became a requirement for all regions in England. This was a response to various lobbies pressing at national level for a more coherent approach to development-site allocations and infrastructure investments. This guidance was produced in draft by local authorities and then submitted to national government for amendment and approval. The three counties in East Anglia had already been working together for some time on regional issues,[30] and their ideas framed the East Anglia Regional Planning Guidance 6 (RPG6 1991) (DoE 1991). While largely carrying forward the county strategy, RPG6 1991 stressed the economic significance of high-tech development in the Cambridge area. It also emphasised the emerging national practice of a hierarchical allocation of numbers of new dwellings to be accommodated in each region, which the regions and counties then had to distribute to districts. This was combined with the specification of criteria based on environmental sustainability principles intended to govern the choice of development locations. These stressed a 'sequence' of locations, from those already developed and near public transport to those furthest away (Murdoch and Abram 2002; Vigar *et al.* 2000).[31]

RPG6 1991 primarily gives an indication of government policy in the planning field.[32] It was unclear at the time how far national government would back its own sustainability principles. Meanwhile, the combination of the 1989 *Structure Plan* and the policies in RPG6 1991 had encouraged even more development sites to come forward in the Cambridge area. This was exacerbated by the lack of site specificity in the structure plan. Given the uncertainty of the national commitment to a new, more sustainable policy direction, it was left to the arenas of major public inquiries to determine how far policy had really changed.[33] These inquiries focused on a group of new settlement proposals around the A10, another group around the A14, and on major retail projects. Arguments raged about how far it was wise to emphasise increasing road provision, rather than managing demand by limiting provision for cars and expanding public transport. Eventually, with one exception, all the projects were refused permission, indicating a firmer stance on sustainability principles at national level. One new settlement was allowed, however, which became the new village of Cambourne, on the A429, not far from the A14 and the end of the M11, given planning permission in 1993. This was a relatively low-density proposal for a development of 3,000 dwellings, with poor potential for developing light rail or guided bus provision, and was not one of those put forward by the county planners.

These developments all required amendments to precise land allocations in local plans and to the county structure plan, although, due to the property slump of the early 1990s, the pressure on decision-making was not quite so great in the early 1990s. The Green Belt Local Plan was finally adopted in 1992 after its 11-year gestation period. As the inquiries proceeded into the major development proposals, the outer boundary was determined in such a way that any new development likely to be approved would be beyond the green belt. With this fixed, South Cambridgeshire could proceed to approve its own first district-wide local plan, in 1993, as required by national legislation in 1990. Cambridge City, meanwhile, started preparing its district-wide local plan in 1990, in order to develop the policies indicated for the city in the structure plan and to incorporate the small area local plans prepared in the 1980s. However, its preparation was caught up in the uncertainties generated by the major inquiries. The City Council's Labour and Liberal Democrat councillors sought a stronger emphasis on environmental sustainability and social equity issues across the whole of the council's work, with more emphasis on the provision of 'affordable housing'. The resultant plan, adopted in 1996 (CCityC 1996), was carefully linked to the City Council's community strategy, a statement of values and a concept of 'urban structure', with long-standing concepts of green wedges re-expressed as 'green corridors' and 'structurally important open spaces'.[34]

The 1989 County Structure Plan and the 1991 East Anglia Regional Planning Guidance 6 had left questions hanging over the 'sustainability' and practicality of continuing with the dispersal strategy. The outcome of the major inquiries, coinciding with a weak property market and an economic downturn, suggested also that the pace of housing development in the Cambridge area could be reduced from expectations of the late 1980s. National government was also defining the 'sustainable development' agenda more emphatically (Owens and Cowell 2002). In this context, in 1992, the county embarked on yet another structure plan revision. Political control at county level had swung to a Labour/Liberal Democrat majority, and councillors were keen to promote more environmentally sustainable land-allocation strategies and to resist development pressures from Hertfordshire and London seeping into the area south of Cambridge. The 1995 Structure Plan, approved after modifications through the relevant inquiry process, clearly signalled that the policy of restraining Cambridge's growth, and the dispersal of economic pressures further north and east was no longer viable, although reiterating many earlier policies.[35]

The new county councillors wanted to initiate a move away from the development opportunities created in the 1989 plan, but did not have time to revise the draft structure plan prior to the inquiry. The inquiry inspector recognised the policy shift among councillors and at national government level, and removed two controversial sites from the revised structure plan, arguing that a strategic assessment of development needs and green belt allocations was needed. The first was for a major housing project near Trumpington. Cambridge City Council strongly favoured the release of this site for housing, having resisted its allocation as a science park. The second was for an additional new settlement. Development conditions in the 1990s did not seem to justify this project, with all the controversy and uncertainty this would generate.

Thus, by the mid-1990s, the overall justification for the policy of selective restraint and growth containment had shifted from a conservation and landscape argument for protecting the 'special character' of Cambridge and its setting, to arguments grounded in the co-evolving agendas of the economic dynamic of a high-technology growth node and principles of 'sustainable development'. This latter argument was being elaborated nationally into an emphasis on developing on brownfield rather than greenfield sites, reducing the need to travel by locating jobs and homes nearer together and protecting environmental resources. Meanwhile, there were continuing concerns about economic growth and the strains this put on the housing market (with serious affordability problems building up and fuelling outmigration and hence commuting), on the labour market and on traffic congestion (Morrison 1998). Sustainability, in this context, also called up an earlier planning idea of 'balanced' development – of housing, work opportunities and transport.

This period illustrates very well the nature of the relations between levels of government in the planning system at this time. Although the counties and districts undertook much of the work of developing the details of planning policies, this was performed in the consciousness that national policies and inquiry decisions could overtake and derail their strategies at any time, and that these national initiatives were by no means consistent with each other. The instability of national policy created continual problems for the county and districts as they struggled to keep up with the twists and turns of a national orientation to both sustainable development and the reduction of regulatory pressure on developers. To retain their regulatory power over the location of development and their ability to negotiate with developers for contributions to mitigate the impact of growth, they needed firm and clear county and district policies expressed in plans vertically-aligned with national policies. Any uncertainty and inconsistency were liable to be exploited by developers and their legal advisers. In southern Britain, the planning system as a whole was becoming increasingly legalised and shaped by inquiry decisions and legal challenges in the courts. Counties and districts needed robust and legitimised strategy, to ground their positions and to reduce market uncertainty about the scale and location of development. But producing formal development plans inevitably took time, as procedures allowing consultation, objection and inquiry/review had to be followed and the different levels of government had to somehow co-align their shifting positions. These practices consolidated a regulatory planning policy community of civil servants, local planners, consultants, developers and lobby groups, with a distinctive vocabulary and techniques (Murdoch and Abram 2002).

But in this process of vertical alignment in Cambridgeshire, although there was careful consideration of the Cambridge Sub-Region, there was no real debate about the qualities of the area. Cambridge was portrayed as a university town, a market town, a regional centre, a dynamic economic cluster, a tourist destination, an accessible city, a green and civilised one, a sustainable one, offering life opportunities for all. But these different identities fluttered across the policy landscape without providing clear directions for the kind of place that the city and the surrounding area might be evolving into. Councillors at both county and district level were against 'sprawl'. Cambridge was envisaged as a compact settlement, surrounded by its green belt, despite its internal structure of major open spaces. The city was presented as situated in a landscape of villages, with market towns beyond. Ideas about development corridors sometimes emerged, with a proposal for an M11 growth corridor spinning off from a study of growth corridors in the South East in the 1980s (Crang and Martin 1991), but these remained muted and primarily linked to arguments about locating development near public transport routes.

In this context, some local actors began to press for a more coherent assessment of the city's future. As the principal policy and plans officer with Cambridge City Council wrote after the inquiry into the Structure Plan:

> The result [of past policies] has been that dispersal has been of housing, not jobs. Cambridge's commuting hinterland has expanded rapidly and commuters overwhelmingly arrive by car, causing the city's notorious congestion. The new structure plan pays lip service to sustainability, but perpetuates the split between jobs and homes.... [it is time for] a clearing of the decks for a major strategic review of the Cambridge area. [In this review, we need to consider] what sort of city [. . .] we want Cambridge to be and how ... urban expansion [is] to be accommodated.... Holford's maxim that one cannot make a good expanding plan for Cambridge seems increasingly untenable. The [structure plan inquiry] panel has given Cambridge's planners the challenge to work out how it might be done. (Hargreaves 1995: 1112).

To mobilise to meet this challenge, some city councillors and officers made approaches to the university to press for a more positive strategic approach to growth in the sub-region. This led, in 1996, to the creation of the informal growth-promotion network, Cambridge Futures.

MOBILISING FOR 'SUSTAINABLE GROWTH'

CREATING THE 'CAMBRIDGE SUB-REGION'

By the mid-1990s, the policy of restraining the growth of the immediate Cambridge area and dispersing its development energy to more distant areas, always contested, had finally lost its support at both county and national level. Without such backing, the regulatory power to restrain development around Cambridge was weakened. At national level, the potential of the Cambridge high-technology economic 'cluster' had become too important to be endangered by restraint policy. Further, the concept of sustainable development, which had significant leverage locally and was also strongly supported in the national-level department responsible for planning, housing and urban policy, undermined the planning arguments for dispersal if the result was to increase car-based commuting. The national government's modifications to the 1995 Cambridgeshire Structure Plan, while backing the plan, set the stage for a major review of spatial strategy, a position that was not affected by the change of national government in 1997 to a large Labour majority. In fact, this accelerated the momentum for a review, by strengthening and altering the regional-level government agencies, and, in 2003, introducing a de facto, if partial, national spatial strategy for England in the Sustainable Communities 'programme for action' (ODPM 2003).

The story of strategic planning in the Cambridge area between 1995 and 2005 is one that moves up and down the levels of formal government and in and out of an array of actors – public, private, formal and informal (see Table 5.2). By the early 2000s, the importance of promoting the various high-technology economic activities, and of their location in the immediate Cambridge area, was widely accepted (While *et al.* 2004). Substantial economic growth in Cambridge was in the interest of promoting national economic competitiveness. Buttressed by sustainability arguments, this implied major housing and retail development in and around Cambridge, with major infrastructure investments, particularly 'high quality public transport', to support such growth. Locally, key stakeholders arguing for substantial growth agreed on the importance of demanding more infrastructure investment from national government. Local politicians, well-aware of citizens' concerns about environmental qualities, struggled to promote a full range of sustainability arguments (from reducing resource use and the impact of climate change, to provision for walking and cycling, and an emphasis on high-quality design). They also sought to ensure a strong connection between the allocation of sites for development and the provision of physical and community infrastructure. To some extent, this could be accommodated by speeding up the post-1950s practice of selective release of sites for development on the boundary of the Cambridge built-up area and the allocation of development sites in the villages or in new settlements beyond the wide green belt. But sustainability arguments encouraged the concentration of development along public transport 'corridors'. This corridor concept, however, challenged the well-established imagery in the Cambridge area of a city in a rural setting of open land and villages with market towns beyond.

This nexus of political and planning actors around local development issues, however, carried insufficient weight to influence national government and lever in the resources needed to tackle the growing infrastructure deficit, let alone cope with substantial growth. Moves were therefore made to connect with other networks to raise the profile at national level of the growth needs of the area. Cambridge has always had a dense array of networks, many well-connected to influential groups nationally and internationally (Keeble *et al.* 1999; SQW 2000). The university had already in the 1990s set up the Cambridge Network and CULIL (Cambridge University Local Industrial Links) to promote local linkages. The Chamber of Commerce was also involved in considering growth needs.[36] The task now was to make a link from these university–business

Table 5.2 Chronology of planning strategies, 1995–2005

Levels of government	*1995*			*2000*		*2005*
National		Labour government elected 1997		Reform of Planning System proposed 2001	Sustainable Communities Action Plan 2003	Barker Report 2004 GAF and CIF created 2004 2004 P&CP Act Revised PPG3 Draft 2005
Regional	Government Offices in regions created (1994)		EEDA created 1998	RPG6 2000 approved	Enlarged East of England region created	Regional Spatial Strategy in preparation ....
County		Cambridge Capacity Study (Chesterton)		Cambridge Sub-Region Study (Buchanan)	Third Structure Plan review (appr. 2003)	
District: city	Local Plan adopted 1996				Local plan under revision	
District: SCDC	Local Plan in preparation				adopted	Local Development Framework in preparation ....
Other major events and decisions		Cambridge Futures formed Cambridge Network formed	Cambridge Futures Study Greater Cambridge Partnership formed			Cambridgeshire Horizons established

concerns to a regional development focus. This was partly encouraged by the creation of the Greater Cambridge Partnership in 1998, under the auspices of the East of England Development Agency (EEDA).[37] A more important arena for a while was the Cambridge Futures group. This arose as a joint initiative between the university, via Vice-Chancellor Alec Broers, and the City Council (via Mayor John Durrant). For city councillors and officers, an alliance with the university and business interests linked them to the promotion of Cambridge's national economic significance.

There was significant overlap between the various groups. Cambridge Futures quickly drew in the county level and focused on studies to develop options for future growth. A key resource was the University School of Architecture, whose head, Marcial Echenique, led a study of future development options. A further study, undertaken for the Cambridge Network by a team including Echenique and the Vice-Chancellor, Alec Broers, looked at the role of information and communications technology in minimising the impacts of growth. Meanwhile, the county had commissioned a study of development options, undertaken by consultants Chesterton (Chesterton Planning and Consulting 1997). This explored the relative merits of concentrating or dispersing new development, and the potential for public transport corridors and the idea of a further major new settlement. These initiatives gradually came together, united by the recognition that any growth strategy needed to be inserted into revised regional planning guidance and into the priorities of the ministries controlling funds for transport, health and education expenditure. The problem for a rapidly growing area in England was that core funding for local government was based on existing populations and their relative prosperity, not on future growth. This implied that provision for growth could only be made through normal local expenditure once it had occurred. Yet, locally, people resisted growth if it was accompanied by such lag effects. The struggle was to get access to investment funding for urban growth, which for many years had been targeted to areas in need of regeneration. In the Cambridge area, the resistance to growth was quickly articulated around the agenda of house-price rises, damage to the historic city centre and traffic congestion (Dawe and Martin 2001; Kratz 1997). In this context, those arguing for a growth strategy had to show that planning regulation and infrastructure investment could proceed in a coordinated way, a difficult enterprise in a centralised and functionally divided government context.

Until the 1990s, the Cambridge 'area' was contained reasonably well within the boundaries of Cambridgeshire County. In previous strategies, it was referred to sometimes as a sub-area, or a sub-division, usually encompassing the city and South Cambridgeshire District Council areas. By the late 1990s, however, it was clear from the various studies that the functional linkages of the area where wider, at least as far as commuting flows were concerned. This lead to the definition of the 'Cambridge Sub-Region' as a unit for statistical purposes that included parts of surrounding counties, some of them not even in the East Anglia region. The task of the promoters of planned growth in the sub-region was to get their ideas incorporated into the emerging revision of Regional Planning Guidance 6. This would not only frame and legitimise the strategies in a new county structure plan. It would, they hoped, help to lever in the needed infrastructure investment.

There followed another period of policy development across all the levels of government, during which the East Anglia area was incorporated into a larger East of England region, including Hertfordshire and Essex. New planning instruments were also created that were proclaimed as a radical reform of the planning system (DTLR 2001). The revision of RPG6 1991 was underway by 1997. The first draft acknowledged the importance of sustainable development principles, and suggested that development could be located in public transport corridors. Despite the reference to corridors, a planning idea that was gathering momentum in debates on spatial strategy in the neighbouring South East region, the core spatial organising principle in the East Anglia draft strategy centred on sub-regions, reflecting earlier city region ideas. Following a national emphasis on increasing housing output to meet projections of increasing population in southern England, the amount of new housing to be accommodated in the Cambridge area was increased significantly (see Figure 5.6).

The revision of RPG6 was then overtaken by the re-organisation of regional planning procedure. Until the late 1990s, local government conferences had prepared the guidance and submitted it to national government for approval. The nationally approved statement then became the basis on which local planning authorities prepared their structure and local plans, in a clear hierarchical relationship. However, there was no public consultation or formal inquiry into these policy statements, nor were they debated in the national parliament. This raised difficult problems of legitimacy. The procedure was therefore amended, in the future expectation of the emergence of elected regional authorities, to require an inquiry.[38] The East Anglia Regional Planning Guidance revision became the first to be subject to such a procedure. But, by 2000, when RPG6 was approved, it was clear that new regional guidance would need to be prepared for the newly expanded East of England area, and that the Cambridge Sub-Region would be relocated in a much wider context. By this time (2002), the emerging planning legislation proposed transforming 'guidance' into a Regional Spatial Strategy, designed to replace structure plans. The advocates of the Cambridge Sub-Region as a growth node had therefore to lock their ideas into RPG6, into the Cambridgeshire and Peterborough Structure Plan that followed, and into the emerging regional spatial strategy for the new, wider region.[39]

The inquiry into the draft Regional Planning Guidance in 1999 provided the opportunity for the presentation of clearer ideas about the scale and location of growth in the Cambridge Sub-Region. However, its primary leverage was on allocating funds for urban regeneration and transport projects, in deciding the amount of development to be accommodated in different parts of the region, and in establishing criteria for development locations. The final guidance, RPG6 2000 (DTLR 2000), combines economic competitiveness arguments with environmental sustainability considerations.[40] It incorporates many of the ideas developed in the Cambridge Futures studies and positions the 'Cambridge Sub-Region' as of key regional and national significance. The region is acclaimed in the guidance as a world class and an innovation capital of Europe. Many of the 1995 Cambridgeshire Structure Plan policies are re-stated, but with more emphasis on sustainable development. Some 50 per cent of development is to be on already developed ('brownfield') sites. A 'sequential approach', as advocated in national policy, was used to define criteria for the location of housing sites; this embodied a conception of a compact city, with higher-density sites nearer the city centre. But a study of growth locations and urban capacity was still awaited. This was eventually undertaken by a team led by consultants, Colin Buchanan, along with a further study of how the ideas of the Cambridge Sub-Region Strategy could be implemented, undertaken by consultants Roger Tym and Partners.[41]

RPG6 2000 thus envisages substantial growth, based on expectations about the growth impetus of the 'Cambridge Phenomenon'. By this time, such an impetus was taken as a given force, to be nurtured and promoted by all levels of government, as well as by key players such as the university. The hyping up of science-based high-technology growth in the 1980s (Crang and Martin 1991) had been moderated since then by the property slump of the early 1990s and by the 'dot com' bust of the early 2000s. Cambridge companies survived the former quite well, but were inevitably hit by the latter, though without serious consequences for the area's long-term growth trajectory. The area's economic dynamism has been increasingly based on a large number of small enterprises, spinning out from one company to another, and overlapping with each other. It is supported by a very large number of consultancies, again overlapping, and by some significant venture capital financiers. The result is a cultural ambience which supports new enterprises and expects considerable redundancy (i.e. firms fail as well as succeed).[42] Commentators and academic analysts were all according credibility to claims for the national and international significance of the economic cluster in Cambridge.[43]

But the Cambridge Sub-Region local economy was much more than just a high-technology cluster. It was a major centre of public administration and of all levels of education. The city attracted large numbers of tourists, overlapping with education in the large number of language schools. The area's economic dynamics were by this time part of the complex nexus of the overall London metropolitan economy. It had become increasingly attractive to enterprises linked to the development of Stansted airport, which by 2000 was expanding rapidly. Cambridge had therefore become not only the major economic growth node in East Anglia, but a major city within eastern England, a trajectory that the Holford plan could hardly have imagined. The struggle for growth in the Cambridge area was thus not only about what scale of growth could be accommodated with adequate infrastructure and sufficient quality; it was also about how to discriminate between the demands of the 'world class, high technology cluster' and those of all kinds of other economic activities for which Cambridge was a prestigious and preferred location. The idea of selective restraint and managed growth thus still permeated the strategies for the Cambridge Sub-Region.

RPG6 2000 had reversed the overall strategy of dispersing growth pressures away from Cambridge. The promoters of the Cambridge Sub-Region had argued that it was desirable to accommodate more housing in the region. The County Structure Plan now needed to be revised to provide a strategy for the location of this development. Rather than the in-house assessments of trends and options that were undertaken in the 1970s and 1980s, the technical knowledge base for the Structure Plan was provided by consultancy studies. During the period between 1997–2004, a large number of studies of development issues in and around the Cambridge Sub-Region had been undertaken, by a number of different consultants (see Table 5.3), reflecting a strong national tendency to contract out the production of research and intelligence for public-policy purposes. Two studies had already been commissioned, on growth locations, from Colin Buchanan and Partners, and on how to implement major developments, from Roger Tym and Partners. A further study was undertaken of the prospects for developing 'multi-modal' transport provision along the Cambridge to Huntingdon 'corridor'. The purpose of the Buchanan study (Buchanan and Partners 2001) was to assess the 'capacity' of Cambridge City to absorb more development and to undertake a strategic review of the green belt, both issues left 'hanging in the air' from the 1995 structure plan. The main focus was on the allocation of housing numbers.[44]

The outcome of these studies, as expressed in the Cambridge and Peterborough Structure Plan 2003 (CCC 2003) was a reiterated version of the locational criteria that had been evolving since RPG6 1991. The critical shift was to emphasise that both Cambridge and Peterborough were key locations for new development. Another new settlement location was also specified. Once again, a choice was made between a location on the A10 going north, and on the A14 to the north-west. The latter, Longstanton/Oakington (now Northstowe), was selected in part because it was on the hoped-for light-rail rapid transit route The Structure Plan maintained the strategy of selective employment restraint, now expressed in terms of the selective promotion of specific clusters, a position likely to be helpful to most existing companies in a tight labour market. The regional dispersal policy remains in the attempt to link Cambridge's growth dynamic to Peterborough, and in the continuing promotion of market towns and some larger rural centres on good public transport routes as locations for economic development projects. Reflecting national policy, housing development in the county area as a whole was to be at higher densities, with a larger provision of affordable housing, to meet local housing needs and to redress the imbalance in the housing market. This also implied less pressure to allocate potentially controversial greenfield sites.

New development was only to be allowed where 'the additional infrastructure and community requirements can be secured, which may be by condition or legal agreement or undertaking' (Policy P6/1, CCC 2003: 57). The text of the plan emphasises the enhancement of the landscape and the provision of cycleways, walking and waste recycling. In relation to the boundary between urban and rural landscapes, the plan calls for districts in their local plans to "maintain a clear transition between settlements and the

Table 5.3 Planning studies undertaken by consultants, 1997–2004

Date	*Title*	*Consultancy*	*Client*
1997	Cambridge Capacity Study	Local Authorities, and Chesterton Planning and Consulting	County Council
1998	Cambridge 2020: Meeting the Challenge of Growth	A working group including Alec Broers and Marcial Echenique	Cambridge Network
1999	Cambridge Futures	Marcial Echenique, Department of Architecture, CambridgeUniversity	Cambridge Futures
2000	The Cambridge Phenonomen Revisited	Segal, Quince Wicksteed/PACEC	Cambridge Futures
2001	Cambridge Sub-Regional Study	Colin Buchanan and others	SCEALA
2001	Implementing the Cambridge Sub-Region study	Roger Tym and Partners	County, EEDA and GO-East
2001	London–Stansted–Cambridge Study	ECOTEC Research and Consultancy, and others	GO-East
2001	Cambridge–Huntingdon Multi-modal Study	A consortium of consultants led by Mouchel	GO-East
2002	Key Worker and Affordable Housing	Department of Land Economy, Cambridge University	Cambridge City Council and SCDC
2003	Employment Growth Scenarios	Experian business strategies	EEDA
2003	Stansted/M11 Development Options Study	Colin Buchanan and Partners	EERA
2004	A Study of the Relationship Between Transport and Development in the London, Stansted, Cambridge, Peterborough Growth Area	Colin Buchanan and Partners with GVA Grimley	ODPM

Note
Demographic studies were also undertaken by Dave King at East Anglia Polytechnic University, for the Local Authorities

countryside" (CCC 2003: 70). There is also a strong emphasis on conserving the landscape qualities of the countryside and promoting access to it from urban areas. This was a clear signal for the Cambridge area that a strong green belt should be maintained, but with the old idea of 'wedges' or corridors of green penetrating into the heart of the city. On the transport front, the structure plan continued to emphasise the development of public transport and the exercise of strong 'demand management' measures to reduce congestion in cities such as Cambridge. The county hoped to promote more frequent public transport in accessible and environmentally friendly vehicles, and to promote two long-standing rail proposals – the rapid transit route and an east–west rail route connecting through Cambridge to Oxford in the west and the East Anglian ports to the east.

For the Cambridge Sub-Region itself, these ideas were translated into a clear policy argument explaining that the strategy of limiting Cambridge's growth was no longer sustainable (see the CCC quotation that starts this chapter). This argument expresses the consensus arrived at among politicians in the county, in the city, and, by this time, in South Cambridgeshire District Council.[45] The planning framework of selective restraint that 'nurtured' the development of the Cambridge high-tech cluster was now to be directed at nurturing a sustainable strategy for a more 'balanced' form of development, with more mixed uses on major development sites, combined with an emphasis on reducing commuting, increasing housing supply and affordability, enhancing the quality of the built environment and ensuring a vision of 'Cambridge as a compact, dynamic city with a thriving historic centre . . . framed by its Green Belt setting' (CCC 2003: 106). The new strategy is expressed in an overall vision statement and diagram (see Figure 5.7).

The plan then allocates 'housing numbers' to each location type within the subregion and specifies the locations where sites on the inner boundary of the green belt are to be released to allow more development. Once again, sites previously considered are brought back into play, following the assessment by Buchanan and Partners, notably in the east and south, along with a new site in the north-west, promoted by the university for its long-term expansion needs.[46] Major retail expansion was firmly resisted except in Cambridge city centre. These proposals were presented in a key diagram (Figure 5.8). But while the CPSP 2003 reflects a major shift in strategy for the location of new development, and asserts the importance of attention to environmental qualities and the integration between development and infrastructure, it has little to say about what kind of place the Cambridge area was to become as a city, rather than merely an economic cluster in an attractive environment. Its place-framing attention is limited to the specification of criteria to shape new development projects and the definition of the morphological boundary between built-up and 'green' areas.

It was left to the districts to locate development sites precisely in their local plans and to produce 'master plans' as planning briefs for each site. Although many of the arguments at the structure plan inquiry were about sites around Cambridge itself, South Cambridgeshire had to accommodate more of the housing total, with the new settlement

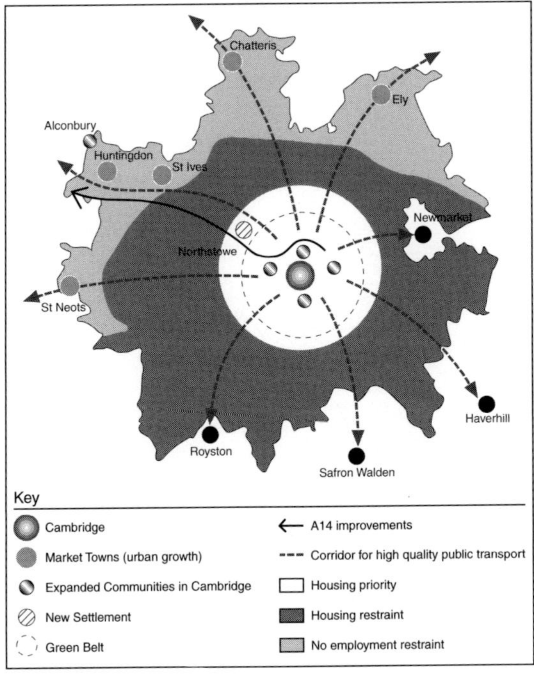

Figure 5.7 The Cambridge Sub-Region vision
Source: Cambridgeshire County Council 2003: 100, with permission

Vision for the Cambridge Sub-Region

'... it will continue to develop as a world leader in the fields of higher education and research, and it will foster dynamism, prosperity and further expansion of the knowledge-based economy spreading out from Cambridge, whilst protecting and enhancing the historic character and the settng of Cambridge as a compact city.... Sustainable and spatially concentrated in patterns of high quality, socially inclusive development will be focused in Cambridge ... to provide a more sustainable balance between jobs and homes ...'

of Northstowe accommodating only 6,000 of the 20,000 that had to be provided for. However, since many sites had been proposed for development and evaluated over the years, selecting sites was not so problematic. It was clear that large sites were available and in suitable locations to accommodate more development than even the targets allocated in RPG6 2000. But for local stakeholders, the availability of sites as such was hardly the issue. Well before the Structure Plan was approved in 2003, the county, the districts and other promoters of the 'Cambridge Sub-Region' were arguing that it would not be possible to proceed without major funding for infrastructure. This could only be obtained from national government. There was also an issue about how to coordinate and manage so much development, with several sites overlapping the boundary between Cambridge City and SCDC. The position of the county as a possible coordinator was by this time seriously weakened as most of their strategic planning powers were to be taken from them by the emerging planning legislation, which relocated the strategic spatial planning instrument to the regional level. The production of a Regional Spatial Strategy for the new East of England region was required, the preparation for which was initiated in 2002. In this context, key stakeholders, under the umbrella of the Greater Cambridge Partnership, produced in 2002 a list of key investments to underpin the growth strategy that they presented to national government (While *et al.* 2004).

The districts, Cambridge City and South Cambridgeshire District Council, had now

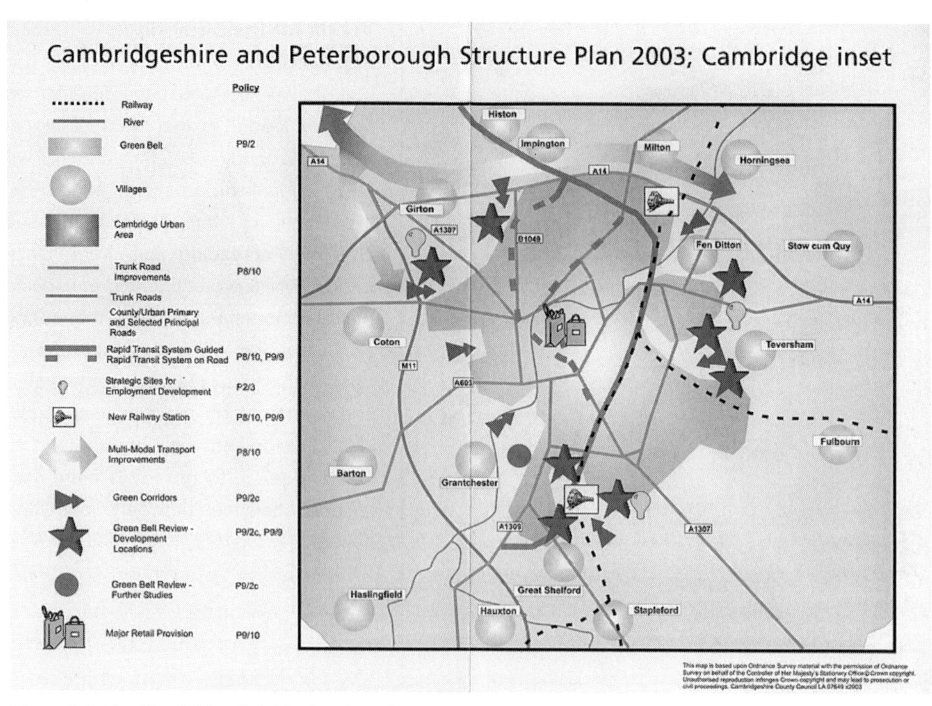

Figure 5.8 The Cambridge Sub-Region: key diagram
Source: Cambridgeshire County Council 2003, end pocket, with permission

to revise their own plans, in the context of the changing approach to development plans introduced by the new national planning act, passed in 2004.[47] By 2003, Cambridge City was already revising its local plan. This was adjusted after a period of public consultation to bring it in line with the approved structure plan. It repeats many of the criteria for new development, emphasising achieving design quality, compactness, environmental benefits and the relation of physical and social infrastructure in all-new developments. This provides an agenda for negotiations about design and about developer contributions with project developers. In terms of urban structure, the city's plan maintains the long-standing concept of landscape corridors stretching through the city, and introduces the idea of 'peripheral mixed-use centres' in the new development areas, to provide future employment-generating development as well as housing. A strong emphasis is given to accessibility, enabled by 'high quality public transport' (CCityC 2004).[48]

The struggle for multi-level co-alignment of spatial development policy thus remained very complex while, at the same time, key stakeholders needed to turn their attention to the implementation of major development projects. This underlined the need to obtain funding for infrastructure investment, not merely for transport, but for community services such as health and education, and for affordable housing. Since the early 1980s, the expectation had been that such funding could be obtained from developers' contributions through the power to demand 'obligations' when planning permission was given. But this inevitably meant that infrastructure lagged behind development, which was not only against the new Structure Plan policy, but extremely unpopular locally. Therefore, the sub-region stakeholders had little option but to campaign hard for more 'upfront' investment, although this meant that their local initiatives were repositioned within a larger and much more complex regional and national governance landscape.

LOCAL SUCCESS IN AN UNSTABLE GOVERNANCE CONTEXT

British local government has traditionally had substantial capacities but limited autonomous resources and virtually no formal power to legislate. This situation was reinforced from the 1980s and continued with the Labour government of 1997, although the latter promoted building a stronger regional tier of government and in theory a 'new localism' (Corry and Stoker 2002). To obtain investment funding, the promoters of growth in the Cambridge Sub-Region had to target the national level, to gain attention to their needs. Such attention built on a well-developed base, through the traditionally strong networks between the university and the civil service and the government's enthusiasm for the 'Cambridge Sub-Region'. The national Treasury was also increasingly concerned about the link between the supply of new housing and macro-economic stability, with a house-price boom in full swing, with rising numbers of households combined with investment shifts from equities to property following the 'dot com' bust, but yet new house-building was not expanding. This led to an interest at national government level in the relation between economic performance, housing prices and production, and the planning system's role in allocating land for housing development. But critical for the Cambridge area was the link between new housing and infrastructure, which meant that the attention of the government departments responsible for transport, education and health was also needed. For the promoters of the 'Cambridge Phenomenon', central government was the 'problem' (SQW 2000).

The main national department dealing with planning and urban development issues, by this time called the Office of the Deputy Prime Minister (ODPM[49]), was responsible for local government, the regional agenda, urban policy, housing policy and the planning system. With respect to this range of functions, the ODPM and its predecessors under the Labour government had been pursuing an often-conflicting agenda of managerial improvement, promoting 'modernisation', 'holistic' or 'joined-up' government, 'urban renaissance', sustainable development and quality environments, as well as 'new localism' ('6' *et al.* 2002; Corry and Stoker 2002; Johnstone and Whitehead 2004), though with a very weak conception of how different policies and initiatives connected together in cities (Marvin and May 2003). In the planning field, the emphasis was on expanding house-building, particularly of affordable housing negotiated through the planning system via developers' contributions, and 'modernising' the planning system into a more flexible, proactive and responsive form of 'spatial planning', rather than time-consuming bureaucratic land-use regulation[50] (Tewdwr-Jones and Allmendinger 2006).

The Cambridge Sub-Region experience was very attractive in this context. Here was a part of the congested outer Metropolitan area accepting growth and not fighting it, showing how a locally articulated strategic approach could succeed in a very sensitive local environment, with an integrated, 'spatial' approach to managing development. As a planning strategy, the ODPM had no problems in backing the planning framework. Providing resources was another matter, however. Through its urban policy remit, the department provided investment funds for urban regeneration. It had not been providing resources for areas of growth, which, it was assumed, could 'pay their way'. Lobbies promoting increased investment in northern England, and in the major cities outside London, well-connected to Labour ministers, sought to protect and enhance these allocations (Jonas *et al.* 2005; Marvin and May 2003). Within the growing areas of southern Britain, many stakeholders were clamouring for more infrastructure investment, responding to the same kind of citizen complaints as in the Cambridge area. Appealing to the national level therefore catapulted the Cambridge Sub-Region promoters into a highly competitive governance landscape.

In this context, the ODPM took the bold step of producing the nearest thing to a national spatial plan for England achieved since the 1960s[51] (see Figure 5.9). Sustainable Communities: Building for the Future (ODPM, 2003) was an attempt to balance the various claims for development investment between the regeneration and growth lobbies. Its focus was primarily on how to produce quality living environments and on the housing production agenda, but it was intended to provide a framework for coordinating the inputs of other departments to the areas targeted for investment. In addition to the well-established urban areas where regeneration funds had been focused for some time, the Sustainable Communities 'Plan' (actually called an 'Action Programme'), identified four 'growth areas' around London. Two remain from the 1970 Strategic Plan for the South East, the Milton Keynes/South Midlands area, and Ashford, in Kent, on the eventually appearing Channel Tunnel high-speed rail line. The third, Thames Gateway, was prefigured in 1970 but evolved in the regeneration strategies for the London docklands and industrial areas along the mouth of the Thames, and had been strongly promoted as a way of counterbalancing growth in London's western M4 corridor, which was no longer considered a growth area. The fourth, the 'London–Stansted–Cambridge' area, is much more recent, deriving from the debates on the Regional Planning Guidance 9 for the South East at the end of the 1990s.[52] The Sustainable Communities 'plan' mentions the possibility of another new settlement in the Cambridge area. The 'Plan' was followed up in 2004 with two new funding mechanisms, the Growth Area Fund (GAF)[53] and a Community Infrastructure Fund. John Prescott, Deputy Prime Minister, also worked hard to persuade his Cabinet colleagues to allocate funds from transport, education and health budgets to the growth areas.[54]

By 2004, therefore, the Cambridge Sub-Region promoters had acquired a new national status as a major growth node in a newly defined growth area, access to a new investment fund, and some priority in other government departments for infrastructure

Actions to Achieve and Maintain Sustainability

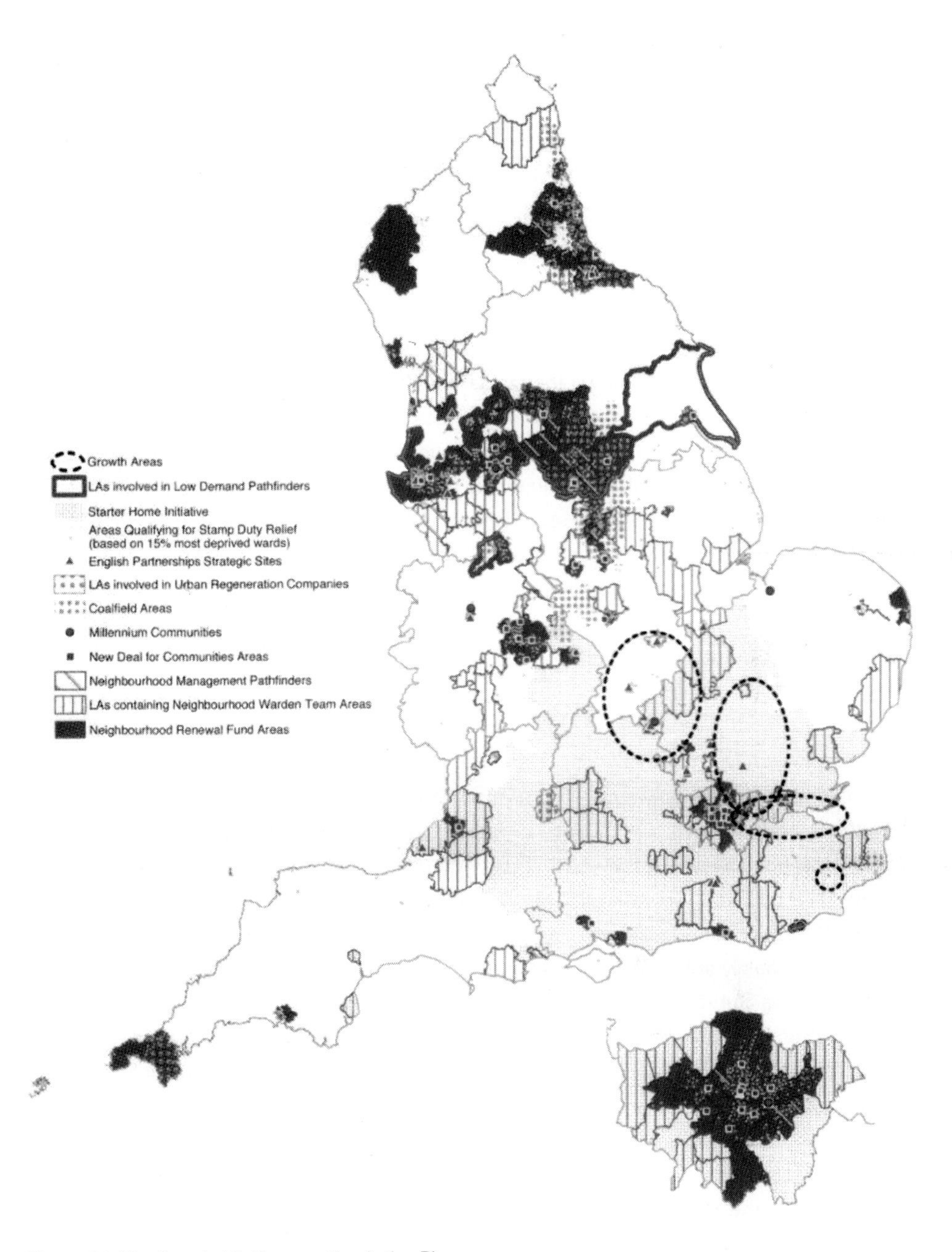

Figure 5.9 The Sustainable Communities Action Plan

Source: ODPM 2003: 67, © 2003

Note

The dotted ellipses are the growth areas

spending in their area. But it still remained to work out where and how the available funding would be disbursed, and how major developments would be managed. On the planning front, despite general support for the strategy embedded in the Cambridgeshire and Peterborough Structure Plan 2003, two factors at national level generated uncertainties. One was the new Planning Act, eventually approved in May 2004 after a two-year gestation. This relocated the structure planning role of counties to the regional level, weakening the coordinative power of the county, and required districts to convert their local plans into 'local development frameworks', which meant considerable extra work. The other destabilising factor was a review sponsored by the Treasury of the relation between the house-building industry and the planning system, the Barker review (Barker 2004).[55] This opened up the policy debate about how much developers should contribute to the direct costs that their projects imposed on a locality and how much they should pay as a tax on the value generated by the conversion of 'greenfields' into urban development. Developers also tended to adopt a 'wait and see' attitude in negotiations over contributions, until there was more clarity about what infrastructure the new government funds would provide. As local actors commented in 2005, this uncertainty had the perverse effect of slowing down rather than speeding up the delivery of new housing development.

With all these changes in government responsibilities and in policies, the critical arenas for urban development strategy had moved from the County Planning Department and the informal networks promoting the Cambridge Sub-Region to the national level and the wider regional level, with the Government Office for the East of England (GO-East),[56] and the East of England Regional Assembly (EERA), consisting of appointed local politicians and other stakeholders, as key players, along with the regional development agency (EEDA).

The Cambridgeshire and Peterborough Structure Plan 2003 was largely absorbed into the emerging Regional Spatial Strategy (RSS). This was partly because Cambridgeshire politicians felt they had already made their contribution to accommodating growth.[57] In addition, however, it was known that there was capacity in sites already approved to accommodate further development, so that a roll-forward in housing targets to 2021 was easily achievable if development on these sites got underway. The issue for the Cambridge Sub-Region stakeholders by this time was not capacity but delivery. The big struggles over the RSS focused further south, in Hertfordshire and Essex, where county/district relations had been riven by conflicts for some time. National government once again felt that a major 'strategic assessment' of the location and form of development in the London–Stansted–Cambridge growth area, now stretched to reach Peterborough, should be undertaken. ECOTEC consultants had already carried out an assessment of the Harlow area, complementing that undertaken by Colin Buchanan and Partners for the Cambridge Sub-Region. The draft of the EERA Regional Spatial Strategy (RSS), available in February 2004, was put on hold at the request of the government, while this new study was undertaken, again by Colin Buchanan and Partners (Buchanan and Partners 2004).

The impetus for this study was the problems now being experienced by government in persuading local stakeholders in Hertfordshire and Essex to accept further growth in the Stansted-to-London part of what was increasingly referred to as 'the M11 corridor'. Into this study was lobbed the requirement that the east of England should accommodate a further 18,000 dwellings by 2016. The study largely left the established proposals for the Cambridge Sub-Region intact, as did the draft RSS, which focused on identifying development nodes to the south. It makes much use of the corridor concept. In contrast, the RSS had extended the well-established concept of regional sub-areas, each with a predominant urban node, into a general idea of a region composed of city regions. Both studies sought to develop a stronger concept of the spatiality of territorial organisation than had been common in structure plans in the previous 20 years (see Figure 5.10).

The Regional Spatial Strategy was then reassessed in the light of the Buchanan study and the increased housing numbers, and approved in November 2004 by the East of England Regional Assembly (EERA) to go out for consultation, prior to amendment and an inquiry due in Autumn 2005. Almost immediately, however, the Conservative councillors on EERA, with a national election due the following May, withdrew their

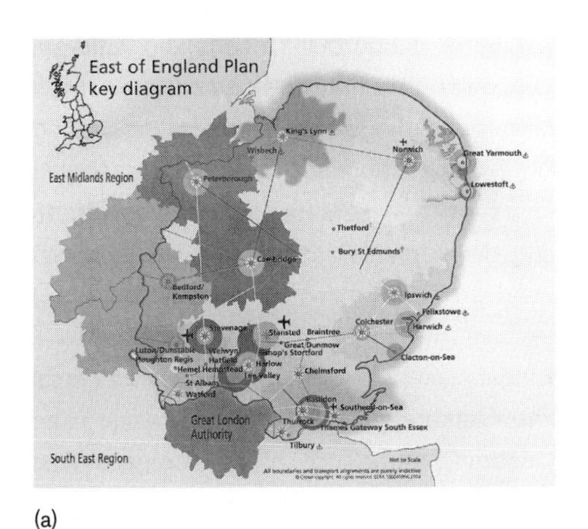

(a)

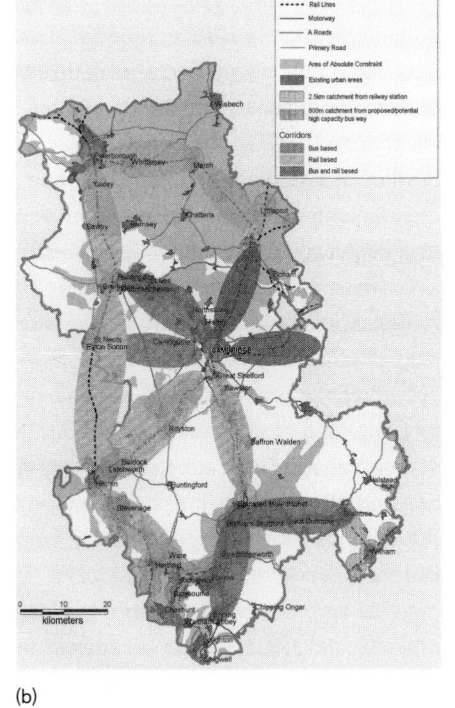

(b)

Figure 5.10 Corridors and sub-regions in the East of England (a) East of England RSS Key Diagram 2004; (b) potential growth corridors suggested by Buchanan and Partners 2004

Sources: (a) EERA 2004: 301, with permission of the East of England Regional Assembly; (b) Buchanan and Partners 2004: 86, crown copyright 2004

support for the strategy, on the grounds that the infrastructure to support such development was not available.[58] However, by March 2005, the councillors came back into the discussion arena, while maintaining their concerns about infrastructure provision. By this time, the consultation period had ended, generating 26,000 objections to the strategy. Few of these related to the Cambridge Sub-Area. However, by early 2006, in response to new demographic projections, the pressure to increase still further the numbers of dwelling units to be accommodated was revived.

Locally, the focus of attention had shifted to getting the major development sites in the Cambridge Sub-Area underway. During 2004, there was much discussion in the nationally-designated 'Growth Areas' about the appropriate 'delivery mechanisms' for managing growth. In the Cambridge area, there was little support for an agency with powers taken from the local authorities, yet it was recognised that some agency was needed to manage development sites and to make bids to the national 'growth area' funds. The outcome was the creation of Cambridgeshire Horizons (CH) in 2004, a semi-independent agency, funded for three years principally by the ODPM, to act as a development facilitator and coordinator for the major sites. All bids for the growth area investment funds were to be channelled through CH before being sent on to GO-East. By 2005, CH had become not merely a development coordinator, bringing different parties together, smoothing out difficulties and speeding up the development process. It also had a networking and knowledge-circulating role, acting as an arena for discussing common problems, for open debate about project priorities and for exchanging experience.[59] While each major project had its own stakeholder group seeking to develop masterplans, coordinating and phasing development, and linking public realm requirements with development practicalities and profitability considerations, CH acted as a useful arena in which they all met.[60]

Thus, in the decade from 1995–2005, a local stakeholder coalition was successful in mobilising behind a growth strategy for the Cambridge Sub-Region, breaking the long-standing hold of the 'Holford' plan, and inserting an agenda of development sites and infrastructure needs into regional and national policy arenas. The practice of regulatory drip-feed of sites released for development had not been changed, but the quantity of land released was substantially increased, and connections between planning regulation and infrastructure investment made in a more coordinated and strategic way. By late 2005, funds for the guided busway from Northstowe to Cambridge had been approved. But there were costs to this success. The local coalition lost some of its powers to the regional and national level, and, in focusing on the implementation of the major growth projects, did not continue to sustain strategic debate about the nature, qualities and urban morphology of an expanded Cambridge.

In addition, the coalition's efforts were forced into the procedures and discourses established by national government. Although, by the early 2000s, a vigorous movement had developed within the national planning policy community to transform the practice of planning from its narrow emphasis on regulatory issues to a more strategic 'spatial planning' focus (RTPI 2001), the politics of vertical mobilisation reduced engagement with citizens in debating urban region futures. Instead, key actors in planning and development arenas depended on their various networks to connect to local concerns. Many citizens remained continually active in their interest in the qualities of the area, and councillors were largely responsive to their viewpoints. What was missing in all this strategy-making work was a wide-ranging debate to 'summon up' any new orienting idea of what the rapidly changing urban region emerging around the old city of Cambridge might become, beyond the initial work on 'futures' in the late 1990s.

Citizens and other stakeholders in the Cambridge area were typically both proud of the recognition of Cambridge as a 'special place' and very protective of its particular qualities. But the 'special place' of Cambridge by 2005 was very different to that of the 'university and market town' of the mid-twentieth century. It had become a key economic driver in a regional and national context, an important locale in the wider economic nexus of southern England, positioned in a group of locales globally significant for a particular industrial form that emerged around new technologies in the late twentieth century. But it is also a small historic town, cut into by green tongues of landscape, and surrounded by villages in a rural landscape, accessible by cycle and on foot. Those seeking the space for growth also know that all these special qualities have to be respected. They live with multiple views of the city's identities and continual contest about changes. They also know that economic growth on its own will not be tolerated by a vocal and well-informed citizenry, who care about environmental conditions and are aware of all kinds of ethical responsibilities to the environment and to fairer and more equitable forms of development. In this climate, it might be possible for an exemplary case of multi-level political and administrative co-operation to manage growth within a perceived territory which itself escapes formal administrative boundaries.

But this potential is undermined by the weakness of the formal governance capacity for a locally driven development strategy. Informally, the various parties over the years have managed to work out how to 'balance' steady expansion with keeping hold of key qualities of place. The county and the city have been critical arenas in achieving this balance, with the university a powerful third party, with its own contradictory interests in both conservation and growth promotion. Now the university is only one party among many, the county's planning powers have been curtailed, and the key formal arena for arguing over planning strategies and investment priorities is at the much wider level of the region, which is itself merely a slice of the London and South East metropolitan complex. Because of the difficulties of managing growth in such a large urban agglomeration, and because this complex is so important and near to national government, any conflicts are played out up and down all the levels of government, and encounter the contradictions over planning, development and infrastructure policy at interregional and national levels. The result is an unstable wider governance context, with the potential to undermine the stability and local support that the growth coalition in the Cambridge area has sought to achieve around a new development trajectory.

CONCLUDING COMMENTS

The story of planning and development in the Cambridge area over the past half-century illustrates the power of local forces, first, to limit development pressures and then accommodate substantial growth according to locally articulated principles. But it also shows that this only succeeds in the UK context if continual efforts are made to gain support nationally and to express local concerns in the discourses and procedures articulated at national level. Although the Cambridge area has been an exemplar of how a new planning system should be practised, the relations between levels of government in the system have been hierarchical and sectoral rather than multi-level and integrative. Co-alignment between the regulation of development and development investment is particularly difficult in such a context. The Cambridge politicians, planners and other stakeholders succeeded because they continually took initiatives to articulate their position in relation to evolving economic and social conditions and in awareness of how planning ideas nationally were evolving.

The major material outcome of this effort is visible in the protection of the valued landscape – of villages in their rural settings, of small market towns and of a still-small city connected to its green surroundings by 'wedges' of attractive undeveloped landscape. But it is also apparent in the increasing levels of traffic moving along the expanding road network, and in the very high house prices with consequent effects on labour costs and commuting levels, a phenomenon evident across affluent southern England and particularly visible in periods of national economic growth and housing market boom. Despite much local concern about the environmental and social costs of selective growth, these costs threaten to undermine the idea of a 'balanced', environmentally sustainable and socially equitable growth strategy.

The planning system, with its procedures for plan-making and for inquiries into plans and specific development proposals, has provided the critical arenas for both articulating and legitimating growth-management strategy. In contrast to many other parts of the country, spatial strategy-making legitimated through the planning system has been a key tool through which local actors have sought to realise a 'selective' approach to growth. Plans have been continually adjusted to retain strategies in good currency and appropriately aligned with the discourses and techniques advocated in national policy. This has been necessary to defend planning decisions against strong developer challenges whenever weaknesses or uncertainties in strategy have been perceived. These strategies have typically been carefully crafted, and focused on key strategic issues, with arguments directed at both local audiences and national government expectations. They have also been infused with a strong perception of the local landscape. The idea of a city region, in a terrain of small market towns and villages in a rural setting, has retained its imaginative pull on planning strategies, despite a shift from a strategy of growth dispersal to one of compact urban expansion. This, in turn, is justified by concepts of city region housing and labour markets that are relatively self-contained, even as

Cambridge itself grows in its impacts across the wider region, and the area becomes absorbed into the complex geographies of the vast metropolitan region of southern England.

A rather traditional geography continues to pervade spatial strategy-making in the Cambridge Sub-Region. This is partly related to the continued significance of the old 'small university town' identity that still has substantial meaning for powerful local actors. But there has also been a shift in the knowledge resources mobilised to underpin strategy-making. Whereas, in the mid-twentieth century, the emphasis was on developing a careful account of the local territory, which could inform strategy-making, an approach still manifest in the technical survey work undertaken by the county planners in the 1970s and 1980s, by the 1990s there was no longer the staff for such work. Instead, knowledge has been provided through discussion in the various network arenas, through occasional special studies carried out by the university's Departments of Architecture and Land Economy, but most particularly by consultancy companies skilled in addressing policy issues structured by national government preoccupations. This material is often not easily accessible to the general public.

The Cambridge Sub-Region story thus illustrates a situation with substantial local capacity to manage development processes in a situation where there are always conflicting values and claims about development options and trajectories. This capacity uses formal government arenas, but activates these through the informal networks that connect different groups to politicians and officials, and link local actors to national politicians and civil servants. In this way, the 'Cambridge Sub-Region' has been brought into existence, even though there is no formal organisation to represent it, and even though it extends across several administrative jurisdictions. However, these networks are still largely those of an 'establishment elite'. How well they will survive in an enlarged Cambridge and in a governance landscape in which they are more exposed to the highly conflictual politics of other parts of southern England, remains to be seen. As throughout this story, much depends on the capacity of national government both to encourage integration between land-allocation strategies through the planning system and investment in infrastructure and services in areas of substantial change, and to decentralise itself, to give institutional space for the development of local capacities for the governance of place.

NOTES

1 A phrase used by a key actor in the 2000s.

2 Road numbers in the area have changed. The current A14 was formerly the A45. I have used current numbers.

3 The report was written in the format of a survey, followed by analysis and then the strategic spatial conception. A major theme was to prevent industrial development in the city, with echoes of the fate of Oxford and the expansion of the Morris car plant there reinforcing a preservationist stance.

4 Cambridge was designated a city in 1951.
5 Local government departments in the UK are headed by full-time appointed professionals, answerable to council committees, or (since the late 1990s) to a Cabinet of elected members.
6 The Ministry had a special interest in Cambridge during the war, due to the opposition of the University to planning controls. A study by Dykes Bower for the Ministry suggested a cap of 100,000 on the growth of Cambridge and the separation of settlements (Cooper 2000).
7 Holford held the Chair of Planning at University College London.
8 Holford and Wright rejected the bypass proposal of the 1934 Davidge report, in favour of an inner ring road largely using existing roads.
9 The *Distribution of Industry Act* 1945 required all firms proposing industrial developments over 3,000 sq ft (27 m^2) in building area in congested areas such as London to seek a permit (called an *Industrial Development Certificate* in 1947), providing the power to force firms to seek locations elsewhere.
10 These New Town and Town Expansion Schemes allowed urban councils seeking to expand to import firms and households from the London area, as part of the London decentralisation strategy.
11 This was finally approved in 1965.
12 The County also designated some villages as 'growth villages', an idea that evolved in the 1950s (Morrison 1998).
13 The Third London Airport Commission was appointed in 1968.
14 See Royal Commission on Local Government in England, Vol. 2, *Memorandum of Dissent by Mr D. Senior*, Cmnd 4040–1, HMSO, London, 1969.
15 The study emphasised the contribution that Cambridge made to the national economy and its important role as a centre for the wider region. It noted that the University needed more space for expansion and that there was an urgent need for more housing in or near the city.
16 The 1968 Town and Country Planning Act.
17 Growth was welcomed in the north and east of the region which were suffering from a fall in employment in farming.
18 During the 1970s, with the development industry in recession and a Labour government, attempts were made to create a more structured process for capturing and distributing land values generated by urban development between public and private interests (the 1975 Community Land Act), but this foundered on developer opposition and implementation problems. Instead, 'planning gain' or developers' contributions attached to each planning permission became the main tool for redistributing value. For a summary of this complex story, see Healey *et al.* (1995).
19 The new village of Bar Hill suffered from this, with the quality of development delivered ending up well below that which the initial design brief had hoped for.
20 The R. Travers Morgan Study proposed a range of measures to reduce congestion within Cambridge, including a new eastern bypass. Most of these measures were incorporated into county transport policies (CCPO 1977). The retail study was undertaken by J. Parry Lewis, a Manchester University professor who argued strongly for the creation of a new major subregional centre outside the city, with a substantial growth in population of the area to support such a centre at an appropriate scale (DoE 1974). This attracted hostility even during the preparation of the study, and the idea was definitively rejected, thereby confirming that the historic central area would remain the core of retail provision.

21 There was consultation on initial problems and issues, on the county's technical analyses and on the issues arising from the public consultation. This led to the production of a draft 'Written Statement' of the structure plan, which was given widespread publicity. Finally, and accompanied by a report on the public participation and consultations, the text was submitted to the national government Secretary of State for approval, following a public inquiry. The county planners worried about the breadth and depth of voices expressed through their consultation processes, but noted the diversity of views on some issues, particularly over the scale and location of employment growth.

22 The inner boundary had been defined in the 1960s, but now needed amending. The outer boundaries had never been approved in a statutory plan.

23 There was little understanding of the extent to which the development industry, especially housebuilders, benefited from a clear regulatory framework.

24 In the UK, it was positioned with a group of high-tech clusters that were seen to be emerging in 'Silicon Fen' (the Cambridge area), 'Silicon Glen' (in the Scottish central lowlands) and the 'M4 corridor' to the west of London (Haugh 1986).

25 The report also notes the value of the strategy of selective restraint in protecting both sites and labour supply from competition from other economic growth pressures (SQW 1985).

26 Changes in 1987 in the national 'Use Classes Order', a regulatory tool that defined land-use categories, created a 'business' class, which allowed companies to shift easily between light industry and office uses.

27 However, the planners were able to draw on expertise available in other county departments.

28 Until 1987, these were called 'circulars'. Then they were given a more explicit status, and referred to as *Planning Policy Guidance* (*PPGs*). In 2004, this term was replaced by *Planning Policy Statement* (*PPS*).

29 Cambridge City Local Plan 1996 makes frequent reference to 'interests of acknowledged importance' in justifying a whole raft of limitations on development.

30 In the arena of a Standing Conference of East Anglia Local Authorities (SCEALA).

31 In this way, the 'drip-feed' approach had become a demand-driven 'housing numbers game', which in turn was challenged by the concept of 'plan–monitor–manage' and the assessment of urban capacity (Gunn 2006; Wenban-Smith 2002).

32 By this time, although other government departments had a regional presence and concerns, there was little coordination between them. It was therefore civil servants in the regional offices of the national Department of the Environment who examined the draft RPG6 1991 to assess how far it was consistent with, and developed, national policy. In 1994, government offices in the regions were created to improve coordination. In East Anglia, the relevant government office was GO-East.

33 The planning system allows for four opportunities for a review of local planning decisions. Such a review involves some form of public inquiry, all held under the authority of the Secretary of State for the Environment (i.e. national government). Those who are refused planning permission by a Local Planning Authority are entitled to lodge an objection, which will trigger some form of inquiry. Or the Secretary of State may 'call in' a major planning application for national level determination. A structure plan (and now a regional spatial strategy) is examined through an 'Examination-in-Public'. A local plan or other local development planning document is examined by a local public inquiry. These inquiries are organised through the well-regarded

national Inspectorate. The consistency of local decisions with national policy is a key criterion for Inspectorate judgements.

34 An important role for the plan was to provide procedural grounding for the demands on developers for contributions to infrastructure, for compensation for loss to environmental assets due to development and for contributions to providing more affordable housing.

35 The county nevertheless maintained the 1989 policy of continuing to attempt to slow the rate of growth in the Cambridge Sub-Area, reiterated the policy for compact settlements, classified into types, and affirmed the need for selective restraint on economic activity in Cambridge itself, while accommodating high-technology firms that 'needed' to be in Cambridge. Particular attention was given to achieving a better integration between new development and transport provision, with an emphasis on the promotion of public transport, including some form of 'advanced' public transport system. Concerns about water supply were also creeping into the discussion about accommodating development. South of Cambridge, large areas were included in a designated 'area of restraint'.

36 The Cambridge Preservation Society, so active in mid-century, was in contrast little in evidence at this time.

37 The Labour government established development agencies in each English region, with powers to distribute regeneration funds and grants for local economic development, in the Regional Development Act 1998.

38 In the form of an Examination-in-Public.

39 The situation was made even more complicated by local government re-organisation in 1998, which had sliced off Peterborough from Cambridgeshire County, although working relations among politicians and planning staff at the strategic planning level remained good.

40 Its economic analysis draws on the Regional Economic Development Strategy, produced by EEDA in 1999.

41 The key issues in these studies centred on which sites should be released from planning constraints for housing and other development, and what conditions and obligations should be imposed on developers, the location of major retail investment, the possibilities for public transport investment and the priorities for highway improvements.

42 See Cooke 2002; Crouch *et al.* 2001; Garnsey and Lawton Smith 1998; Keeble *et al.* 1999; SQW 2000.

43 It was referred to as a 'maturing milieu' (Garnsey and Lawton Smith 1998); as demonstrating 'institutional thickness', with strong external linkages (Keeble *et al.* 1999). In 2000, comparisons were no longer made with other UK areas, but with a few international clusters (Crouch *et al.* 2001). According to *The Economist* (21 February 2004), the area was 'Europe's nearest equivalent to Silicon Valley'.

44 The consultants took the housing figures from the RPG, developed a database of possible development sites to get an idea of 'capacity', and then arranged these into various options that could be evaluated according to a range of criteria, broadly intended to measure various dimensions of 'sustainability'. The various options were structured in terms of ideas about transport corridors and the possibility of new settlements (Buchanan and Partners 2001). The data used was primarily supplied by the county, and most of the sites assessed had already been the subject of development interest over the previous 20 years.

45 Although SCDC politicians did feel that they had been dragged into a scale of growth they would have preferred not to accept.

46 Cambridge City had hoped that development could be extended in the east across the Cambridge airport site into the villages of Teversham and Fullbourn, as an alternative to a new settlement beyond the green belt, but South Cambridgeshire District Council argued successfully at the EIP against this, on the grounds that the latter were villages that needed to be kept separate from the built-up area, in line with green belt policy.

47 This required local authorities to produce a Local Development Framework, consisting of a Core Strategy, a Statement of Community Involvement, and a suite of Local Development Documents relevant to particular projects and issues. A district-wide Local Plan was no longer required.

48 South Cambridgeshire District Council waited until 2004 before starting to revise its local plan, and therefore had to follow the formal requirements of the new Act.

49 There have been continual recombinations and separation of functions in this national department. The most recent reorganisation relevant to this study was in 2001, when transport was allocated to a separate department, and 'environment' was moved to a new department where it combined with agriculture and rural development. There was a further reorganisation in 2006.

50 That regional planning as promoted in the Cambridge area might have positive benefits in protecting landscape, maintaining high land and property values and excluding some competition for sites and labour for favoured industries was not often recognised.

51 However, as some have pointed out, this plan just emerged, without any public debate or any approval in an elected chamber of government (see letter to *Planning* 2 July 2004: 10).

52 The M11 growth idea had been revived from the 1980s during the EiP on RPG9 for the South East (S. Crow, pers. comm., July 2005).

53 £164 m was allocated in the SCP for the growth areas outside the Thames Gateway. This was divided up to give the M11 corridor £40 million in March 2004, with further amounts later, for which bids to the GAF had to be made. The GAF was originally called the Growth Areas Development Grant.

54 In this, he was backed by Gordon Brown, Chancellor of the Exchequer, who wanted to see a substantial increase in housing provision in southern Britain, to cool the housing market.

55 This had been set up by the Treasury to investigate why the supply of new houses was not responding to demand.

56 The initial regional idea had been to create elected assemblies at the regional level, but this collapsed when voters in the north-east of England overwhelmingly rejected such a proposal.

57 Cambridgeshire councillor John Reynolds, who had steered the structure plan through to approval, took on a similar position in EERA.

58 This was a political position that the Conservative Party were to take nationally in the May 2005 general election, with the cancelling of the Sustainable Communities Plan announced in their election manifesto, but also reflected real local concerns.

59 Two of the top management team of Cambridgeshire Horizons had previously worked for the County and the City Council and were well-networked into Cambridge planning and development arenas.

60 However, its funding was short-term and its support vulnerable to shifts in the local political climate as development emerged into physical form and generated new pressures on transport and social infrastructures.

CHAPTER 6

STRATEGY-MAKING IN A RELATIONAL WORLD

> Strategic planning is selective and oriented to issues that really matter. As it is impossible to do everything that needs to be done, 'strategic' implies that some decisions and actions are considered more important than others and that much of the process lies in making the tough decisions about what is most important for the purpose of producing fair, structural responses to problems, challenges, aspirations, and diversity (Albrechts 2004: 751–752).

> It is strategic *thinking and acting* that are important, not strategic planning. Indeed, if any particular approach to strategic planning gets in the way of strategic thought and action, that planning approach should be scrapped (Bryson 1995: 2).

THE 'RESTLESS SEARCH' FOR A GOVERNANCE OF PLACE

The three accounts in the preceding chapters illustrate the complexity of the challenge of strategy-making for urban areas. It was never an easy task, but for planners and urban development managers in the early twentieth-first century, it seems even more challenging than it was to their predecessors. Attempts at developing strategies that have the power to shape subsequent events involve an expanding range of parties. They draw in several levels of government. They often draw together different policy communities and government sectors to address issues to do with the qualities of places and the coordination of state and private action. Those involved in spatial strategy-making have to think ever more carefully about who they should build relations with, and how this should be done. The strategic interventions they generate potentially impact on people's everyday rhythms and ways of using space. Strategies may have effects on property rights, business interests, daily life movement patterns and come to touch deeply-held values about places and environments. As a result of this range and complexity, conflicts over strategies or elements of strategies can be intense and long-lasting. Those involved in strategy-making work find themselves in the midst of whirls of complexity and conflict, performing difficult institutional work in building new policy perspectives and ideas through which to attempt to shape key aspects of urban region development. They continually have to consider the potency and legitimacy of their activities. Their strategy-formation initiatives may fail to accumulate shaping power. If their activities succeed in acquiring force, the projects and regulatory interventions they promote may fail, be overtaken by events or have unexpected adverse impacts. Strategy-making is a terrain full of the 'tragic choices' that confront those involved in collective action oriented by some conception of a 'collective interest' (Forester 1993).

Yet, despite the challenges and complexities, the practice of spatial strategy-making has persisted in European urban areas since the mid-twentieth century. The three accounts in the previous chapters illustrate repeated efforts to articulate strategies to guide and give justifications for interventions in urban development investment, area management and development regulation. These efforts have had effects, material and immaterial, beneficial in some ways, constraining and harmful in other ways. This repeated recurrence of governance concern for spatial strategies suggests that the demands for a strategic approach to the spatial development of urban areas is neither tied to a particular political or economic configuration nor is it merely an epiphenomenal gloss on more fundamental processes. It arises from inherent tensions in the complex conjunctions of urban dynamics. The various forms and foci of spatial strategy-making episodes reflect a continual search for a strategically focused way of responding to these demands. The accounts illustrate a 'restless search' (Offe 1977) for appropriate policy discourses and for governance practices to meet the challenge of specific spatial conjunctions.

The accounts also underline how difficult it is to generalise about the relation between governance capacities, approaches to spatial strategy-making and outcomes. They highlight the variable fertility of specific governance contexts, between and within countries, and in different time periods, for an explicitly strategic approach to the governance of place. Each story has its own trajectory, related to its own particular evolving interaction between institutional sites and a wider context of political, economic, social and environmental forces. Each illustrates the power of the constraints of these wider forces to shape local opportunities and the force which particular actors and groups of actors have, in turn, to open up opportunities and to influence the dynamics affecting them.

The accounts in themselves thus make the case for a fine-grained understanding of the institutional contexts and situated trajectories (pathways) of efforts at spatial strategy-making for urban areas. This is one of the critical insights of a 'sociological institutionalist' understanding of governance processes (see Chapter 2). But they also illustrate important commonalities. Some of these arise from the dynamics of what we now refer to as a 'global economy', particularly the changes in the technology and location of manufacturing production that have pushed into redundancy industrial sites in Amsterdam and Milan, as well as much of the workforce once employed there. Commercial activities, cultural and design industries and highly skilled new technology activities have created new kinds of work opportunities, cultures and relationships. Other common threads arise from socio-cultural shifts, in material welfare, in demographic movements, in lifestyle expectations and attitudes to the city, to nature, to diversity and difference, to materiality and identity, and to the relation between the individual and governance. These economic and social shifts are drawn into the sphere of the state, of governance, politics and administration, by elite networks, lobby groups and social movements, which stretch between and across cities, regions and countries. These create the momentum for critiques of the European welfare state, or for the adoption of environmental policies by all levels of government, or the articulation of urban social movements that promote cosmopolitan understandings of social justice in the multicultural and diverse city. They show the influence of political responses with similar political concerns; from the building of welfare-state capacities in the mid-twentieth century, to the opening up to more participative influences in later decades; from the concern with the fiscal crisis of rising demands on the state to ideas about reducing the costs of formal government through efficiency drives, contracting out of services, changing the distribution of effort between levels of government and between public and private actors, and, in urban development in particular, much greater reliance on private-sector investment initiatives. The accounts also show the way similar planning and management ideas circulate between cities and countries, shaping how key actors, and particularly urban planners, have responded in particular periods to the demands and opportunities for spatial strategy-making. The 'restless search' for a way to manage the governance of place is thus structured by dynamic configurations of broader forces, as the regulation theorists argue (see Chapter 2), but in complex and subtle ways in which space for innovation with potential wider consequences is always present.

In all three cases, those involved at the turn into the twenty-first century found themselves in governance contexts of change and instability. Established policy discourses were being challenged. Traditional arenas and governance networks were being displaced and actors were searching for new relations with other levels of government to make linkages with actors in economic and socio-cultural spheres. Those involved often sought out new ways of understanding urban dynamics and new ways of thinking about governance and how to do it. Actors at the core of governance activity experienced directly the momentum of 'restlessly searching', in struggles both to innovate and change trajectories and to hold on to valued qualities or institutional positions. The outcome, in the hindsight of history, may be new stability, through a new, hegemonic 'mode of regulation' in European urban regions, or, less dramatically, new 'urban regimes' in specific cities. Or, a more likely outcome, a recognition of the condition of dynamic instability in governance processes may come to shape how people think about and perform urban governance processes.

The institutionalist perspective articulated in Chapters 1 and 2 is a particularly relevant basis on which to develop an understanding of the potential of spatial strategy-making for urban areas as a way of building governance capacities, of assembling knowledge and developing understanding and of conceptualising the meaning of place and spatiality. It helps in grasping how strategic mobilisation develops in unstable conditions, with multiple claims for governance attention and continual multi-vocal pressure on governance actors to demonstrate their legitimacy. It emphasises that specific episodes of urban governance activity need to be understood in relation to their broader governance context, their interactions with ongoing shifts in the wider economy and society and the trajectories of governance discourses and practices that have developed in particular places. It focuses attention on the relations through which governance activity is performed and the conceptions of urban conditions and potentialities that are mobilised in such work.

In the next three chapters, I return to the themes introduced in Chapters 1 and 2, to develop a relational approach to the governance of the 'places' of urban areas. I draw on both academic literature relevant to each theme and the experiences of the practices presented in this book. Having set up the approach to each theme, I outline its implications for the concepts and practices of spatial strategy-making focused on the place qualities of urban areas. As I do this, I draw out the aspects of such concepts and practices that are likely to promote or inhibit those governance processes which encourage a rich and diverse conception of urban dynamics and hold considerations of distributive justice, environmental well-being and economic vitality in critical conjunction. Each chapter has a similar structure, with the themes presented in reverse order to their introduction in Chapter 2. Finally, in Chapter 9, I draw the arguments together around the overarching theme of governance capacity. In this chapter, I turn to the nature of strategy-making in a dynamic, relational urban context. First, I briefly review the experience of the three urban areas and comment on the contribution of a key 'planning policy community' evident in all three cases. I then explore the meanings of 'strategy' in key traditions of thought, before developing a relational approach to strategy-making. I conclude with a comment on the power of strategy.

THE PRACTICES OF STRATEGY-MAKING

In Chapter 1 and 2, I raised questions about what strategies are, about the institutional work they perform, and about how they come to have effects. The case accounts provide a rich empirical source for addressing these questions, as each contains several episodes of strategic effort to shape the development trajectories of their urban areas. Some of these efforts have had significant material effects, shaping the location of major urban extensions, the positioning of transport infrastructure, the manner of redevelopment and conservation of urban cores, and the promotion of new spatial development nodes in the widening daily life networks of urban areas. The debates mobilised in some of the efforts, and the strategic statements made in the plans, have sometimes been important in building and reinforcing conceptions of the identity and qualities of the respective 'places'. In such periods, strategic thinking and acting both inhabits and gives life to the arenas and processes of formal planning systems.

At other times, plan-making and agreed strategies of one period have been pushed to the sidelines or deliberately overridden by shifts in political priorities or by the force of particular interests – by commercial interests and especially the growth dynamic around

Schiphol airport and other transport nodes in Amsterdam; by party networks and real-estate investment interests in Milan; by university interests, conservation lobbies and research and development companies in Cambridge. In these periods, strategic thinking may move away from formal arenas, into particular alliances and the strategic actions of key actors. Formally produced strategic plans of earlier periods may then be experienced as irritating constraints, carrying the remnants of past strategies to interfere with the freedom to manoeuvre of new initiatives, or, as in Milan in the 1980s, providing an administrative 'gate' to trigger a hidden practice of developers' contributions to 'ease' progress through development regulations. In these circumstances, formal plans may become merely statements of already well-established strategic principles, necessary to give legal legitimacy to regulatory and investment decisions. Alternatively, they are subject to continuous adjustment, as in Milan, creating severe inequalities between those who know how to negotiate adjustments and those who do not.

These varied experiences of strategic planning efforts through time emphasise that strategies do not necessarily reside in formally prepared plans, nor are formal plan-preparation processes necessarily the main arenas through which strategies are formulated. The significance of formal procedures for making strategies and plans depends on institutional specifics, such as the construction of public and private rights and obligations in urban development processes, on the powers and relations of different levels of government, on the range and depth of networks which connect actors together to mobilise attention to urban issues, and on the scale and nature of demands for the legitimacy of strategy-making activities. Each episode in each of the case accounts has been positioned differently, confronting both a different urban geography and history, and a different institutional conjuncture. The trajectory of each episode combines both 'path dependency' and generative force, shifting discourses and practices, and reshaping policy communities and institutional arenas. In Amsterdam, a formal strategic plan is an expectation embedded not just in the mindsets of government actors but more deeply in the city's governance culture. In the Cambridge area, the formal plan-making arenas are needed to give political and judicial legitimacy to a strategy that is likely to generate substantial challenge and conflict. In Milan, the formal process is necessary in some form to allocate legal rights to develop sites in particular ways.

These experiences underline that strategic planning activity must be understood as a situated practice with its effects deeply structured by the specificities of time and place. In the enthusiasm for planning in the early post-war period, some of this specificity can be seen in the plans produced, where imaginative planning experts of the early days of the 'planning movement' took up the challenge of producing planning ideas for particular cities. In Cambridge, the key planning issue at the time was 'roads', the central issue of Holford's study. In Amsterdam, it was building quality urban neighbourhoods on reclaimed land. In Milan, it was producing a regional transport system and building well-designed housing schemes. Spatial strategies focused on urban conditions are thus not formulae, to be taken from a management textbook, government guidance or planners' kitbag of plans. They grow out of specific situations, both as regards development trajectories and institutional contexts.

Yet the cases also illustrate that ideas about ways of making spatial strategies diffuse from one place to another. This is not a direct process, although sometimes a particular city's practice becomes iconic for many others. Dutch planning, as exemplified in the Amsterdam case in the mid-twentieth century, became a beacon for those developing planning approaches in other European countries, just as Barcelona became a symbol of a strategic approach to 'urban renaissance' in Europe in the 1990s. More usually, ideas about the process and content of urban strategies diffuse within and between national policy cultures, through direct contact in arenas of professional and policy exchange and through professional and academic literatures (see Chapter 8). These then interact with pressures experienced locally, which may in turn express wider economic, social and political dynamics. Strategies shaped by the modern movement and the political and economic pressure to provide mass low-cost housing in the post-war period were challenged in the 1970s by social movements that highlighted their disruptive and socially unjust consequences. Amsterdam and Milan both experienced influential urban protest movements, in touch with each other, and with movements elsewhere (Mayer 2000). Across Europe and North America, these movements helped to shape a critical mood that demanded a stronger citizen voice in government and a less paternalistic attitude from politicians and government officials.[1] Even in Cambridge, the echoes of this concern encouraged the county planners to give a great deal of attention to ways of consulting with citizens. In Amsterdam, these initiatives left a legacy of consultation mechanisms and city centralisation that is now embedded in current governance practices. In contrast, in Milan, the heritage of these movements exists only as a memory among some political groups and in the planning community.

In the 1980s and 1990s, all three cases were influenced by wider European debates about the need to make urban areas 'economically competitive' by developing their 'assets' of various kinds. In Amsterdam, this discourse co-existed uneasily in the 1990s with the well-established social welfare emphasis on urban neighbourhood quality and on environmentally focused agendas that had become embedded in the 1980s in national and local policy discourses about place-management. The environmental agenda in Cambridge, as in the UK generally, both refurbished a culturally grounded resistance to urban sprawl, and also linked to questions about the environmental costs of the growth in road traffic. Through the issue of increased commuting and traffic congestion, common cause could be found with the 'competitiveness' agenda. In Milan, the challenge to the promotion of city assets for economic positioning purposes was carried through arguments about the social justice of the distribution of benefits and the accessibility of the assets. By the 1990s, in all three cases, as elsewhere in Europe, the ability to 'balance' these competing discourses within an overall urban strategy became an ever-more-complex task, both intellectually and politically. Meanwhile, citizens in all three places increasingly demanded attention to the everyday 'liveability' of the built environment.

These multiple experiences of strategy-making for urban areas thus emphasise the importance of grasping the specificity of situations in developing an understanding of why strategies take the form that they do, the institutional work that they perform at particular conjunctures and the effects that they have, both anticipated and as they unfold through time. This does not mean that the wider context of a particular urban governance nexus can be ignored. While a trajectory is achieved in a specific institutional setting, it is continually shaped by influences connecting that setting not merely to its own evolutionary story, but to other forces that reposition a local story in wider debates and conjunctures. Thus, while key actors in all three cases had a sense of arriving, in the early twenty-first century, in some kind of new configuration of local trajectories and wider forces bearing in on the dynamics of their urban context, the history and potentialities of each situation were very different.

'PLANNING POLICY COMMUNITIES'

Who then were the key actors in these various episodes of spatial strategy-making? The accounts make reference to politicians, party networks, lobby groups, business interests, landowners, developers and citizens/residents. They refer to many different agents within the formal sphere of government – water managers, highway engineers, the providers of public facilities, council managers and chief executives, national-level civil servants. But above all, they highlight the role of those identified as 'planners', either through their professional or academic affiliations, or their position in a governance context. It is not just that these 'planners' took on the mantle of orchestrating efforts in spatial strategy-making. Others expected them to do so. Politicians seeking some kind of spatial strategy expected their planners to produce one. Citizens and businesses imagined that planners made strategies in plans, even though they experienced planners often as regulatory bureaucrats. In the mid-twentieth century, the protagonists of the 'planning movement' confidently responded to these challenges, as the Amsterdam and Cambridge Sub-Region accounts illustrate. By the end of the century, this identity was under challenge, both by critics who complained of planners' ways of thinking and acting, and by planners themselves, many of whom saw that developing a strategic approach involved complex interactions with many other 'stakeholders' in urban region dynamics.

Those identifying themselves as planners were, in any case, not merely sitting in a municipal planning office, although such an institutional site linked to the regulatory power over property rights was always an important arena of strategic planning activity. In the cases presented in this book, those trained as planners or acknowledging themselves as planners were to be found as planning officials in a range of government agencies, as members of companies and of lobby groups, as academics, as members of consultancy firms, and sometimes as politicians. But other stakeholders have also been important, particularly politicians at all levels of government, party networks, other national, regional and local public agencies, special development companies and partnerships, lobby groups of residents, or special issue associations, property interest groups and business groups. Also important, but not so visible in the cases, are the expressions of general 'public opinion', as witnessed in the Amsterdam referendum on the creation of the Amsterdam metropolitan region, and in the continual commentary by the media on all kinds of issues concerned with urban conditions and the actions of 'the planners'. So the governance landscape of any city at any time is in some way positioned in relation to a wider 'community of interest and activity' around issues to do with urban planning and development. In the context of the formation and development of European welfare states, such 'policy communities' have typically evolved around particular government functions. They have developed institutional forms expressed in established relations, in discursive histories and practice trajectories that hold members into common, if fluid, frames of reference through time.

A policy community may be understood as networks of relations and frames of reference that develop among those actors interlinked through regular relations around the articulation and operationalisation of a particular set of policy issues, and from which a shared understanding of issues and debates evolves.[2] It is bound both by recurring interactions, and by common reference points and knowledge, and thus is also an 'epistemic community' (Haas 1992). A 'planning policy community' may thus be understood as the nexus of relations and debates that flow among those regularly involved in the practice of managing urban development projects and the regulation of urban development. Within this loose association, the self-recognising 'communities of planners' are those especially trained and experienced in the practice of planning tasks. These communities have their own national specificities (Sanyal 2005). A planner trained in Italy and working in Milan, coming from a broad architectural tradition, has a different conception of the planning task to one trained in the Netherlands or in the UK, where a social-scientific orientation is much stronger. Each is organised professionally in different ways, and each have different traditions of linkage with other stakeholders. But there are significant overlaps and exchanges that allow ideas, and, increasingly, personnel, to flow between them (see Chapter 8).

This movement of planning ideas emerges from the case experiences. In Amsterdam and Milan in the 1950s, the strategic planning task was seen as the production of a general 'urban region' plan, expressing a comprehensive morphology within which the preparation of specific projects and regulatory instruments could be located, even though the real action and focus was on urban extension. The Holford plan for the Cambridge area is interesting because it is deliberately much more selective in focus, concentrating on the politically contentious issue of new road proposals. Planners in Milan in the 1970s and in Amsterdam in the 1980s once again attempted a comprehensive urban strategy for their cities, set in a sub-regional context. By this time, the British structure plan had become a different kind of instrument. Rather than a spatially configured morphology, it had turned into a set of policy principles for negotiating development projects initiated by private actors. The Milan *Documento di Inquadrimento* was an attempt to move Italian practice in this direction in the late 1990s, drawing explicitly on British experience. A similar development was also happening in Amsterdam in the 1990s, as the role of public investment in urban development began to diminish in favour of a greater reliance on the private real-estate sector.

In Amsterdam and the Netherlands, the shift to a greater reliance on private investment in development raised questions about the City Council's well-established practices of planning and managing major urban development projects, and the way they were situated in 'structure plans'. In Britain and Italy, in contrast, the momentum in the 1990s among those mobilising to change planning practices focused on reviving a more strategic approach in which to situate the practice of negotiating over development projects initiated by the private sector. In developing conceptions of what 'strategic planning' involves, ideas from within planning communities have mingled with those from the management and marketing literatures. Experiences in different parts of Europe have been exchanged in debates at the European level about strategic spatial planning and development, particularly in relation to the development of the *European Spatial Development Perspective* (CSD 1999), on which ministers and planning officers from across Europe worked together (Faludi 2000, 2002; Faludi and Waterhout 2002). A key element of this emerging approach has been the production of a 'strategic vision' as an orienting device to inspire multiple actors, that can be translated into a framework within which development projects and development regulations can be located. National legislation in England in the mid-2000s, and in the Netherlands too, now gives much more attention to such a 'vision' or 'core strategy', so consolidating a practice already appearing in the Cambridge case. But the relation between visions/strategies and development projects remained uncertain. The drafters of new legislation and the promoters of strategic-visioning exercises tended to assume a linear relation, with project proposals and development criteria developing from the strategy. In Milan, in contrast, an astute grasp of the governance situation emphasised how strategies might evolve from innovative project negotiation and assessment practices.

By the end of the twentieth century, both advocates of a new strategic thrust from within the planning policy community, and external critics of planning concepts, were making demands for a 'culture change' within the planning community. In England, this demand was expressed in exhortations and national-policy statements, aimed not only at changing the mindsets of planners involved in land-use regulation, but in shifting the established practice of contestation between developers and anti-development lobbies, mediated through the planning system. In the Netherlands and the UK, planners were being encouraged to build arenas and institutional practices more capable of horizontal, intersectoral co-alignment. A similar movement was underway in Italy, finding expression in the Milan area in the work of the Province, as well as several coalitions of municipalities. A strategy was presented in these ideas about culture change and coalition-building as a critical way of changing policy frames and creating momentum around new agendas. What conceptions of strategy were being mobilised in this movement?

MEANINGS OF 'STRATEGY'

In Chapters 1 and 2, I emphasised that strategies are complex social constructions. They involve difficult institutional work in drawing together sets of actors and their relational networks and creating new policy communities and networks that can act as carriers of strategic ideas across governance landscapes and through time. In the planning histories of the three cases, it is evident that what is 'strategic' and 'structural' has been understood in different ways. This is not a phenomenon specific to these cases. There have been major movements in academic thought and in the practices of planning and management that have carried different meanings of strategy. In Table 6.1, I draw out four such understandings, and link them to ways they are manifest in the three cases.

The idea that an urban development plan should embody a 'strategy' flowed into the planning epistemology in the 1960s, from management science and business practice (Albrechts 2004; Mintzberg 1994). Before that, the planners of the first part of the twentieth century, deep in architectural and engineering traditions, emphasised the concept of 'structure', as in the structure of a building. Urban areas were conceived as

Table 6.1 Meanings of strategy

Strategy as . . .	*Underpinned by . . .*	*Expressed through . . .*	*Illustrated by . . .*
Physical structure	Morphological analysis	Plans as maps and designs	1953 Milan PRG 1935 Amsterdam GE Plan 1985 Amsterdam *structuurplan* (in part)
Orienting goals	Socio-spatial analysis to identify threats to goals	Policy statements about programmes of action to achieve goals	1970s facet plans in Amsterdam 1980 Cambridgeshire structure plan 1980 Milan PRG
A framework of principles	Systematic technical and interactive search procedures to reduce uncertainty	Framing concepts, projects and programmes; policy criteria	Cambridgeshire structure plans 1985 Amsterdam *structuurplan* (in part)
An inspirational vision	Interactive processes to imagine futures and mobilise attention	Metaphors, storylines and manifestos	Futures exercises in Cambridge and Amsterdam in the 1990s

having 'structures' that created frameworks to be filled in by detailed area-development schemes and specific building projects (Burtenshaw *et al.* 1991; Hall 1998; Webber 1964). This concept of structure lives on within the British structure plan (now superseded) and the Dutch *structuurplan*. It remains in use when referring to the structuring of spatial patterns. For example, in relation to his work on the Flanders Structure Plan, Albrechts writes that the plan provides:

> structuring principles which will be capable of imposing some order on the current chaos and to introduce new spatial orders on Flanders with the aim of meeting the needs of sustainable development (Albrechts 2001: 87).

The management concept of strategy came with a different imagery, that of the battlefield, and the distinction between 'strategy' and 'tactics' (Mintzberg 1994; Solesbury 1974). This was linked to the idea that a strategy could be expressed in organisational goals, which expressed core values. These could then be developed into trajectories and principles to guide specific actions through careful analysis. Planning, in this conception, involved the working through of strategic goals into specific action programmes, following a linear process, underpinned by technical analysis and evaluation of alternative actions (Mintzberg 1994). These ideas flowed from management science into policy analysis and into the planning field as the 'rationalist' paradigm, centred on logical–deductive reasoning (Breheny and Hooper 1985). In these ideas, the concept of strategy shifted from a morphological form to a step-by-step planning process through which strategy, programmes and projects could be arrived at. Rich debates emerged in the 1960s about the initial impetus for strategy-making. Was it the existence of a problem or the search for achieving a goal? How closely were strategies, action programmes and projects actually related? Etzioni famously reframed the distinction between strategic and detailed planning in the metaphor of a camera, with its wide-angle lens and its zoom lens. Effective planning, he argued, needed a careful combination of both perspectives, in a 'mixed scanning' approach (Etzioni 1973). Implicitly, strategy was associated with the wide-angle lens.

By the 1970s, several European countries, including the UK and the Netherlands, had changed their planning systems to create requirements for both broad, strategic or structural plans and detailed area and topic (facet) plans. Area plans could be specific design master plans, or development briefs, to frame detailed development negotiations, or general principles to guide area management. An important issue in these strategic plans was the relation between the statement of strategy and the allocation of property-development rights. In the British case, all plans are advisory, rights being established when a development permit is granted. In the Netherlands, rights are allocated in a statutory local area zoning plan, the '*bestemmingsplan*'. In Milan, as in Italy generally, the city-wide general plan still carries the formal function of allocating property rights, a connection that leading Italian planners have been seeking to uncouple for some time.

Despite these alternative tools, the dominant conception of a strategic plan, illustrated in the three cases, was of urban strategy embodied in comprehensive, spatially specific 'plans-as-maps', with the capacity to mobilise the provision of land and finance for physical development and to govern the exercise of public regulation over private property rights. But the conversion of ideas about strategy into planning instruments required the linking of strategic planning activity to specific institutional arenas. The reality of the complex practices of struggle between levels of government and between factions and interest groups over the content of strategies, as illustrated in the three accounts, gradually encouraged an understanding within the planning field of strategy as a political process of focusing attention among the many parties whose activities collectively shape urban dynamics (Bryson 1995). This recognises that the state no longer has direct 'steering capacity' over the investment strategies of a landowner or financial investor. Nor does it have the controlling power to demand others to deploy their resources in particular ways. Instead, the state has to 'steer' by persuasion, seduction and inducements. In the terms of Allen's conceptions of power relationships, the power of legal authority gives way to the power of persuasion and inspirational seduction (Allen 2003, 2004) (see also Chapter 2).

In the management field, in policy analysis and planning, conceptions of the nature of strategy and models of strategy-formation processes began to shift in response to the recognition of the socio-political dynamics through which emergent strategies are produced. As is evident in the three cases in this book, and despite the conceptions still being mobilised in the revisions of planning systems, strategy formation does not proceed in an orderly way through specified technical and bureaucratic procedures. It is a messy, back-and-forth process, with multiple layers of contestation and struggle. Strategies emerge from these processes as socially-constructed frames, or discourses. Strategy formation is not just about the articulation of strategic ideas, but about persuading and inspiring many different actors, in different positions in a governance landscape, that particular ideas carry power, to generate and to regulate ideas for projects.[3]

Within the planning field, there were some precursors to what has become a major paradigm shift in thinking about strategy-formation processes. In the 1970s, Friend and colleagues developed a concept of policy-formation processes centred on strategy-makers operating through a range of networks, with the objective of reducing 'uncertainty' in the surrounding environment (Friend *et al.* 1974; Friend and Hickling 1987). These ideas emphasised that strategies were social constructions and involved collective learning processes, but they centred on technically managed processes within contained 'action spaces' for planning activity. In contrast, the emphasis in later work has been on the way technical analysis, political debate and multi-vocal understandings of issues combine and clash in multi-actor and multi-valent contexts. Faludi, drawing on the work of Friend and colleagues, also made a clear distinction between strategies and 'strategic plans' as frames of reference, and 'project plans' related to specific actions in the urban environment. This distinction seemed to describe well the Dutch practice of situating development projects in a spatially specified strategic framework (Faludi and van der Valk 1994; Mastop and Faludi 1947). In Italy, in contrast, Secchi made an influential argument in the 1980s that strategies could be carried in the work of shaping projects. The conception of the city and its strategically important spatialities were, he argued, brought into conscious attention as major projects were imagined, debated and shaped into what materialised as physical form (Secchi 1986). This idea influenced how the strategic effort in Milan in 2000 was conceptualised. Such concepts helped to justify the 1990s 'turn to projects' across Europe to which the new 'turn to strategy' was a response (Healey *et al.* 1997; Salet and Faludi 2000).

STRATEGIES AS EMERGENT FRAMING DISCOURSES

The notion of strategy as reference frame grew in part through these ideas. But concepts of 'framing' emerged separately through the recognition that what gives a strategy focus and leverage is some kind of synthetic integration. A frame is an 'organising principle that transforms fragmentary information into a structured and meaningful whole' (van Gorp 2001: 5, in Fischer 2003: 144). A frame provides 'conceptual coherence, a direction for action, a basis for persuasion, and a framework for the collection and analysis of data' (Rein and Schon 1993: 153). A strategy is thus more than a framework of principles. It has the quality of an inspirational, motivating 'vision', supported by a way of 'seeing'. It offers a direction and provides some parameters within which specific actions can be set. It creates ideas about how future opportunities may be grasped and threats avoided. It is supported by storylines and metaphors that create meanings and provide foci of attention (Hajer 1995).

Such frames do not necessarily precede action. They are continually shaping and being shaped by the flow of action, in recursive rather than linear processes, as Barrett and Fudge (1981) argued many years ago. Strategies are as much 'found' as explicitly created (Mintzberg 1994). They become 'recognised' as emergent issues capture attention. A 'vision' may be articulated at any time (Bryson 1995). It may emerge through reflection on the justification for a project or a regulatory norm, as Secchi argued. Yet it is also a creative product, made by imaginative endeavour and intellectual development. The examples of strategy-formation processes in the various episodes described in the cases suggest that strategies that have been developed with careful attention to institutional specificities, which combine a grasp of such specificities with a conception (or 'vision') of the identity and trajectory of a city's development, are more likely to mobilise enduring attention than those that invoke a generalised rhetoric drawn from a store of planning concepts circulating in national and international professional communities.[4] But even very well-developed and locally situated episodes may be unable to withstand strong challenges from a different political and economic nexus, as happened to the Milan *PRG* strategy of the 1970s, to the Amsterdam 1985 Plan, and, finally, to the Holford plan. Thus the social processes through which frames are produced are deeply affected by the institutional contexts in which they are located and the history of struggles over meanings and values which have preceded them. Explicit strategy-formation processes may be significant initiators of new discourses that flow into and transform practices. At other times, new discourses take slow form as the parameters of action shift, a process emphasised in the institutional design of recent interventions in Milan. These different trajectories through which strategies may emerge is nicely illustrated by Mintzberg (1994) (see Figure 6.1).

To have long-lasting effects, strategies need to move from the stage of frame construction or discourse structuration, to discourse institutionalisation in Hajer's terms (Hajer 1995); that is, to the routines of practices. Given the complex nature of urban governance landscapes, this means that strategic frames producing significant effects must have the capacity to travel and to be translated into all kinds of institutional arenas through time, without losing their core ideas and motivational capacity. They need persuasive and seductive properties. In complex institutional landscapes, not only are many groups of actors and actor networks likely to be involved in and affected by strategy formation processes. To have effects, strategic frames need to flow through these networks and into the routines of 'communities of practice' (Wenger 1998) through which material changes to urban development are produced. Strategy-formation processes involve learning through networks, and may lead to the creation of new communities of practice around strategic frames, generating what some have called 'network power' (Hajer and Wagenaar 2003; Innes and Booher 2000).

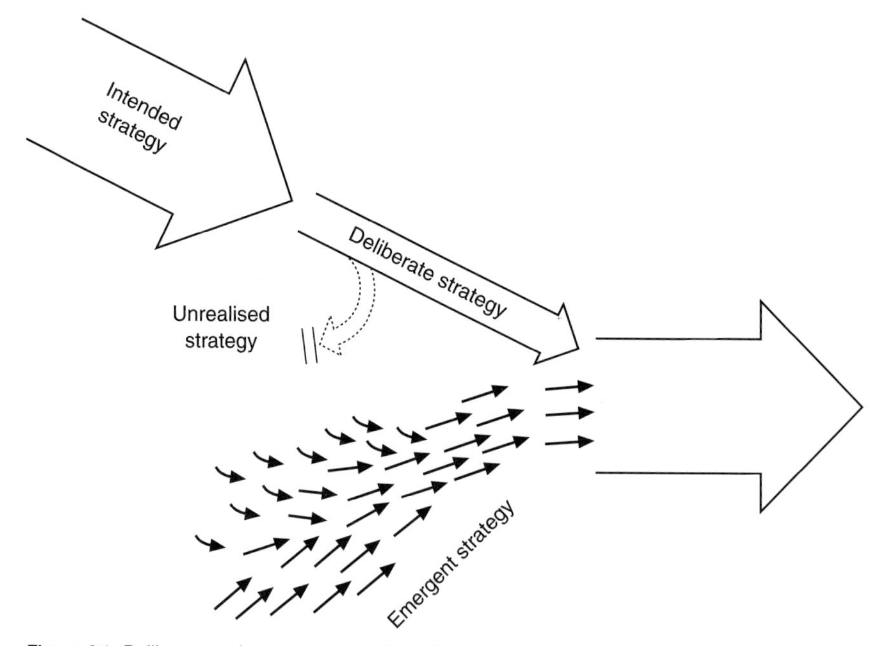

Figure 6.1 Deliberate and emergent routes to finding strategies
Source: Mintzberg 1994/2000: 24

Strategies, in this conception, are emergent social products in complex governance contexts, with the power to 'frame' discourses and shape action through the persuasive power of their core concepts. If new strategic frames accumulate sufficient power to enrol others, to travel across significant institutional sites of urban governance and to endure through time, then they are likely to have significant effects in shaping the future. They have transformative potential.[5] The existence of a strategy, understood in this way, is thus not to be found merely in rhetorical invocation through the use of the word 'strategy' or 'vision' or the production of some kind of image. It is to be found in the way a discursive frame is used in generative, coordinative and justificatory work in governance contexts.

Such a conception of strategy arises from a relational and interpretive perspective on governance processes. This emphasises two dimensions of the relations, or connections, which strategy-making involves. The first is the way a strategic frame imagines connections between phenomena, highlighting where critical attention and interventions may be needed. The second is the nexus of relations through which a force builds up behind a strategic frame, sufficient for it not merely to attain some priority in governance attention, but also to endure and flow to influence critical arenas where action is shaped and which a strategic initiative seeks to influence. In these processes of constructing intellectual capital and socio-political force, a strategy may be continually re-imagined, with meanings and priorities shifted. A powerful strategy is one that has interpretive flexibility but which retains and focuses on key parameters as it travels among governance arenas through time. Such an understanding of strategy has a particular resonance with contemporary urban governance conditions, which are widely perceived by academics and practitioners as characterised by transitions, transformations, uncertainties and instabilities. In such conditions, social-learning processes become more important than bureaucratic procedure, rationalist scientific management or pluralist politics as modes of strategy formation (Christensen 1999).

In summary, then, what are the critical dimensions of 'strategy' that arise from the above relational and interpretive conception? Strategies are selective constructions, 'sense-making' devices, created from a mass of material. Their formation occurs through time, but not necessarily in defined stages and steps.[6] They are created through processes of filtering and focusing attention, highlighting some issues and pushing other issues to the sidelines. The formation of new strategies changes the dynamics of inclusion and exclusion of those interests and issues favoured by a strategy. Persuasive strategies orient and inspire activity, through motivating people with future hopes (Albrechts 2004; Friedmann 2004), and through giving some actors an idea of what other actors may be up to. They mobilise intellectual and social resources to create the power to carry a strategic frame forwards, just as they may also mobilise resistances. This is, of course, one reason why some strategically acute actors may resist a demand or fashion to produce a strategy.

Strategies that emerge in urban contexts are therefore complex social constructions, which are both institutionally embedded and transformative. They are efforts in collective sense-making. Strategies accumulate power by grasping moments of institutional opportunity in imaginative and politically astute ways. If they accumulate power through mobilisation and persuasive processes, they generate the political force to shape resource flows, structure regulatory norms and arguments, and inspire the invention of new projects and interventions in urban dynamics. Strategy-formation processes involve mobilising actors in many different social networks, drawing on their knowledge and resources. Such processes create knowledge and re-order values which, in turn, feed back into networks and may create new networks and 'communities of practice' around a new strategic discourse. In this way, strategy-formation processes are both dynamic, emergent social constructions and also contribute to stabilising and ordering complex realities. A strategy that accumulates substantial persuasive power becomes a part of the structuring dynamics within which subsequent actions are embedded.

KEY DIMENSIONS OF URBAN REGION STRATEGY-FORMATION PROCESSES

I have so far argued that spatial strategies can exert a powerful force in shaping interventions in urban dynamics, though they may not always achieve this. It is therefore important to recognise how and where this power is exercised. This is particularly important if our concern is with the ability of an urban spatial strategy to keep multiple issues in conjunction and to reflect a rich and inclusive array of experiences of urban conditions, rather than the narrow pursuit of a single dominant understanding, such as economic competitiveness. I now examine four key dimensions of strategy-formation processes that arise from the perspective outlined above: the filtering of ideas; the framing of strategies; the generation of mobilising force; and the potential for transformative force. I draw out the institutional work involved in each dimension, the power dynamics of inclusion and exclusion associated with such work and the potential for creative, generative power to be released. Through these dimensions, I highlight critical questions that those evaluating and designing strategy-formation processes focused on urban dynamics should consider.

FILTERING PROCESSES

In a relational understanding, urban areas are understood as geographical spaces transected by very many webs of relations that weave across, in and around each other, generating nodes of activity and identifiable places with distinctive social and physical qualities (see Chapters 2 and 7). Grasping some kind of understanding of these relations is a challenging imaginative and intellectual task. Assessing whether, when, where and how to intervene in these relations in an attempt to make a significant difference to trajectories and outcomes is, in turn, a complex political task. Many issues struggle for attention and many stakeholders struggle to get their understanding and prioritising (their 'rationalities' and frames) at the centre of governance initiatives (Albrechts 2004). Any strategic frame arises from some kind of filtering and sorting process among these understandings. These processes are not just about arriving at a robust way of understanding an urban region. They involve struggles about the prioritising of interests, rights and claims for policy attention. Although the rhetoric about urban strategies often refers to words such as 'comprehensive', 'balanced', 'integrative' and 'holistic', and strategies when selected may be justified in relation to some overarching principles, such as their contribution to 'welfare', 'sustainable development' or 'well-being', the filtering processes that underpin strategy formation are the first of two critical points where the exclusions and inclusions of a strategic orientation are determined.

The image of how this filtering was done in the mid-twentieth century emphasises the role of the planner, drawing on expert judgement and various studies to determine the key parameters to give focus to a strategic frame. The Amsterdam and Cambridge cases suggest that the reality at the time was more interactive than this, with skilled and respected planners working hard with politicians and key government actors to get support and precedence for their understandings. In Holford's plan for the Cambridge Sub-Region, the agenda was in any case pre-set by the high profile of the 'roads' issue in local politics and the close connection between local activists and national politicians and civil servants/officials. In Amsterdam, the emphasis on urban extension and transport reflected pre-existing national government funding programmes.

By the late twentieth century, filtering processes were much more obviously situated in an ongoing multidimensional flow of studies, debates, challenges, claims and counter-claims. Key 'sorting moments' may happen at the start of a strategy formation. But they may also occur during a process. In the Cambridge Sub-Region case, the formation of a sub-regional lobby group in the mid-1990s, which included some economic actors along with county and district planners, provided an initial arena that excluded many other groups with a stake in the area. While the momentum of the lobby group flowed into the formal processes of constructing county and regional strategies, this came up against challenges of technical knowledge (particularly about the 'feasibility' of transport options) and against other issues that surfaced as the formal processes of structure plan preparation and approval proceeded. In Amsterdam, as the power of the national spatial planning ministry to set national development investment priorities ebbed away in the 1990s, and as more conflicts surfaced about what kind of urban area was emerging and where the priorities for investment should be located, the planning department and key politicians were continually reformulating their intellectual understanding and reshaping their relations with other actors. Elaborate consultation processes and orchestrated debates played an important role in the filtering processes that underpinned what became strategic frames for urban development. In Milan, many in the planning community would like to see such processes develop. But the political commitment to such practices has been limited and largely rhetorical, with little tradition of connecting such broad debates with specific action programmes. And even in

Filtering processes
When in a strategy-formation process does issue filtering occur? (a potential throughout)
Where does it occur – in what institutional sites/arenas?
How does such filtering take place and through what practices and mediums?
Who is filtered in and who excluded through such processes?

Box 6.1 Filtering processes

Amsterdam, some significant stakeholders have had little voice in recent strategic discussions about their locale of living and working.

If strategies for urban areas are to be understood as efforts in collective sense-making, then a key quality of strategies lies in the nature of the 'sense' being made. This sense can partly be seen in the way key actors 'read' the emergent reality of an urban area and grasp the governance landscape of stakes, claims and interests struggling for voice and presence in the construction of the collective sense. In Chapters 7 and 8, I look more closely at these 'readings' through a discussion of spatial concepts and forms of knowledge that feed into strategic planning processes. But, in evaluating strategies and in designing institutional processes to encourage strategy formation, it is also important to consider when the filtering of issues and claims may occur, in which institutional arenas it is likely to occur, how such settings may shape filtering processes, how filtering actually takes place and the impact of all these on whose stakes and interests get prioritised and whose ignored (see Box 6.1). The case accounts emphasise the point made by a social constructivist/interpretive theory of strategy-making that there are few general recipes. The when, where, how and who of filtering processes is deeply contingent on historical and geographical specificities. This is an important reason why such processes need to be kept in the foreground of evaluative attention.

FOCUSING AND FRAMING

Mintzberg (1994: 272) argues that 'great strategies' are constructed in 'fertile minds' and from 'myriads of small details'. Constructing them 'requires a mental capacity for synthesis, with imagination'. Such strategies create and give sense to a mass of confusing signals and challenges. Any strategy, in this perspective, involves a selective focus, a way through the morass of issues, ideas, claims and arguments to identify one or more concepts, images and/or principles that are both 'meaningfull' and orienting. Such strategies may be arrived at by systematic search procedures, such as carefully evaluated assessments of alternative directions, and/or the construction of future scenarios and efforts at 'backcasting' to see how ideas about alternative futures might affect present decisions (Albrechts 2005; Secchi 2002). Or they may be the result of active campaigning by groups promoting particular interpretative frames, as in the Dutch discussions of spatial concepts such as the *Randstad*, *Deltametropool*, corridors and '*vleugels*' ('wings') (van Duinen 2004), or the growth-promotion lobby in Cambridgeshire in the 1990s. Or they may arise in an imaginative leap, by a strategic thinker, as traditionally planning consultants claimed to be able to do, or through the creative discovery processes of collaborative encounter (Innes 2004; Innes and Booher 2003). However they appear, strategies come into existence through an act of 'recognition' in which they are 'summoned up', 'seen', 'named' and 'framed' (Schon and Rein 1994; van Duinen 2004).

The formation of new strategy is more than merely an aggregation of issues and claims that have survived prior filtering processes. Many statements of strategy in formal planning documents are little more than this. But such aggregations do not have the quality of a frame, which can perform institutional work in focusing and making sense of diverse indications and activities. Such aggregations in formal plan statements may meet procedural requirements to have a strategy, to release funds from some other agency or to meet a regulatory procedure. But they do not have the quality that creates a strategy with long-lasting effects. Strategy formation that has effects involves the generation and consolidation of a new frame, a new discourse with its supportive storylines and metaphors.[7] This involves not only 'naming' the discourse but also the reframing of many issues within the perspective of the new 'sense' (Fischer 2003; Schon and Rein 1994). Such a strategic frame is highly selective, foregrounding some issues and backgrounding others. It also has integrative properties, drawing diverse issues under the shelter of the frame. A strategic frame is synthetic. It promises to sustain these sheltering and synthetic properties through time, as its persuasive power carries the frame forward. It has seductive, attractive properties, an inspirational 'vision' to motivate a range of actors through time. Such a strategic force both transforms and reorders. But it also structures the future, fixing new understandings and creating new categories (Mintzberg 1994). A strategy with such framing capacity is thus potentially a very powerful governance instrument. It is perhaps unsurprising that government actors in the present era champion the idea of strategy and include requirements for a strategy statement in policy instruments, but are very cautious when it comes to realising such an idea. This in part explains the rhetorical invocation of the terms 'strategy' and 'vision' in much contemporary policy talk, attached to very little meaningful or persuasive content (Shipley and Newkirk 1999; Shipley 2002).

The cases in this book illustrate different ways in which strategic frames are arrived at. The Cambridge Sub-Region story is of a strong frame imposed on a dynamic urban development reality and sustained over 50 years. The frame existed in the minds of some influential stakeholders prior to a major planning effort, and then persisted against all attempts to change it. What the Holford strategy did was to 'name' it, to give it a planning logic. The growth of Cambridge was to be capped, development pressure was to be dispersed to the surrounding settlements and beyond, and modest measures were to be taken to deal with the increase in road traffic, within the limits of conserving the distinctive heritage of ancient colleges and romantic water meadows. The frame of contained growth was continually reinforced by a much broader implicit spatial strategy of urban 'containment' and compact urban growth in England, underpinned by strong societal support (Hall *et al.* 1973). This was reinforced in the Cambridge area by the university's own interest in restricting everyone else's development but its own. Changing this strategy required a direct struggle, inspired by an alternative, nationally significant frame, that of the 'Cambridge Phenomenon' and its storyline of an internationally important cluster of economic innovation, which could be positioned within a much wider policy discourse of 'economic competitiveness'.

In the Amsterdam area, Schiphol's development was also justified through the invocation of the global economic significance of its 'mainport' function. But the strategic frames governing urban development evolved more steadily than in the Cambridge case, through continual interaction with national, municipal, inter-municipal and more recently, multi-stakeholder discussion arenas. From an unchallenged position as a well-defined city with strong influence and investment support from national government, delivering a welfare-state strategy of residential neighbourhoods and industrial areas, new framing ideas have evolved to emphasise the complex mix of diversity, heritage and commerce, distinctive locales and wide-ranging connectivities of a cosmopolitan locale within the expanding urban region. Planning efforts have attempted to mobilise concepts of networks and nodes, of openness and urbanity, accessibility and urban quality. But it has not been easy to infuse these ideas with inspirational energy. Instead, the emphasis has been on continual re-alignments needed to make strategic conceptions catch up with the major projects, which are much more clearly framed as inspirational visions of how particular locales could develop.

In Milan, architectural design traditions foster a local culture in which visions about projects, particular locales and urban futures are continually being produced. But the difficulty since the 1950s has been to create framing concepts that sustain a strategic conception of the city while performing the institutional work of shaping public and private investment projects. Such frames do exist, such as the sacrosanct protection of the old core from significant redevelopment and long-standing ideas about development axes. But these concepts frame implicitly. This encouraged those involved in the work on the *Documento di Inquadrimento* to emphasise that strategies should evolve from processes of recognising emergent potentialities. Their intervention sought to create the conditions to encourage such recognitions. They appreciated the significance of Mintzberg's emergent way in which strategies were 'discovered' (see Figure 6.1). But the cases also show that the discursive terrain was too full of history for emergent concepts to proceed independently of previous attempts to shape urban evolutions. The above four ways in which strategic frames may arise are summarised in Box 6.2.

All the cases illustrate the intensely political nature of producing strategic frames that have a clear spatial content. It is perhaps no surprise that many vigorous efforts to

Focusing and framing
Challenging a well-established frame in a direct discursive struggle (e.g.: Cambridge area).
Evolving an established frame by continual adjustment, co-alignment and re-consolidation (e.g.: Amsterdam).
Creating conditions in which explicit strategic framing work can eventually develop (e.g.: Milan).
Discovering strategies through 'recognising' emergent conditions.

Box 6.2 Focusing and framing

reframe urban spatial strategies may not succeed. New conceptions repeatedly challenged the Holford plan in Cambridge, until a new growth strategy finally broke through in the 1990s. The Amsterdam case shows how new ideas were often reworked back into older frames, or failed to get leverage. But when a strategic frame survives, it carries both generative and constraining power. It opens up new connectivities and conjunctions which, in turn, may lead to new creative synergies. It fixes and focuses. It draws attention away from some issues in order to concentrate around new ways of seeing a situation. As it integrates a new set of relations, so it may also disintegrate older relations (Healey 2006b). Using a more relational vocabulary, a new strategic frame that carries power enables decoupling from established governance discourses and practices, and either the formation of new practice arenas or the recoupling of old practices with new discourses. A new frame changes the boundaries of inclusions and exclusions. For this reason, there is much discussion in recent literature on strategy-formation on how to prevent new strategies from developing hegemonic 'lock-in', from fixing so firmly that adjustments to the threats and opportunities of new situations cannot be 'seen' and adapted to.[8] Within the planning field, such 'locked-in' frames can be found in concepts such as the *Randstad* in the Netherlands, and the English 'green belt' concept, both of which are so deeply embedded in popular and political understandings and practices that they appear resistant to repeated efforts to supplant them.

This suggests that those evaluating the emergence of strategic frames and those designing processes to encourage their emergence need to give particular attention to where framing work is likely to take place. When within a process of articulating strategy the definition of a frame is emerging, whose efforts are critical to its formulation and how it is 'named' and consolidated to carry significant meanings and legitimacy? Because the generation of a new strategic frame is a creative process of imagination and discovery, it cannot be confined into a step-by-step technical process of formulation. It will inevitably destabilise and perturb established frames and conceptions. Building new strategic frames thus both needs mobilising power to help the process along and creates mobilising power once a frame begins to have attractive and seductive force.

GENERATING MOBILISING FORCE

Strategy formation is institutionally complex work. It involves capturing ideas, issues, tensions and understandings from 'above', 'below' and from round and about. The force behind what become strategic concepts may 'bubble up' from all kinds of positions in a governance landscape (Mintzberg 1994).[9] Explicit endeavours in strategy formation may 'invent' strategies, but often such endeavours merely 'recognise' and 'name' strategies, giving them a more concentrated force by filling out implicit orienting concepts with meanings and justifications that resonate with a felt need for some form of strategic direction by one or more key actors. In part, strategy formation is an imaginative activity, depending on creative efforts to visualise future possibilities. But it is much more than this. Strategies with depth connect imaginations about the future to the potentialities and constraints embodied in all kinds of evidence about the ongoing flow of events and meanings, of connectivities and impacts (Mintzberg's 'myriad of small details'). Strategic thinking involves a capacity to see the wider significance of details and an ability to position and give them meaning in the context of a strategic frame. It is in the work of positioning and the giving of meaning that the political work of strategy formation develops. Strategies that have effects are not just abstract concepts, floating in the ether of design and planning discourses. They gather force because they resonate with the values, perceptions and particular needs of key actors. They develop energy as they are positioned in critical governance arenas. They answer to the sense that some kind of strategic orientation is needed to give meaning, justification and legitimacy to a stream of activity.

'Sensing' that a strategic effort may be helpful in an urban context itself reflects certain assumptions and predispositions about how governance works. As Whittington (1993: 30) notes, 'the very notion of "strategy" may be culturally peculiar'. Some governance cultures, as in the Netherlands, demand clear strategic formulations by government agencies as justifications for actions. In other situations, articulating strategies openly may seem politically dangerous, constraining the autonomy of powerful actors to operate according to their own strategic sensibilities, judgements and interests. Even where key actors in a governance landscape feel the need for an explicit strategy, developing this around some concept of the 'place' of an urban area may seem unusual, difficult, alien. How is 'strategic force' around such a focus generated?

There are many candidates in the literature from which an answer to this question can be developed (see Box 6.3). It is often argued by planning officers in government positions that the key is to legally enforce the formation of a strategy as a statutory duty. In England and the Netherlands, there was continual pressure in the later twentieth century to have 'up-to-date' plans to set the framework within which more detailed plans and regulatory decisions could be made. Similarly, in Milan, the invention of a new instrument grounded in regional statute, the *Documento di Inquadrimento*, was intended to provide a justificatory argument within which specific decisions could be located. The hope is that embodying a strategy in such statutory documents will give formal legitimacy to specific actions. Many planners, and often those who look to planners for help in strategy

Mobilising power in strategy-formation
Statutory duty – for example, to have a strategy or strategic plan. **Intellectual and imaginative power** – of skilled planners, for example. **The forces of economic interest** – of land and property developers and major industrialists, for example, driven by capitalist logic. **Advocacy coalitions and networks** – around urban qualities. **Strategic actors and leaders** – individuals, groups, and sometimes an established governance capacity.

Box 6.3 Mobilising power in strategy-formation processes

formation, emphasise the intellectual and imaginative power of individual planners and a skilled planning team. This justifies the hiring of consultancies to produce strategies, or the building up of special strategic planning teams. In Amsterdam, the preference was for a strong in-house planning team. In Milan, regular use was made of academic consultants. In the Cambridge area, respected planners were used as consultants in the 1940s and 1950s, while consultancy firms were regularly used in the 1990s and 2000s. But a close look at the skill of these planning teams shows that the force they are able to generate lies not just in their intellectual analysis and synthesis. It lies in the capacity to position and craft intellectual arguments so that they have resonance in political contexts.

This practice led many strident critics of planning activity in the 1970s and 1980s to accuse planners of being merely a technical arm of an oppressive state that served the forces of capitalist interest under the mask of promoting the 'public interest' (Castells 1977; Cockburn 1977). This shifted the origin of the force behind strategy-formation for urban areas to some logic arising in the sphere of (capitalist) economic development. Planners, politicians and other lobby groups in Amsterdam, Milan and the Cambridge area have been repeatedly accused of being driven by the logic of capital accumulation through property development or, more recently, the logic of developing urban assets to promote economical competitiveness. The cases show, however, that although such exogenous pressures were present, in the flow of economic activity and in the rhetoric orienting policy, other pressures were also evident, such as the defence of landscapes of cultural identity, or the challenges of addressing different conceptions of urban quality. The force that creates moments of opportunity for strategy formation is more to do with the perception of 'deficiency', either through a general unease with established approaches, as in the concern for some kind of strategic frame within which to locate the proliferating project proposals in Milan in the 1990s, or the perception of a crisis, as in 1990s Cambridge. These perceptions may be shaped by exogenous economic or political forces. But arguments about the power of these forces are also harnessed by actors to buttress their case for greater strategic attention to local experiences of urban dynamics.

Since the 1980s, in recognition that the power to make a difference to what happens within urban areas is more widely diffused than allowed for by simple notions of government control or economic determinism, a much greater emphasis has been placed on creating force through coalition building. The advocacy and formation of 'partnerships', 'platforms', 'round-tables', and new arenas of all kind, was a major phenomenon in 1990s urban governance across Europe (Cars *et al.* 2002; Pierre 1998). Such coalitions can be built around mutual interests and interdependencies (Booher and Innes 2002), but also through recalling past identities, as Le Galès (2002) suggests. Once coalitions begin to form, they may develop positions, values and discourses. They become advocacy coalitions (Sabatier and Jenkins-Smith 1993) and discourse coalitions (Hajer 1995).[10] Over time, such coalitions may develop some of the qualities of a policy network (Klijn 1997). The building of such a coalition and emergent policy network around the idea of the 'Cambridge Phenomenon' was clearly a critical move in the Cambridge area case. The difficulty of building a powerful coalition in Milan in the 1990s was one reason for the very targeted approach to a strategic intervention there in the late 1990s. In Amsterdam, coalition-building had become a constant process by the 1980s, as the City Council continually reworked its relations with national government, the province, other municipalities and key actors in the economy and civil society. A strategy that could express a coalition position, as in the various spatial concepts mobilised in Dutch spatial planning, was an important (if not always successful) tool in sustaining coalition focus (van Duinen 2004; Zonneveld 2005a).

However, coalitions may not always agree, still less develop a shared understanding. The difficulties in Amsterdam in 2002/2003 over longer-term development allocations in the Haarlemmermeer–Amsterdam–Almere area provide one such example. The withdrawal of Conservative politicians from their prior agreement to the East of England Regional Spatial Strategy in 2004 is another. Key actors may leave a coalition, as the city of Almere did in the Amsterdam region in the 1990s. This suggests that there are two kinds of coalition to be considered. One consists of those actors necessary for a strategy to get significant leverage. The other consists of the advocates of a strategy. A great deal of development work may occur in the arenas of the latter. But until these overlap with the former, an 'advocacy' coalition lacks the strategic force to build a new policy discourse. The story of the Dutch 'Deltametropool' concept provides an interesting example of where an advocacy coalition carefully switched arenas, to position its ideas within a politically powerful arena (van Duinen 2004), but then lost momentum as national politics changed again.

Finally, some argue that strategic force is generated by the power of specific 'strategic actors', relentlessly pursuing specific agendas (Flyvbjerg 1998), and of 'leaders' with particular skills in driving ideas forward into practices. Bryson (1995) argues that there is little point in embarking on a strategic planning effort without sufficient leadership in place. The role of 'strategic actors' tends to be stressed in stories of contested urban politics. The public management literature, in contrast, stresses leadership capacities. The two come together in the focus on the qualities of 'mayors' in urban-government systems that grant significant executive roles to elected mayors. The importance of the interests and political leadership of mayors emerges from accounts of urban-strategy formation in France (Motte 1995, 1997, 2001) and more recently in Italy (Magnier 2004). But the cases used in this book suggest that strategic leadership is not necessarily embodied in a single person (Bryson 1995), still less a formal political role. It may be found in a group, or be diffused as 'network power' (Hajer and Versteeg 2005; Innes and Booher 2000). Or, as in Amsterdam, it may exist as a general governance capacity for articulating the strategic focus behind particular actions and positions, underpinned by societal expectations. In Cambridge and Amsterdam, the initiative in maintaining momentum moved from one person and group to another. Those who kept the strategy together as it travelled from one institutional site and time to another were not necessarily the most visible advocates of a strategic frame.

Strategic mobilisation thus involves a process of coalescence of intellectual and political forces through which strategies are 'recognised', given names and positioned in specific institutional contexts. Such mobilisation exploits moments of opportunity, where having a strategy responds to some felt need among key actors. Skilled strategic work involves understanding the nature of such moments and the opportunities to 'capture' them in particular directions. Successful strategy formation widens the opportunity space for particular discourses and practices, opening out the 'cracks' in the existing governance landscape to enable different frames to develop leverage (Healey 1997; Tarrow 1994), thereby creating its own structuring force on that landscape. Achieving such power involves intellectual work, the mobilisation of imagination and knowledge, and the mobilisation of political power, through coalition formation and leadership, and through moving around among different arenas. In the case of the formation of urban strategies, both challenges are particularly complex and are often only partially realised. Only some key actors come to 'see' the place of an urban area as a critical focus of attention and only some ways of seeing are prioritised.

GENERATING TRANSFORMATIVE FORCE

Strategy formation processes that have structuring effects succeed not merely by creating convincing interlinked policy narratives and discourses. Strategies succeed to the extent that they shape subsequent events. This happens not through a simple linear process from strategy formation to formal approval in a legitimate political arena and then 'application' to the framing of specific investment projects and regulatory decisions. Strategies with shaping power re-order categories and positions (Mintzberg 1994). They destabilise established routines and practices and generate new ones. They shape practices rather than specific decisions, through providing a different way of 'making sense'. They create momentum around which new policy networks and 'communities of practice' may form. They institutionalise, as Hajer (1995) argues. To achieve this in complex urban region governance contexts, they need the capacity to travel from one arena to another.

In each new arena where they arrive, a strategic discourse will be interpreted in some combination of old understandings and new possibilities. A framing discourse with strong structuring power is able to carry core ideas from one arena to another. In the language of actor-network theory (Latour 1987), those pushing out a strategy across a governance landscape may seek to wrap its core ideas into some kind of synthetic 'black box', so that its key elements cannot be driven apart. However, discourses without interpretive flexibility are unlikely to survive as they permeate across a governance landscape. If they do, they may over-stabilise governance responses to dynamic urban conditions, as in the 'green belt' concept in British planning practice. Effective institutionalisation therefore means that strategies not only survive continual reinterpretation but are enriched by such processes. Strategies are thus orientations, not precise programmes.

Strategies that have transformative power to reshape some aspect of urban governance landscapes therefore need not only discourse coalitions behind their formation but also the capacity to travel across a landscape, and endure in a continually changing form across practices and through time. They need to be able to encourage the formation of new discourses and networks through which the 'ecology of existing games' can be transformed (Klijn and Teisman 1997: 106).

The travelling capacity of a strategy is partly held in place by recognition across different arenas of its functional utility. It serves particular interests. But, more widely than this, it is held in place by the legitimacy of a particular strategic orientation. Legitimacy is not a one-dimensional governance property. The cases show that many different forms of legitimising strategies are in play at different times. Table 6.2 links the source of the power of a strategy to the way legitimacy is established, through the testing and challenging of a strategy's claims. This in turn is linked to the power of a strategy to travel and to achieve transformative effects.

This assessment suggests that strategic discourses are at their most powerful where they become embedded in legal practices, in the routine conventions of relevant communities of practice, and where they are pushed along by tacit understandings. Those mobilising to create new strategic discourses are often seeking to destabilise and transform such embedded strategies. Transformative force can be given momentum by seductive frames and images, by scientific knowledge, by expert judgement, by interactive learning arenas, created by such arrangements as round-tables and partnerships, and by the politics of interest pursued by strategic actors. Such force is potentially resisted by electoral mandates, legal authority, conventional practices, established knowledge, expert judgement, and by strategic actors. A critical task for initiatives in generating transformative strategies with the force to make a difference is therefore the mapping of the institutional terrain, in terms of actors, arenas, networks. Those embarking on transformative initiatives need to know how these link resources, regulatory power and discursive power, taking account of the potential to move with emerging strategic ideas, to mobilise against them or just resist or ignore them.

Table 6.2 Institutionalising urban region strategies

Source of power	*Legitimising processes*	*Travelling capacity*	*Transformative power*
Electoral mandate mandate	Votes; party organisation	Legitimises other sources; weak unless concepts grounded in the public realm already	Gives generalised authority to new strategies
Legal rules, principles and contracts	Law courts	Legal judgements can set limits, and modify principles within frames	Gives generalised authority to new strategies; judges the political arena by its formally agreed strategies
Science; formalised knowledge	Scientific community Respect for science	May buttress a frame or undermine it over time	Contingent on the quality and utility of the science
Conventional frames and practices	Communities of practice	Can resist new frames or convert into conventional terms	Full of resistances that need to be overcome
Expert practical judgement	Personal integrity and respect for experts	Can buttress a frame, or undermine it if experts are not respected	Contingent on the quality and utility of of the judgement
Strategic actors and the politics of interest	Fear; acceptance of domination; calculation of interest	Can reinforce seduction of a discourse (and resist it)	Contingent on the trust of key partners in the power and legitimacy of key actors
Experiential and tacit knowledge of key actors	Interactive arenas; respect for multiple forms of knowledge	Harnesses agencies to act as carriers of new ideas	Potentially strong in reinforcing the 'truth' of a strategy
Local, situated knowledge	Respect for citizen's knowledge and values, and for 'street level' bureaucrats	Helps to relate new ideas to the public realm; and to reach operational 'communities of practice'	Contingent on the significance of citizens and frontline staff in local governance cultures
Discursive seduction	The seductive power of the frame or vision to convince	Strong rhetorical power, but can easily be traduced if the imagery lacks depth	Contingent on quality of the frames and the attitudes of the 'seduced'

The Power of Strategies

Strategies with transformative potential are thus not arrived at by linear step-by-step processes, even though similar steps may be found in many examples. Episodes of transformative strategic potential rise up when moments of opportunity appear. These moments have different potentials and capacities for enlargement. Once a strategic frame is formed and 'named', it may travel. But the distance and manner of such travelling across a governance landscape will be contingent on specific institutional circumstances. Those frames that have long-term, wide-ranging impacts are likely to be slow in evolution and in institutionalisation. They need to accumulate mobilising power. This means that they need to pass through arenas that offer institutional spaces in which many parties learn what it means to 'see' the issues of concern to them in new ways. Probing, contestation and challenge in accessible arenas encourages such recognition by many parties, and tests the utility and legitimacy of strategic concepts from many directions.

Strategy-making, understood relationally, involves connecting knowledge resources and relational resources (intellectual and social capital) to generate mobilisation force (political capital) (Healey 1998c; Innes and Gruber 2005). Such resources (capital) form in institutional sites in governance landscapes which, if a strategy develops mobilisation power, become nodes in networks from which a strategic framing discourse diffuses outwards. The strategic frame travels as an orientation, a sensibility, a focus for new debates and struggles, performing different kinds of institutional work in the different arenas in which it arrives. Efforts in strategy-making may be initiated in many different institutional sites, but to have significant effects, the mobilisation dynamic, with the knowledge and relational resources embodied within it, has to move towards arenas that are central to accessing the resources (over which a strategy needs) to gain influence and to have effects (it also needs to accumulate) sufficient legitimacy to survive in governance landscapes where power is diffused and attention continually shifting.

Although the concept of strategy and plan, strategy-making and plan-making, are so closely connected in planning and management thought, it helps to separate them conceptually. The term 'plan', in the context of attempts to shape urban development in some explicit way, is perhaps best understood as either, or both, a development investment programme in which funds are allocated to specific projects, or a specification of land-use and development rights and obligations, the rules governing physical transformations. Urban strategies need to frame both development and regulatory activities to have effects, while investment projects and regulatory principles obtain greater legitimacy through being grounded in a broader strategic frame. But plans can proceed without strategies, or with merely a vague rhetorical invocation of a strategy. Or spatial strategy-making may have another focus than on urban areas. Its 'place' of attention may be the nation, a wider region, an urban node, a neighbourhood, a new development or redevelopment area where a new 'piece of city' is proposed. Or a strategic conception of an urban area is articulated in the design and justification of a major project. The key to the active presence of an urban strategy lies thus not in a specific arena, or type of policy statement, but in the 'summoning up', in the flow of critical activities, of an explicit idea of an 'urban region' that embodies some collective sense and is recognised in many of the institutional sites of urban governance.

Deciding to embark on an explicit effort to articulate an urban spatial strategy is a challenging choice. Because planning systems typically contain a requirement for the existence of some kind of urban plan, and because the planning tradition has stressed the importance of a 'comprehensive' and 'strategic' content for so long, on many occasions planning 'strategies' are produced to fulfil statutory requirements and professional expectations. Or 'strategic visions' may be required to fulfil criteria for accessing specific funding streams. But these are not necessarily strategies filled with the kind of content elaborated in this chapter. As in many of the Amsterdam and Cambridgeshire plans, strategies may largely be rolled forward from one period to the next with minor adjustments. A critical judgement for those concerned about urban futures concerns when to attempt the creation of a new strategic frame, when just to prepare the ground for such a frame to emerge, and when to merely move along with the flow of events, allowing the patterns and potentialities of an urban area to emerge without any attempt at deliberate shaping. Such a judgement involves considerations of institutional design at two levels: first, about the potential for designing a strategy-making process and the form this might take; and, second, about the potential effects across a complex governance landscape of having a powerful strategic frame.

How such judgements are made will depend in part on the skill and perceptions of those involved in reading emergent opportunities and challenges relevant to the considerations of urban futures. But they will also depend on a reading of the institutional dynamics of the governance landscape. Is the situation one where there is so much conflict, fragmentation, competition, confusion and uncertainty that innovation is stifled, inequalities magnified and valued resources are being diminished (a zero–minus game, a lose–lose game)? In such situations, would developing a strategic frame at the urban-region level help to reduce the damaging dimensions of the confusion through creating some strategic stability, fixing some of the dispersed and competitive energy as a collective resource? Or is the situation one where the structuring power of past policy frames is so embedded in policy discourses and practices that it inhibits the capacity to evolve in new ways. In such situations, would the discovery of a new strategic frame help to destabilise the governance landscape and release adaptive energy? Figure 6.2 presents this strategic choice in a simplified form. Here power is conceived as a force, or as energy. This figure suggests that there are critical strategic choices to be made about the balance of 'fixing' and 'destabilising'. I return to this issue at the end of the next chapter.

Judgements as to whether it is desirable to embark on a strategy-formation process and whether a resultant strategy is likely to promote a broad range of concerns

Figure 6.2 Fixing or releasing energy through strategies

and experiences of urban life cannot be made in the abstract, or even by some kind of calculation as to the fit between indicators about a particular context and indicators about what a strategy can achieve. There is no calculative route to such judgements. They are deeply political, practical and situated, as I hope to have shown in the three accounts. If strategy-making processes succeed, they have impacts – material, epistemological, ontological and on governance capacity. But how these arise needs a more careful examination of the subject matter of urban strategies and the knowledge resources available to be mobilised in order to recognise, elaborate and develop understandings of this subject matter. These issues are developed in the next two chapters.

NOTES

1 Other examples can be found in Vancouver (Punter 2003) and Portland, Oregon (Abbott 2001).
2 For a more detailed discussion on the definitions of policy communities and policy networks, see Klijn 1997; Rhodes 1997; Vigar *et al.* 2000.
3 See Whittington (1993), Mintzberg (1994), Morgan (1997) and Christensen (1999) for contributions from management science; and Schon and Rein (1994), Bryson (1995), Healey (1997), Kickert *et al.* (1997), Gualini (2001), Hajer and Wagenaar (2003) and Innes (1992), for contributions in planning and policy analysis.
4 Other examples include the Vancouver experience (Punter 2003), Portland, Oregon (Abbott 2001), and, in the mid-twentieth century, the Greater London Plan (Hall 1988) and the Clyde Valley Regional Plan (Wannop 1995).
5 See Albrechts 2004; Bryson 1995; Healey 1997.
6 Note that Healey (1997), Bryson (1995) and Albrechts (2004) all indicate a set of tasks or 'steps' involved in strategy-formation processes, but all have caveats about a set sequence of steps.
7 Eckstein and Throgmorton 2003; Hajer 1995, 2001; Throgmorton 2004.
8 Bryson 1995; Mintzberg 1994; Whittington 1993.
9 Mintzberg stresses 'bubbling up' from below, within an organisation (p. 364).
10 Note the differences between advocacy coalitions and discourse coalitions.

CHAPTER 7

SPATIAL IMAGINATIONS AND URBAN 'REGION' STRATEGIES

> It is hard to produce a plan that at once captures the conditions of the society, city or policy area and also meets the demands of each of the citizens experiencing the problems society is mobilised to process. It's hard to be both scopic (viewing from above) and comprehensive and immediate and individual and responsive (Perry 1995: 210–211).

> Cities and regions possess a distinctive spatiality as agglomerations of heterogeneity locked into a multitude of relational networks of varying geographical reach. As such, they express, perhaps more than other socio-spatial formations (nations, households, organizations, virtual and imagined communities), the most intense manifestations of propinquity and multiple spatial connectivity (Amin 2004: 43).

INTRODUCTION

Strategies, as discussed in the previous chapter, are devices for focusing attention. If they come to shape how interventions in urban areas are designed and materialised, they have real, material impacts on the potentialities afforded by urban areas and the way these are distributed among those with a stake in them. Spatial strategies focus attention on the 'where' of activities and values, on the qualities and meanings of places, on the flows that connect one place to another and on the spatial dimensions of the way activities are organised. In the discussions, analyses and disputes that surround the formation and use of spatial strategies for urban areas, this spatial dimension may not be immediately visible. The emphasis may be on general problems – congestion, pollution, lack of affordable housing, conserving historic buildings, the shortage of sites for new companies. Or they may be on appropriate processes – when and how to organise consultation processes, the nature of formal inquiries, how to reconcile different viewpoints. But what gives spatial strategy its distinctive focus and contribution is the recognition that 'geography matters' (Massey *et al.* 1984). It is not just traffic congestion in general that is a problem, but specifically where this occurs, what the impacts are and how and where they are experienced, and, as a result, who is affected by congestion and its impacts. It is not just the inability of the housing market to produce affordable housing that is the problem, but the way housing markets work to distribute living opportunities for different people within an urban area, so that poorer people may end up facing inequalities not only in access to housing, but to work opportunities, health services, education and leisure opportunities. It is not just the conservation of buildings that is at issue, but the way conservation measures impact on the overall quality of an area, in terms of property values, visits from tourists and traffic flows. Strategies that emphasise the spatiality of activities and relations thus foreground some critical interconnections and qualities arising from the evolving co-existence and juxtaposition of multiple activities and webs of relations in particular areas, locales and territories.

But what is this 'geography' and how does it 'matter'? There can be no doubt that specific physical qualities of Amsterdam, Milan and Cambridge were powerfully present in the material experiences of people living in, working in, visiting in and making policies for these places. Perhaps even more so, they existed as places in the imagination of residents, business groups, elites, tourists and policy-makers. Analysts of territories, the core concern of urban and regional geography, have long debated the relation between the experienced city and the imagined city.[1] Is an urban area a bundle of property rights, a landscape, a set of activities, a collection of networks, a jurisdiction, a symbol? Does it exist objectively, to be 'found' by appropriate analysis, or is it a social construct, to be discovered and imagined by some kind of creative process? And, however it exists, is an urban area just a 'presence', a 'place-in-itself', or can a place 'act', to become a 'place-for-itself'?[2] The answers to such questions are important because they frame political initiatives and policy programmes. These initiatives and programmes in turn affect the daily life experience of those living in, working in, visiting and passing through urban areas.

In this chapter, therefore, I enter into the realm of geographical debate about materiality, identity, perception and representation. These debates are full of difficult, apparently abstract, issues. But the issues they raise enter into practical strategy-making, as planners in Amsterdam argue over 'contours', 'layers', 'networks and nodes', or in Milan over a 'strategy of relations' and 'polycentric' urban patterns, or in Cambridge over 'green belts', 'corridors' and 'village landscapes'. In developing the capacity for strategic focus, I argue that those involved in strategy-formation need to give careful attention both to the geographical dimensions of urban dynamics and strategies and to the kind of geography they are using. This matters because spatial conceptions translated into government interventions have material, ontological and epistemological effects. They affect the physicalities of the experience of place, the meanings that attach to them and the knowledge developed about them. Strategies focused on urban areas carry within them particular spatialised ways of 'seeing like a state' (Scott 1998).

The three case accounts highlight many dimensions of the geographies of spatial strategies. But, by the end of the twentieth century, all were struggling to come to terms with a context of exploding spatial connectivities, in which physical proximity was not necessarily the primary determinant of how one activity affected another. In this chapter, I argue that a relational geography, rather than a geography of simple physical proximity, has the potential to open up a productive way, both intellectually and politically, through this struggle. I build up this argument in stages. I first introduce the tensions between representations of space, and the spaces of material existence and cultural identity. I then look more closely at various dimensions and vocabularies used in developing representations of urban areas, drawing out the tensions between a geography of physical patterns and a geography of relational dynamics. I next delineate the perspective of a relational geography, focused on multiple juxtapositions of proximities and connectivities (Amin 2004). Finally, I bring the discussion back to the implications of such a geography for the focus and content of spatial strategies for urban areas.

REPRESENTATIONS OF SPACE – POLICY CONCEPTIONS, DAILY EXPERIENCES AND CULTURAL SYMBOLS

What *is* an urban 'region'? Is it a clearly defined object, some kind of 'thing'? Is it just an idea we have of an ambience, or a history? Does it exist within some kind of hierarchy of spaces – the world, continents, nations and districts and neighbourhoods? Or is it formed in a landscape of horizontal networks or adjacent urban regions? Is it a set of clearly definable relations that interlock more within a specified area than with areas outside? Does it have a clear core? Are its boundaries fixed or indeterminate? Is it a pattern, a structured order, with socio-physical expression, or a continually emergent assemblage of potentialities? Is it an active subject that can *do* things or is it just a passive analytical or symbolic construct? If so, who constructs it and for what purposes? Because of these definitional questions, I have been careful in this book to avoid using the terms 'city' and 'urban region' as if their meaning was known. Instead, I have used the terms 'urban' and 'urban areas' to direct attention in a general way to the phenomena I am referring to.

Within urban and regional geography, different perspectives have developed to organise answers to these definitional questions. These give different meanings to words such as 'place', 'space', 'spatiality' and even 'geography' itself.[3] In the geographical debates, there are conflicts between perspectives that view space as a surface on which objective patterns can be discerned, and those that understand spatiality as an inherent property of any social and natural relation; between place as an objectively discerned focus of activities and qualities, and place as actively produced through experiences and meanings, as an 'event'; between spatial dynamics as composed of physical patterns succeeding each other, or as a process of continually intersecting, transecting, conflicting and synergetically innovating interactions between multiple trajectories, space as a 'simultaneity of multiple trajectories', to use Doreen Massey's expressive phrase (Massey 2005: 61).

If spatial dynamics are a process and place an event, as Massey insists, then to understand 'geography' requires attention to the production of space. In an influential contribution to analysing the social processes through which 'space' and 'spatiality' are produced, Henri Lefebvre (1991) proposed the distinctions in Table 7.1. As his terms are

Table 7.1 Three ways in which 'space' is produced

Lefebvre's labels	*His definition*	*'Translation'*
Perceived (spatial practices)	Daily routines and interactions with the routes and networks of of 'urban reality'	Routine material engagement and experience of being in and moving around urban areas
Conceived (representations of space)	As in the conceptions of 'scientists, planners, urbanists, technocratic subdividers and social engineers, as of a certain type of artist with a scientific bent' (p. 38)	Intellectual conceptions of urban areas, produced for analytical and administrative purposes
Lived (representational spaces)	'As directly *lived* through . . . images and symbols' expressed in symbols and signs (p. 39)	Cultural expressions of place qualities and spatial meanings

Note
After Lefebvre 1991: 38–40

not always easy to grasp, in this table I have expressed them in several ways, including my own 'translation'. Lefebvre is interested in the complex interactions between the material world and the way we live in it, systematically attempt to conceive it, and give emotive expression to our experiences. He argues that 'space' is continually being produced by human processes of routine material engagement, of intellectual conception and of cultural expression. He proposes a triad to describe this range of understandings of space.

Lefebvre stresses that these distinctions are analytical. In the flow of life, all are present, inscribed in any thought or action and interacting with each other. He is particularly concerned that those who 'conceive' space, in which he would surely put those who are involved in spatial strategy formation, tend to get leverage on the power to shape the material 'urban reality' to which the rest of us then have to adjust. In this, he parallels the concerns of others – philosophers such as Habermas, sociologists such as Foucault – that the world of 'systems', of government and the corporate economy, seem to penetrate and dominate the way we live today. Echoing this sentiment, the anthropologist James Scott writes: 'State simplifications . . . strip down reality to the bare bones so that the rules will in fact explain more of the situation' (Scott 1998: 303).

These rules, Scott argues, then turn onto the reality, organising it, shifting power to experts and state functionaries and away from those with 'local knowledge'. Scott, however, does not therefore seek to remove the state. He recognises that some governance activity and the systematic simplifications that go with it are inevitable in complex social formations, as discussed in the previous chapter. Instead, he demands that this activity should be challenged to adopt richer and more varied organising concepts, infused by knowledge from outside itself. This has become a key concern of those developing the political implications of a relational geography, who are searching for an approach to urban development that cultivates mobility rather than stasis, openness rather than closure, richness and mixity rather than homogeneity, and which centres policy attention on improving the daily life experience of urban areas (Amin and Thrift 2002; Massey 2005).

For Lefebvre, the lived spaces of images and symbols can provide a route, through imagination, to 'change and appropriate' the domination of those producing 'conceived' spaces. More generally, the case accounts in this book show that the conceptions of strategy-makers are not autonomous constructs, although it sometimes seems that planning policy communities operate in rather introverted thoughtworlds. To gain significant leverage, the conceptions of professionals and policy-makers have to have some kind of resonance with material experiences and cultural imageries. Liggett (1995) links Lefebvre's notion of representational spaces to meanings within a cultural memory. She provides an example of the way Lefebvre's three spaces link together through an example of a planning document, the *Cleveland Civic Vision 1991*. This contains a map of housing sites, but also:

> a handsome colour photograph of single-family residential units in an older neighbourhood. The Central Business District (CBD) and Lake Erie and a beautiful blue sky form the background of the picture. In this setting, the houses [to go on the sites] are not being represented as single-family dwellings, but rather as well-kept homes in an older neighbourhood – that is, as an urban community. . . . The memory of the imagined community [cover] is projected onto the figure for a potential future [the map] (Liggett 1995: 252–253).

In a similar way, the cover of this book projects a polycentric Amsterdam, yet still as a built form on a continuous surface viewed from the city core looking outwards. Liggett emphasises in her example how the planners, following the advice of the marketing world, have learned how to manipulate and appropriate cultural imagery to buttress other arguments and intentions developed through analytical 'representations of space'. But the appropriation is a narrow one and hardly meets Scott's demands for greater richness. My case accounts underline the significance of cultural memory in shaping the conceptions of urban 'regions' mobilised in spatial strategies, such as the deeply embedded notion of the Milan *cuore*, or the ideal of contained towns surrounded by villages in a rural landscape in Cambridge. Notions of physical axes and corridors mobilised by policy-makers gained little real leverage against these conceptions. But the planners in Amsterdam, working with academics, were searching for inspiration from the experiential realm of living in a diverse, open and mobile urban reality, struggling to develop a policy conception around notions of urbanity and accessibility (Bertolini and Dijst 2003; Bertolini and Salet 2003).

Lefebvre emphasises that his different ways of understanding space are in continual dialectical interaction with each other, producing challenges, dominations and resistance. What is 'real', materially and ontologically, is produced through these interactions.

His 'scientists, planners, urbanists, technocratic subdividers and social engineers' (Lefebvre 1991, page 38) and (certain) artists are continually being challenged by the experience of material realities and by cultural images. Thus, as I will argue later on in this chapter, such actors are not outside these realities and images but draw on them selectively to create and reinforce the conceptions they develop in their role as experts and policy-makers. Translated into planning concepts, these images recursively feed back to reinforce the cultural associations, and embed a particular cultural imagination still further into the politics of place qualities. As Van Eeten (1999) shows nicely in relation to the *Randstad*, no amount of analysis of linkages and flows that deny any 'objective coherence' to culturally embedded iconographic representations of place or spatial patterns seem capable of displacing such politically and culturally embedded planning concepts. Thus 'conceived space' is not apart from 'perceived' and 'lived' space, but evolves interactively with these other spaces.

If this is so, then we may expect a strategy for an urban 'region' to reflect all these three dimensions of space. A strategy may acknowledge everyday space–time routines, such as the commuting flows produced by the growth limit and dispersal strategy in the Cambridge area. It may express implicitly culturally embedded ideas about how places should be (the value of Milan's city core, the images of neighbourhood life in Amsterdam's city centre, or of a Cambridge interpenetrated by water meadows and surrounded by a rural idyll of village church spires in a landscape of fields and woodlands). But the main institutional work of an urban strategy lies in its attempt to represent an urban area. It involves 'summoning up' some conception of an 'urban region', which indicates both its internal differentiations and its external positioning.

Many of those involved in the production of spatial strategies are very aware of the relation between their conceptions, material experiences and cultural imageries. Over the past decade in England, there has been much debate in the professional press over the continuing relevance of the 'green belt' as a spatial organising idea, but planners no longer 'own' this idea, as it has been appropriated into a broader popular consciousness (Elson 1986; Rydin 2003b). In the Netherlands, planners talk about finding an 'appropriate fit' between the experiential reality of urban dynamics and the spatial conceptions they mobilise.[4] But the challenge of 'co-aligning' policy-oriented conceptions with both material experience and cultural imaginations of the place and spatiality of urban is fraught with difficulty. This partly arises from the multiplicity and complexity of the social relations of urban areas and partly from the inherent difficulties of capturing these 'comprehensively' and 'objectively'. Following Scott's argument, the necessary simplifications of the conceptions mobilised in urban strategies need to be carefully examined, not just to determine whether they are simplifications of a complex reality, since they inherently will be. They need to be probed to assess what the implications of such simplifications may be. In the next section of this chapter, I offer some suggestions for a critical examination of the dimensions and vocabularies of space and place mobilised in the discourses and practices of spatial strategy-making.

DIFFERENTIATING 'CONCEIVED SPACES' IN SPATIAL STRATEGIES

DIFFERENT GEOGRAPHIES

To summarise the argument so far, intellectual conceptions of the space of urban regions may be derived from many sources and traditions of analysis and design. These, however vaguely formed, underpin the framing discourses of any spatial strategy. Following Lefebvre, these may be built up through an effort to analyse the routines of material engagement (perceived space), for rules and indications upon which to build an idea of what an urban area is now and could be. In doing this analytical work, all kinds of intellectual constructs will then be used as lenses through which to observe the routines and capture in some way their essence and dynamic, for example: labour markets, land markets, river basin flows, journey-to-work patterns, traffic flows. Such concepts attempt to give meaning to aspects of the urban through generating an 'objectified' representation of material reality.

Intellectual conceptions may also be built from cultural values, images and symbols to form an intellectual construct, from cultural icon (Lefebvre's lived space) to a concept of what an urban 'region' is and could be. Such symbolic concepts attempt to give meanings through generating an expressive representation of an ontological reality, a kind of identity (for example, concepts of 'modernity', and 'globalisation', or even the 'compact city'). Architectural debate about the form of cities has a long tradition of such imagery, with a tendency for its cultural resonance to get detached from cultural roots as protagonists have searched for imaginative ways of expressing future possibility.[5] The Milanese design elite enjoyed such imageries in the 1980s. But conceptions of the urban arise also through the cultural lenses through which geographers and other social scientists have imagined space and territory. Examples which are clearly rooted in cultural traditions are the Dutch and English preoccupation with sharp boundaries between town and country (green belts, red and green 'contours') and the recurrence of notions of settlement hierarchies that echo patterns of a pre-industrial age. The Amsterdam planners' search to understand the nature of 'urbanity' and 'accessibility' may also be understood as an attempt to create a conceptual structure for a quality of identity, a cosmopolitan cultural ambience and movement, which can then be used as a basis for strategic urban management. As Lefebvre argues, these analytical and symbolic directions through which conceptions of urban areas are constructed recursively feed on each other and with the material realities and meanings they seek to represent.

Thus, these days, most urban strategies are likely to show a complex mingling of concepts that have evolved in the attempt to capture contemporary experiences and concepts that are themselves the product of particular values, such as the 'compact city' and 'urbanity'. These conceptions are unlikely to be coherent and internally consistent. But this jumbling of different inspirations in the formation of a strategy involves not ony a mixing of imageries. It also involves encounters with different epistemologies, different ideas about the construction of knowledge about place and spatiality. Within the field of academic geography, there has been a strong epistemological shift from a physical geography of spatial patterns arranged on 'space-as-a-surface' to a social–relational geography in which space is not a 'continuous material landscape', but a 'momentary coexistence of trajectories, a multiplicity of histories all in the process of being made' (Massey 2000: 129). In the geography of physical patterns, places, peoples, cultures, etc. become objects located on a surface. 'They lie there, in place, without trajectories' (Massey 2000: 128). This is sometimes called a 'Euclidean' or 'cartesian' geography.[6] The patterns affect each other through physical proximity, with effects between one activity and another varying with simple linear distance. In a relational geography, in contrast, 'cities and regions are seen as sites of heterogeneity juxtaposed within close spatial proximity, and as sites of multiple geographies of affiliation, linkage and flow' (Amin 2004: 38).

In the second part of the twentieth century in Europe, the first, 'cartesian'-style, geography dominated spatial planning thinking and, more generally, ideas about space and place used in other policy communities. But, by the end of the century, a relational geography was struggling to find expression. The impact of such a geography is evident in planning thinking in both Amsterdam and Milan. How do the different 'geographies' manifest themselves in the practical work of forming a spatial strategy and translating general ideas about an urban area into specific techniques and practices? In Box 7.1, I present a set of dimensions to allow a more detailed probing of the conceptions of the place, space and territory linked to the area for which an urban 'region' spatial strategy is being produced. As in Chapter 6, these are provided to encourage both critical analysts and those developing policy ideas to look closely at the content of the specific geographical concepts used and their potential effects.

AN URBAN 'REGION': OBJECT OR RELATION?

The first dimension probes the kind of geography being expressed through a spatial concept. In the Cambridge Sub-Region, the image of a regional landscape organised in an integrated way into hierarchies of market towns, interacting with their surrounding rural hinterlands, provided the dominant regional strategic concept from the 1950s onwards. The strategic planning effort in the Cambridge area can be interpreted in part as a struggle to prevent the emergent reality from bursting out of the constraints of this culturally sustained, physicalist conception. Meanwhile, the justification for breaking this conception, to allow Cambridge more room for growth, derives from a different geography, that of a material experience of complex economic and social relations connecting the Cambridge area to all kinds of other locales, near and far, as imagined in the concept of competition in a 'globalising' economy. In Amsterdam in the 1990s, much more attention has been given to the position of the urban area in relation to a variety of human and non-human networks – of water flows, green spaces, infrastructure flows, and economic and social relations. Planning teams and their advisers have been actively

Question	For example . . .
1 What space is being referred to? (What kind of entity?)	A surface or container? An actor? A material set of relations? A set of formal rights and obligations? A place of encounter? An 'event'?
2 How is it *positioned* in relation to other spaces and places? What are its *connectivities* and how are these produced?	In a hierarchy of places? In relation to (diverse) global forces? In relation to natural resource flows? Proximate and distanciated connections?
3 How is it *bounded* and what are its *scales*?	Clear boundaries between inside and outside? Or a more 'porous' treatment? Unified concepts of centre and periphery/edge, or more multiple and fluid concepts? Concepts of networks and nodes? Concepts of layers of networks? Concepts of open networks with multiple nodes, networks, and scales/reach?
4 What are its *'front' and 'back'* regions?	Who and what is 'in focus'? Who is present? How are non-present issues and people brought 'to the front'? Who/what is 'in shadow', in 'back regions'?
5 What are its *key descriptive concepts*, *categories* and *measures*?	Land uses? Property rights? Social groups? Spatial metaphors/concepts? Landscape types? Economic activity systems? Aesthetic qualities?
6 How is the connection between past, present and future established?	Linear, circular or multiply-folded? 'Sliced up' comparative statics or dynamic emergence? Relation between daily, weekly, yearly and generational time? Relation between policy, investment and regulatory time? *continued*

Box 7.1 Dimensions and vocabularies of space and place

Question	For example . . .
7 *Whose* viewpoint and *whose* perceived and lived space is being privileged	Residents in general? Multiple social groupings? Politicians? Policy communities? Businesses? Property developers and investors? Activists? Stakeholders without local citizenship and property ownership rights?

Box 7.1 Continued

seeking what it means to work with these juxtapositions and encounters between multiple flows that arise in a highly urbanised context. In Milan, besides a strong historic sense of a place-in-itself, with a distinctive social, cultural and economic ambience, there is also a day-to-day recognition of the material experience of all kinds of economic and cultural relations connecting different networks (in fashion and design, the media, etc.) to all kinds of different places, people and cultures across the globe. The most recent strategic episode has paid great attention to the urban in another sense – as a collection of sites with property rights within a political–administrative jurisdiction. In both Milan and Amsterdam, the area of the political jurisdiction of the City Council has also been important in defining the 'space' in mind when the area is discussed. In these conceptions, two types of tension find expression. The first is between a geography of physical manifestations and a geography of flows of social connectivity. The second is between a geography of activities (functions) distributed across and through space and of a geography of jurisdictions and accompanying political rights. In both, an urban area can become an 'actor' in its own right, but in a different way in each. In the first, it is the urban as an ambience, a locale of particular juxtapositions, which 'acts' through the co-evolving relations and connectivities of those juxtaposed. In the second, the area 'acts' through the formal political powers of a municipal administration in a hierarchy of territorial sovereignties.

THE POSITION OF AN URBAN 'REGION'

The second dimension provides a clue about the kind of 'world' in which the 'space' of the urban 'region', as represented in a planning concept, is seen to exist, and how it connects to other 'places'. Plan-making practices in Britain in the 1980s and 1990s, with their primary focus on the justifications for limiting the private right to develop land and property, produced 'development plans' that were positioned in a landscape of hierarchically organised administrative jurisdictions and semi-legal argumentation (Rydin 2003b; Tewdwr-Jones 2002). In Europe generally, their mid-century predecessors were positioned in a landscape of functional relations between living and working, sometimes coloured by a sense of the distinctive historical identity of particular cities and their hinterlands. The focus was on the scale of people's daily movements through the urban fabric. Implicitly, this assumed a broader geography of adjacent city regions spread across a national or European 'surface' (Dühr 2005; Kunzmann 1998).

Since the 1980s, urban and regional policy-makers have paid much more attention to the way an urban 'region' is positioned in relation to other 'regions', both within Europe and internationally, with which they are seen to 'compete' (Amin 2002). This conception of 'cities-in-competition', a central idea in the policy discussion of urban 'economic competitiveness', has strongly influenced the policy discourses on spatial development in Europe that emerged in the 1990s. These sought both to equip Europe as a continent with the assets to compete effectively with other global blocs and to enhance the capacities of regions across Europe to compete in a global landscape of competing regions.[7] Jensen and Richardson (2000, 2004) argue that these concepts reflect a tension between a Europe of 'places' and a Europe of 'flows', following the dichotomy set up by Castells (1996). This suggests a place concept embodying physical proximity contrasted with a flow concept expressing a relational geography of social connectivity. Both geographies pay attention to the interplay between places and flows. They differ in their conceptions of that interplay. In the discourse of 'European spatial planning', the concepts of 'place' and 'flow' are perhaps more closely linked to the historical image of a Europe of city regions, internally integrated and connected to each other through transport routes. Jensen and Richardson (2004) argue that European spatial policy initiatives in the 1990s sought to develop, through investments in large-scale transport networks, an 'integrated' European space, reproducing and reinventing a historical image, to serve the demands of an 'integrated' economy. But this 'monotopic' approach to integration ignored the way such infrastructures 'tunnelled' across the places they passed through, reinforcing the centrality of the core economic regions in the zone around Paris, London, Frankfurt, Milan, Brussels and Amsterdam. The struggle in Amsterdam to connect emergent business nodes such as *Zuidas* to the city of liveable neighbourhoods provides a very clear example of this tension.

The rhetorical emphasis in this 'competitiveness' discourse as realised in urban contexts is predominantly on economic relations and place qualities, on the scales and times of economic value-added chains and of the investment decisions of major corporations, even though the actual political work of a strategy may be to capture resources from regional, national or EU governments, as in the struggle to get national recognition for the *Zuidas* project. The Amsterdam case also shows another kind of positioning, with respect to natural resources and processes, in this case the complex water flows of a delta environment, in a transnational river basin and ocean flow system subject to climatic changes which could have major impacts on the fine-grain of local environments. It is perhaps unsurprising that, of the three cases, it is in Amsterdam (and the Netherlands as a whole) that most effort has been made to represent a region as located in a dynamic landscape of networks and nodes, understood relationally, at least in part. In this conception, the critical quality of urbanity lies in the interrelation between proximate juxtapositions of all kinds and sets of relations that connect one firm, or person, or household, or arts centre, or pressure group, with those in all kinds of other, more-distant places 'distanciated' in physical terms. Accessibility and interconnectivity, understood in terms of both physical proximity and in the flows of relationships-at-a-distance, become critical qualities needing cultivation in the space of a dense, cosmopolitan urban area. In this relational context, 'presence in places' is established not only by physical presence, but also by the way non-present people, things (such as websites, laboratories, design catalogues) and images ('Barcelona!', 'urbanity', 'sprawl', etc.) can be called up, and called on, in the flow of face-to-face contact.[8] This way of thinking about connectivities raises complex questions for attempts to analyse and fix the potential impacts of a policy intervention, programme or project.[9]

BOUNDARIES AND SCALES

> The space of the city is shaped by many forms and levels of boundaries, each with multi-level configurations and meanings. It is a process through which space is constantly divided and re-shaped in new forms. A living city witnesses, throughout its history, constant change in its spatial configurations, shaped by changing boundaries which define and redefine areas to have different functions and meanings . . .
> (Madanipour 2003: 60–61).

What is inside and what is outside an urban area? Does an urban area have an 'edge'? Does it have a centre and a periphery, or is it just a collection of overlapping networks and nodes, each with its own particular pattern and dynamic of attractor points, margins and edges? A focus on jurisdictions allows clear boundaries to be set. The policies over development regulation in the *Comune di Milano* do not apply to neighbouring communes unless these other communes formally adopt them. Many intercommunal strategic initiatives in fact arise where there is a mutual interest in adopting a common regulation in an area (for example, with respect to car-parking charges, or the management of public transport routes). The struggle to develop arenas for intermunicipal co-operation in Amsterdam can be seen as attempts to gain some kind of control over key relationships that affect the ability of the *Gemeente Amsterdam* to deliver its objectives for the municipal area itself. It is searching for a way of lining up the space–time dimensions of key functional relations with an arena within which some kind of formal territorial sovereignty can be exercised (Gualini and Woltjer 2004).

The urban planners of the mid-century in Britain and the Netherlands imagined that 'objective' boundaries to urban areas could be found through clear distinctions between 'town' and 'country'. This essentially cultural ideal then became the basis for strategies which had major impacts on how the built landscape evolved. By the 1960s, planning analysts gave more attention to functional areas – journey-to-work areas, labour-market areas, housing-market areas, service areas for commercial and welfare services. It was assumed then that these were primarily contained within urban areas and integrated with each other. Repeated efforts to reorganise local government in Britain have attempted to achieve this lining up of cultural, functional and administrative dimensions of territorial organisation. The Cambridge Sub-Region case suggests the limitations of these attempts. A similar impetus provided the momentum behind attempts to create metropolitan regions in Italy, France and the Netherlands in the late twentieth century (Lefèvre 1998; Salet *et al.* 2003). There has been much discussion in recent literature on urban and regional governance about attempts to 're-scale' the state, 'downscaling' from the nation state and 'upscaling' from the municipality.[10] But, as Amin (2002, page 386) argues, this re-scaling debate tends to see places as 'sites of geographically proximate links or as territorial units'. It promotes an idea of a hierarchy of places (Marston and Jones III 2005; Nielsen and Simonsen 2003) and suggests that an appropriate 'fit' between administrative jurisdictions and functional dynamics can be found. Instead, Amin argues, territorial jurisdictions are sites of complex relational juxtapositions, folding different scales in, around and in-between each other. Relating jurisdictions and functional dynamics in urban areas are better understood as matters of continual tension, requiring complex negotiations and alliances (Gualini 2004a, c), manifestations of the 'restless search' discussed in Chapter 6.

The dynamics of economic and social development of the late twentieth and early twenty-first century have increasingly undermined the idea that economic, social, political and environmental relations can somehow be combined into a common pattern and momentum focused at any particular institutional site, whether this be the neighbourhood, the city, the 'region' or the nation. The justification for an urban 'region' strategy and the design of any specific interventions that rest on such a conception is likely to encounter all kinds of problems of legitimacy and operationalisation, as experienced in both the Amsterdam and Milan cases. Instead, a relational geography suggests that mobilisation around the place qualities of urban areas develops its force through connecting specific relational dynamics with specific qualities of juxtapositions, while at the same time creating a relational layer of its own which in some way adds to, affects and enriches the mixture of networks and juxtapositions already in existence.

In both Milan and Amsterdam, planning practitioners and the academics advising them were conscious of the need to have a different kind of representation to express the multiple space–time dimensions of a relational geography of juxtapositions and connectivities. The Milan *Documento di Inquadrimento* focuses attention on a critical investment zone, without any suggestion of boundaries. From the late 1980s, the Amsterdam strategic planners searched for a way of expressing a dynamic, social concept of urban networks, flowing across, through and around the space of the urban, linked to their awareness of the significance of the economic and cultural relations of a European-scale commercial centre and an international transport hub and tourist destination. Yet the Dutch experience shows how difficult it is to 'escape' a physicalist tradition. The concept of 'network' can all-too-easily be translated into a physical pattern, and the concept of a 'corridor' can be both a physical shape and a frame for thinking about mobilities. Dutch planners and politicians found it difficult to give up their red and green 'contours', which demarcated both landscape types and policy domains.[11]

FOREGROUND AND BACKGROUND

The fourth question shifts attention from concepts of boundaries to concepts of cores, nodes, central places. The sociologist Anthony Giddens emphasises that what is seen as a 'core' is what is positioned in the front of attention, or at the front of the stage of a performance. He makes a contrast between 'front regions', the visible focus of attention (Giddens 1984), and 'back regions', hidden from view, though present. In an urban context, front regions might be particular spaces, such as city centres, certain neighbourhoods, industrial sites, major redevelopment projects, green spaces. Or they might be activities – daily life, economic complexes, property-market transactions, cultural productions, tourist 'hot-spots'. Back regions might be the places people come from when they arrive at the 'front'. 'On stage', actors tell us about other places and actors they have encountered, bringing the non-present into presence. Or the non-present actors and places may be ignored, made present only by their invisibility to the actors declaiming 'on stage'. Or a strategy may avoid any mention of certain key issues, knowing that these will be ever-present in the minds of their audience. The elite nature of the distinctive imagery of Cambridge and the conception of Milan's '*cuore*' have this invisible yet ever-present quality. For this reason, no strategy or plan should be 'read' only in terms of what is said or written. Foregrounds need always to be situated in relation to their backgrounds.

Giddens links this notion of front and back regions to geographical concepts of core and periphery (see Figure 7.1). In his discussion of front and back 'regions', Giddens suggests that social life has the quality of a performance, carried out in a particular setting on which it is 'staged'. By analogy, a spatial strategy may itself be considered as some kind of dramatic event, a performance, telling a 'story' about places and

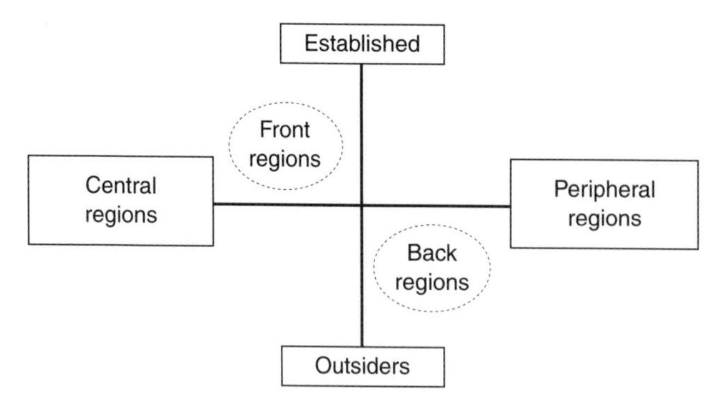

Figure 7.1 Front regions as established, core regions

Source: Adapted from Giddens (1984: 131, Figure 9)

people, in which all kinds of issues and relations are brought into the 'spotlight', emerging from and blending back into the backstage shadows. In the mid-twentieth century, the urban as physical form took centre stage in the strategic representations of all the three cases, reflecting the dominant international planning tradition of the time. Within this, however, different issues were in focus. In Amsterdam, the emphasis was on housing extension on reclaimed lands, with some attention to redevelopment of the urban core. It was not until the 1970s that the city centre itself became a critical 'front region'. In Milan, in contrast, the city centre, as an ambience and as a store of property value, has always been in centre stage, even if sometimes a silent player. The debates over the emphasis in strategic planning initiatives have been over what else would be there as well, a surrounding landscape of neighbourhoods and subcentres in the 1960s and 1970s, or a collage of major development projects in the 1980s and 1990s. In recent years, the neighbourhoods, the old *quartieri*, seem to have faded into the background as far as planning attention is concerned. In Cambridge, the struggles over accommodating growth in relation to the university landscapes of ancient colleges and attractive water meadows have occupied centre stage. Very much in the shadows have been issues about access to housing for those on low incomes.

The metaphor of a 'stage of performance' that lies behind Giddens' concept of front and back regions is helpful in thinking about a 'spatial strategy' as an exercise in story-telling, focusing selectively on particular issues and relations. There is a long tradition of identifying the qualities of policy-making as an exercise in dramaturgy (Hajer 2005; Majone 1987). This metaphor brings into consideration two other properties. First, it shows how non-present issues, objects and actors can be brought into 'presence' through the performance of the drama. Through the performance 'on stage', all kinds of relations, both near and far, and the values and knowledge that flow through them, may be called into presence. Through the drama, an 'urban region' may be 'summoned up' in some form, called into presence (Amin 2004: 34). Second, performances as dramas imply an audience and an interaction between performance and audience. In the theatre, this may be multi-layered. The audience appreciates, situates and criticises the performance – in terms of competence in drama and, in terms of the story, the values and aesthetics of the play. But many plays, and characteristically in classical Greek theatre, include choruses or characters who criticise and comment on the actions and values of key actors in a drama. These commentaries bring into focus the tragedies and comedies of a story, the inherent backcloth of conflict and dilemma against which the main actors make their choices. This hints at a way of thinking about how to maintain in the foreground, in 'presence', people and issues that might otherwise be swept away as 'invisible'. The metaphor of a drama, in which the action and many of the characters may be off-stage, is rich in potential for developing representations of the relational complexity of urban areas. The metaphor also emphasises that developing an account of an urban area with which to infuse a spatial strategy is no smooth process of signing up to a conception. It is filled with 'drama' – struggle, agony, comedy and tragedy – in which different parties 'agonise' over difficult moral and material dilemmas. This suggests that a strategy without a drama may lack persuasive capacity and hence ability to accumulate power.

CONCEPTS, CATEGORIES AND MEASURES

Spatial strategies are full of metaphors and measures that create categories and their constituent boundaries and that generate techniques and their logics. Within the daily flow of planning practices, these are often core 'taken-for-granted', normalised concepts. The language of a strategic performance scripted by members of a planning policy community for each other is typically full of references to these practices. I discuss the knowledge content of such material further in Chapter 8. Such categorisations are used to translate general ideas about the place qualities of urban areas into specific measures that can then be used as principles to guide development investment or land-use regulation. Such ordering devices may come in many forms and often co-exist in a jumble of taken-for-granted categorisations (see Box 7.2). For example, the very visible air and noise pollution generated by industrial activity in urban areas in the nineteenth and

Category	Examples
Land use	Divisions into zoning types in Amsterdam and Milan The English Use Classes Order
Functions/activities	Divisions into topics, 'facets' Divisions into economic, social, environmental and infrastructure issues
Spatial concepts	Concepts of contours, axes, corridors, cores, settlement hierarchies
Environmental relations/ impacts	Measures of noise and air quality Measures of landscape quality Environmental impact assessment analyses and measures
Design qualities	Criteria for major new development projects in structure and local plans in England Design principles in master plans and development briefs
Social groups	Divisions between local and non-local needs Divisions between households Divisions between residents and non-residents Divisions among groups based on age, gender, ethnicity, etc. Social groups as communities in neighbourhoods

Box 7.2 Types of categorisation found in spatial strategies

twentieth centuries encouraged the 'zoning' of land uses to improve conditions in residential areas. Such 'zoning' is often now criticised without reference back to the conditions that led to the practice.

Most formal planning systems work with definitions of land-use categories that give property development rights. Land-use categories typically express conceptions of urban areas as made up of different types of 'activity' on specified plots of land. Classifications developed in the mid-twentieth century tended to draw on the model of an integrated urban region, with an economic base centred in industry and a superstructure of commercial and administrative activities. The organisation of arguments around these categories may then provide a structure for subsequent political and legal argumentation about the legitimacy of a particular intervention, or about its appropriateness. But these categories, and the land and property rights they provide, are continually under pressure as new activities emerge or as new linkages and forms appear, for example 'high-tech' industry, logistics hubs, retail hyperstores and new leisure activities. In Milan, and to an extent in Amsterdam, the response initially was to reduce the range of zones to make them more flexible. In the UK, in contrast, adjustments are continually made to the national Use Classes Order.

Spatial metaphors, such as compact cities, green belts, corridors, centres, gateways, hubs, greenways, deconcentrated concentration, networks, etc., provide principles to justify clear physical boundaries in regulatory policy and the direction of investment to particular locations. The techniques for assessment of environmental costs and benefits may create a language of location, as in the Dutch sequential test for office location (A, B, C) or the British sequential test for housing and retail developments, or the widely-used distinction that emerged in the 1980s between 'brownfield' and 'greenfield' sites. The vocabularies of urban design, sometimes developed into specific design guides, provide all kinds of principles for judging the quality of a proposed new development and how it may 'fit' into its surroundings. Finally, strategies reflect, and these days typically express, an idea of the social groupings into which a society may be organised. These may relate to an idea of the social groupings identified as living in an urban area, or the social groups with which the strategy-makers regularly interact, or to a normative idea about the groups with whom they *should* interact. The importance of these concepts of social groups lies in the way they affect who gets invited to join a strategy-formation effort and how the 'social impacts' of policies and projects are considered. For example, are only physically close neighbours affected by the way Amsterdam's *Zuidas* project develops? And how far should Milan's *Piano di Servizi* focus on the neighbourhood as the locus of the relation between people and service use?

Such categories carry strategic concepts into the fine grain of ongoing practices, often interpreting them in ways that lose the initial strategic meaning, or merely convert a new idea into a vocabulary of well-established categories. All three cases show how certain spatial metaphors become embedded in the consciousness of a planning tradition, yet continually resurface when ways of expressing new strategic ideas are developed. The implication is that any strategy-making effort that seeks to change the way an urban area is conceived needs to pay close attention to these techniques and the practices that build up around their use.

TIME AND SPACE

Albrechts (2005) argues that the focus of strategic spatial planning is on the future. But in what sense is the 'time' of the future being called into presence? Is it as a dream-like imaginary? A prediction? A potential in the present? A spatial strategy focuses on some aspects of place qualities and on the significance of spatial conjunctions and disjunctions as social and natural relations flow in, around and through urban areas. But, as many geographers argue, every relation exists in time as well as space. We are, ontologically, flows as well as entities. Materiality and identities are 'in formation', 'on the way', 'on the move', as well as 'in place' and 'placed'.[12] But the concept of 'time' is as full of difficulty as that of space.

Mid-twentieth-century planners were less bothered by conceptual complications with respect to time. In the modernist conception, time was a simple linear line, connecting past, present and future, on a trajectory of human development. As Massey (2005) argues, spatial arrangements were imagined as patterns occurring in 'slices of time', succeeding each other in a movement of past, present and future. The task of planning thus could be imagined as the production of the spatial arrangement of a future 'slice' of time.[13] The spatial strategies in Amsterdam, Milan and the Cambridge Sub-Region in the mid-twentieth century present the future less as in continual evolution and more as a kind of stasis, a plateau of beneficent development, to be arrived at and then maintained. Even very recent strategic plans have time horizons, although these are often difficult to fix. The Amsterdam *Structuurplan 2003* was supposed to have a time horizon of 2030 in terms of patterns of transport infrastructure, but the difficulties of co-alignment meant that the agreed future 'slice' was pulled back to 2010.

Yet a close reading of the work of even the mid-twentieth-century planners illustrates that 'time' is not a simple succession of periods, each with its own spatial pattern. The Cambridge strategies show a concern for the time of daily, weekly and yearly life, as people move around between home, work, services and leisure opportunities. The plans also speak to political and administrative time, the evolving process of legitimising strategies, the time to organise to make key investments or agree a key regulatory intervention. By the late 1980s, the issue of time had become more complex and contested. The discourse of 'economic competitiveness' created an urgency. Development opportunities need to be created quickly, to capture moments in the flow of decision-making of major economic actors. Urban 'assets' needed to be polished up or produced as soon as possible, to avoid missing the economic boat. Sometimes, too, politicians wanted to leave their mark during their period in office, and urged rapid conversion of idea to realisation, whether in the housing projects of the 1960s or the urban redevelopment projects of the 1990s. But this urgency comes up against other 'times' promoted in political debate. The language of environmental sustainability and sustainable development emphasises the diverse times of environmental systems and human intergenerational time. The language of 'social cohesion' emphasises the time it takes to build relations between and within often disparate and fractured communities of association and of neighbourhood life. Real-estate investors, often portrayed as in a hurry to invest to capture opportunity, may also be found taking a transgenerational view of their properties as long-term investments in a complex portfolio built up over decades or, as in Cambridge (the university) and Milan (city-centre property owners), over centuries. Meanwhile actors involved in governance processes are themselves experiencing the different times of their own daily life encounters with the world as they flow through it. Rather than points and areas in a spatial pattern of a particular slice of time, the 'places' that emerge into recognition are conjunctures of multiple space–times, Massey's 'simultaneity of multiple trajectories'.

If there are so many different 'times' that flow through urban areas, then it is important for analysts and strategy-makers to consider which times they are emphasising and how a strategy is to relate to this complex 'life in place in movement' in multiple times and rhythms in urban areas. One answer, which underpins the strategic interventions of the Milan Commune in the late 1990s, is to be highly selective in focus, leaving most relations to evolve in, around and beyond the specific intervention and its impacts. The argument of the *Documento di Inquadrimento* deliberately steps aside from any attempt to express too much in policy terms and in specific projects. It inserts small interventions in the hope that these will set off future evolutions in various timescales. Yet the '*t-rovesciato*' concept expressed in the *Documento* was being overtaken by the speed of property-market responses to switches of investor preferences between equities and property at the very time that it was being promoted, resulting in new development axes emerging, just as had happened a decade earlier in Amsterdam. In Amsterdam and the Cambridge Sub-Region, the critical time of recent spatial strategies has been political-administrative time, the time over which agreements between government agencies over infrastructure can be expected to hold. But these selective times of spatial strategy-makers are then inserted into the many other space–times flowing around in urban areas. Strategies achieve their effects, both material and imaginative, through impacts on the different space–time horizons of the different relational layers they touch.

WHOSE STRATEGY AND WHOSE CONCEPTS?

Weaving through the above dimensions and vocabularies is the issue of the social consequences and social justice of adopting and positioning particular conceptions and languages. Who is centre stage in a spatial strategy for an urban area and whose space–time is privileged? One of the most attractive features of mid-twentieth-century planning strategies is that they express a real effort to position residents and their material needs and experiences at the core of conceptual attention. How did they live in and move around in a city? How could conditions for them be improved? The conception of a 'resident' was over-generalised but was intended to be inclusive at the time. The contrast with the strategic concern for providing space for business investment and university interests in Cambridge in the 1980s and 1990s could not be stronger. In Cambridge in the 1990s, the privileged viewpoint was that of the economic interests of some sections of a university/business elite, moderated by a concern for environmental sustainability. In many spatial strategies from the 1980s onwards, the dominant discourse of competitiveness in a global economy crowds out the multiplicity of social groups with a stake in what happens in an urban area. The recent Milan strategic planning exercise has been criticised on similar grounds, as privileging the real-estate industry, because so much attention is given to its activities and dynamics (Salzano 2002). But the response of the planning team is that the privileged focus is not the real estate-industry as such, but the position of the 'public interest' in ensuring that real-estate investors respect requirements to contribute to public benefits in return for greater freedom to invest (Mazza 2002).

In Amsterdam for many years, the privileged viewpoint was that of the planning-policy community, and in particular, the politicians and public officials who sought to obtain and fix 'in place' public-sector investment projects. But, over the years, a 'benevolent' concern for providing housing, jobs and a well-designed local living environment for residents has evolved into a much more complex conception. This expresses an appreciation of the multiple social groups within the city, their complex space–time relations within and beyond the space of the Amsterdam municipality, and how this multiplicity contributes to the particular socio-cultural dynamic of the city as an open and cosmopolitan 'special' place. In struggling to give operational meaning to concepts such as 'openness', 'urbanity' and 'accessibility', the city's planning team are searching for ways to give expression to a relational perspective of urban dynamics produced through diverse and dynamic relational networks, transecting and interweaving with each other. They seek, in effect, a strategic approach that will insert an understanding of the multiplicity of the relational networks important for urban life into any particular intervention focus, whether it be the discussion over the form and design of major projects such as the *Zuidas*, or the regeneration of neighbourhoods, or investments in promoting cultural activities in the city. This, in turn enriches the thinking about the social impacts of particular interventions and enlarges the conception of the voices that should be offered space for expression in strategy formation and development. In this way, there is less danger of the 'back regions' of strategy formation being cut off from the front stage. But their critics, equally committed to such a conception of inclusive urbanity, are concerned that the city's planners are proceeding in a too-cerebral way. In their actual interactions, the front stage is still full of members of the planning policy community, with citizens and many kinds of business not even backstage, but in the audience.

A RELATIONAL PERSPECTIVE: IMAGINING THE MANY, ON THE MOVE

TWO GEOGRAPHIES

> The multi-layered city is both a social and a spatial 'coming together' of difference and diversity, chaos and order, fascination and intrigue – a sensual delight, at the same time challenging notions of tolerance and feelings of belonging. The multi-layered city is imagined and real, a creation of our own subjective experiences of the urban landscape as well as a response to the personal – our gender, age, ethnicity, class, physical ability, religious beliefs and sexual orientation (Thompson 2000: 233).

In Chapter 6, I argued that strategies were 'social constructions'. In this chapter, I have examined the dimensions of these constructions. So far, I have emphasised that strategies for urban areas draw on various conceptions of the nature, dimensions and dynamics of the urban. These conceptions, too, are 'social constructions', interrelating with, but not the same as, the experience of the materiality of the flow of living or the 'imaginaries' through which the world around us is valued. These 'conceived spaces' of strategy-makers, to use Lefebvre's term, help to construct the storylines and metaphors which then frame the policy discourses of strategies and feed into the way these are translated across a governance landscape through time. In doing so, strategies not only shape conceptions in institutional sites beyond those in which they were constructed. They also have effects on material realities, as they are used to generate investment projects, to justify resource allocations and regulatory decisions. The discussion in the previous section indicates that the spatial dimension of strategic storylines for urban 'regions' can contain all kinds of assumptions about the relations between people and place, activities and their locational dimensions. Varied 'bits of geography' may turn up in a strategy, often with little awareness of, or attention to, their coherence and consistency, or their compatibility with the apparent intention of a strategy.

In presenting these dimensions, I have highlighted differences between a physicalist geography and a relational one. This raises the question of whether one geographical perspective is better than another. A relational geography is not necessarily in itself the carrier of better values and more effective interventions. Its persuasive power today lies in its resonance with contemporary experiences of multiple mobilities and identities, as the planners in Amsterdam understood. It speaks to a post-'modern' recognition of the dynamic complexity and many contingencies of urban conditions.

The critical shift between the two geographies, emphasised in the previous section, is from one that focuses on physical proximity as its main organising principle to a 'geography of complexity' (Dematteis 1994), in which analysts and strategy-makers are themselves part of the relational reality they are seeking to express (see Table 7.2). Yet, as Dematteis argues, all these geographies are still to be found in scientific analyses, just as they live on in planning concepts. Some analysts of the 'post-modern' condition

Table 7.2 Two geographies

	Geographies of physical proximity (emphasis on physical proximity, sometimes called Euclidean)	*Geographies of connectivity (emphasis on social proximity, connections both physically adjacent and at a distance)*
External analyst (the outsider looking in)	A 'geography' of spatial patterns and physical objects	A geography of overlapping socio-spatial geometries
Internal participant (an actor in a situation)	A geography of local cultures determined by physical morphologies	A geography of complex overlapping connectivities with emergent properties

Note
Adapted from Dematteis 1994: 205

have argued that, in the world of flows and networks, 'places' hardly matter. Others emphasise the 'abstract' dynamics of the 'global' floating above specific places and times, with the 'local' romanticised as the sites of daily life or of radical resistance.[14] In a geography of relational complexity, simple dualisms, such as those between global and local, or place and flow, are replaced by an emphasis on the evolving co-constitution and emergence of potentialities.[15] I now develop the contrast between the two geographies in more detail. I then explore the implications of imagining the analyst and strategy-maker as being 'in-the-world', rather than examining it from outside.

GEOGRAPHIES OF PHYSICAL PROXIMITY

In the mid-twentieth century, the emerging planning community, at that time quite international in its intellectual inspirations, drew on a geography that mixed an appreciation of towns and regions as reflecting some kind of place-related cultural ambience, a 'pre-modern geography' (Dematteis 1994), with a more materialist focus emphasising 'city regions' as integrated local economies, centred on certain productive activities that formed their economic base. Historically, these conceptions preceded the recognition of the impact that transport and communications technology would have on the relation between people separated by physical distance. It also ignored the way people in one place may be linked to other places, both imagined and experienced. It was assumed that people and firms were largely 'rooted' or 'gripped' in place (Dematteis 1994). For some, place was a cultural idea, embodied in history and in the particularity of transgenerational engagement with the local environment. But the dominant conception as translated into planning studies was of space as a surface or container, differentiated by natural features, across which urban settlements connected by transport routes evolved in a hierarchical relationship. 'Places' were constituted by physical proximity. Physical patterns were connected by flows of people and goods along water, rail and road routes and driven by the principle that accessibility varied directly with distance. The patterns were laid out across a physical surface, which 'contained' opportunities (for example, river crossing points) and constraints (for example, coastlines, areas liable to flood, mountains and gorges). Beyond physical determinants, the patterning was driven primarily by economic opportunity, which created a 'productive base', upon which a commercial and administrative service structure would grow. The object of spatial planning efforts was to create and maintain a 'balance' between activities, similar to concepts of equilibrium in economics. For example, the growth of Cambridge was to be held in check by limiting the location of industries there, while the growth of Amsterdam was to be promoted by expanding the industrial and port developments along the port and canal areas.

Overall, it was assumed that these patterning processes were driven by law-like principles. These provided the basis for categorisations into activities (primary, secondary, tertiary industry, services, transport), which became the basis for land-use classifications. For example, planners in Britain in the 1950s were given principles for calculating how much industry was needed to support a new town of a given population size, from which it was then possible to calculate space requirements for different uses – industry, residences and services (Keeble 1952). In the 1960s, similar calculations were being made to determine the space requirements for commercial centres.[16]

This conception, much simplified in the above summary, held great attractions for mid-twentieth-century planners because it seemed to provide a robust scientific basis for drawing up plans to guide future urban growth and development. The 'ideal' city had an appropriate 'balance' of activities and transport links, and adequate spaces for them. 'Balance' implied some kind of equilibrium between supply and demand, the need for work opportunities and work available, the need for housing and support for daily life and availability of housing and local facilities. The ideal city region had a balance of smaller and larger settlements. It had a coherent socio-spatial organisation, with a morphology of places linked together by infrastructures (Graham and Marvin 2001). By the end of the 1960s, the idea had developed that nations should have a balanced system of cities spread across national space, with analysts in the developing nations puzzling about how to build up a settlement structure with more medium-sized towns to 'balance' the emerging hyper-concentration of urbanisation and development in one or two 'primate' cities (Bourne 1975). In this way, a conception of urban area geography built from a pre-industrial European tradition, and, drawing on a deeper European conception of a hierarchically ordered universe, came to 'colonise' development policy in other parts of the world.

By the 1970s, the limits of this conception were becoming obvious in Europe itself. On the one hand, infrastructure investments of all kinds reduced both the costs of distance and provided a more even surface of accessibility to services. People and firms began the steady spread of metropolitan decentralisation which not only increased the space–time dimensions of daily life, but greatly enlarged the spatial reach of weekly and yearly patterns. The material reality of routine engagement with movement through multiple places challenged the conception of integrated settlement systems spread across a city region surface. City centres began to seem less central, as 'edge cities' developed at transport junctions, as in Amsterdam. Industrial areas contained firms with a whole variety of space–time linkages, often little connected to either each other or the rest of an urban area. Transport interchanges such as an airport hub like Schiphol had flows of passengers and goods with very little direct connection to being near Amsterdam. As with any 'paradigm change' in science (Kuhn 1970), the old 'Euclidean' geography did not seem capable of capturing the materiality of what was going on:

> Within this contemporary urban world . . . the modern infrastructural ideal founders. Its essentialist notions of Euclidean space and Newtonian time, of functional planning towards unitary urban order, of single networks mediating some 'coherent' city, are paralysed. It is largely incapable of dealing with the decentred, fragmented and discontinuous worlds of multiple space–times, of multiple connections and disconnection, of superimposed, cyborgian filaments, within the contemporary urban world (Graham and Marvin 2001: 215).

GEOGRAPHIES OF RELATIONAL COMPLEXITY

Whereas the 'old' geography emphasised spatial patterns and general principles of spatial distribution and a balanced 'equilibrium' between the different elements of human activity spread across the surface of a region, the 'new' geography broke with equilibrium models. It has drawn more on Marxist analyses of conflict and struggle between forces within capitalist economic systems. It was inspired by a recognition of the multiplicity, diversity and dynamic mobility of the relations which at any time may weave through, over and under the space of an area, and the variable 'spaces' in which such relations may be integrated. It has more recently been infused with a phenomenological and cultural recognition of the socially constructed nature of perception. What become recognised as urban 'regions', in this 'new' geography, are not objectively identifiable, integrated economic and social systems, but spaces of complex 'layering' of multiple social relations, each with their own space–time dynamics and scalar reach. This focuses the interest for analysts and policy-makers concerned about conditions in particular places on the way these relations pass through a physical area, how and how far they interact as they do so, and how, in these interactions, they both produce qualities of 'place' and 'connectivity' and use qualities and connectivities that already exist:[17]

> the materiality of everyday life is constituted through a very large number of spaces – discursive, emotional, affiliational, physical, natural, organisational, technological, and institutional; . . . these spaces are also recursive spaces . . . carriers of organisation, stability, continuity and change; . . . the geography of these spaces is not reducible to . . . planar (single or multi) or distance-based considerations (Amin 2002: 289).

This conception recasts the relation between flows and places. Instead of a physical patterning of activities rooted in particular pieces of the Earth's surface, with connectivities between activities arising from physical proximity – what is near being more significant than what is further away – a relational geography focuses on relational dynamics that may stretch in all kinds of ways. The metaphor shifts from a 'map' or 'design', to a multiplicity of more-or-less loosely coupled webs, with nodes, links and loose threads. Callon and Law (2004) express this through the image of someone listening to a Walkman or a using a mobile phone as they walk through the city. Massey (2000) uses the experience of travelling from home in London to Milton Keynes where she works, her thoughts sometimes travelling across to all kinds of places in her conversations with the car driver, who happened to be the cultural analyst Stuart Hall, only occasionally noticing the landscape they pass through. Madanipour (2003) illustrates people walking through a public space in their own personal 'bubbles' that link them to ideas, people and places not physically present. Rather than a clear, unified spatial order, with definite hierarchically ordered cores and boundaries, these examples show complex conjunctions, with webs of relations in continual formation, driven by diverse specific driving forces.

In this relational conception, places are materially experienced as significant conjunctions, a mix of concrete objects and material flows, of sensual impressions and emotive memories, to which particular meanings and feelings become attached. Places appear in collective encounters through the way they are recognised in the course of conjunctions and juxtapositions. Places are sites of encounter and attachment. From being just spaces of physical and social co-existence and encounter in the flow of relational interaction, they get called into 'presence' through the accidents of adjacency, through the mapping practices that develop in particular relational systems, through the repeated patterning of daily, weekly or yearly routines, through the way these build up through time to create a 'patina' of history and association that creates a sensibility of 'belonging' to a place. Thus urban regions, cities, metropolitan areas, do not exist objectively, though they arise from very material experiences. As Amin (2004: 34) puts it, cities and regions as spatial formations

> must be summoned up as temporary placements of ever moving material and immanent geographies, as 'hauntings' of things that have moved on but left their mark[18] . . . as situated moments in distanciated networks, as contoured products of the networks that cross a given place.

Once recognised, 'summoned up', places can become 'actors' in their own right, through the recognitions that they call up and the way these recognitions are used, in the same way that machines and techniques have the power to 'act'.[19] Amsterdam, Milan and Cambridge exist as 'places' with an accretion of meanings in the context of many different relational webs, as do Barcelona, Berlin, Birmingham and Budapest, etc. Similarly, places within urban areas can carry a strong recognition – the city-centre neighbourhoods in Amsterdam, the '*cuore*' in Milan, the Cambridge water meadows. Much of the focus on 'projects' in strategic planning activity in the late twentieth century has been about creating 'places' in areas abandoned by industries, buttressing schemes with physical designs and marketing rhetorics to create a different 'sense of place' for a new 'piece of city'.

A relational geography generates metaphors of flow and network more than patterns of settlement and finds expression in icons and sketches, more than in maps and measures. It leads towards a recognition of the complexity of the social processes through which life in movement is experienced. In this view, places and flows are in continual formative encounter through which both evolve. Such evolutions do not tend to a particular 'balance' or equilibrium. This unsettles attempts at prediction and projection from the present to the future. Rather, places and flows develop through complex contingencies that generate potentialities and trajectories which may only be identified as they emerge. This raises the issue of who does the 'summoning up', the recognising of possibilities and potentialities, and where they are positioned in the complexity of relational flows.

EMERGENT COMPLEXITIES AND THE POSITION OF SPATIAL STRATEGY-MAKERS

I now return to the positions of the analyst and strategy-maker in Table 7.2. In geographies of physical proximity, the analyst sits outside the world, looking down upon it, seeking to create a systematic representation of some kind, in Lefebvre's terms. This becomes the 'scopic' view of the 'modernist' planner (Perry 1995), manipulating a landscape of 'things', into a pre-conceived desirable order. Even when the focus shifts to the analysis of relational interactions, there is a tendency for geographers or planning analysts to position themselves outside these interactions, 'looking in' and extracting from this experience an abstract language of description – architectures, geometries, webs, etc. (Simonsen 2004). Yet once we (as analysts and planners) get to see a world of multiple, mobile relations, it is difficult to avoid noticing how we too are situated within the world we are observing and commenting on. This idea gets expanded through the work of cultural anthropologists and sociologists who observe the social worlds and practices of academic production and scientific analysis.[20] Those who observe relational interactions are themselves constituted by such interactions, in settings that provide locales, social worlds, 'lived in' places, a kind of 'place of dwelling' or 'habitus' in which 'communities of practice' develop.[21] Once institutionalised into policy communities, government departments, academic disciplines and departments, these social worlds accumulate power and become hard to shift.

It is all too easy to use notions such as 'habitus' to reinforce the romantic notion of a cohesive community, attached to its particular place of living. A city planner or local politician might, in this conception, see their role as to summon up the 'essence' of the 'community-in-place' and represent it defensively to an outside world. This helps to construct the commonly used dualistic opposition between the 'local community', in its daily life existence, continually threatened by external forces – of global economic forces or of the state. Such conceptions of community are also continually resurrected in the urban planning literature, for example in the 'new urbanist' design ideas developed in the US (Katz 1994), in the ideas for an 'urban renaissance' in the UK (Madanipour 2003; Urban Task Force 1999), and in ideas about 'community involvement' in planning activity.

A relational geography of complexity challenges this conception in three ways. First, people's attachment to a particular place is only one of their many attachments and the relational attachments people have with a place or places are not necessarily coterminous with those of their physically proximate neighbours. Thus people in one place may have all kinds of relations to those in other places, which, as (Massey 2004a) argues, carry responsibilities to people and places elsewhere. Second, these attachments, although shaped by the past, are in continual evolution as people engage in the production of their life trajectories and experiences, in interaction with whatever they encounter. Third, and particularly in urban agglomerations, the very nodality of places is constituted by the complex mix of juxtapositions and connectivities of very many relational dynamics.

Analysts and strategy-makers are inside this mixture. They imagine and operate both in their own communities of practice and in a dynamic landscape of a multiplicity of overlapping relations within and between complex mixtures of physically proximate and distanciated connectivities. From these social worlds, their interpretations, selective attention and ordering devices filter out, sometimes pushed by a deliberate intention to affect others, sometimes by the capacity for ideas and techniques to flow from one institutional site to another, and interlink with other relational dynamics. Understood as an active participant 'in-the-world', those involved in planning work, such as the formation and use of a spatial strategy for an urban area, are deliberately seeking to create effects within the complex relational flows within which they themselves are positioned. They are not outside the 'realities' they are 'planning'. They are positioned within, and continually being shaped by, the ongoing flow of material and ontological 'realities'.

This idea is partly captured by the notion that analysts and planners, politicians, lobby groups, residents, developers, etc., like all of us, are somehow 'embedded' in our particular geographies and histories, in our social worlds, our various 'habituses', our communities of practice. In these various social worlds, we interact with humans and other species, with the animate and the non-animate, continually adjusting and developing our identities and our capacities as we do so. In this way, we are continually 'in formation'. Our 'environments' are thus not 'containers' in which we have existence or 'umbrellas' under which we shelter, but are drawn into our formation processes as we engage with them. To express this, recent interpreters of human–environment relations (see Ingold 2000) and of the ontological geography of a relational world (see Thrift 1996, 2000) seek to move beyond the notion of a place-situated 'habitus' or 'locale of being', to emphasise that life is lived 'on the move', through continual engagement with the material and social world, through which new ways of tackling challenges are developed as well as old ways reasserted to face new experiences. Thus 'living' is experienced in place in movement, a process of encounter and becoming, in an environment of complex interactions between multiple relational systems which are themselves in continuous movement. This is the 'geography' of complexity, a geography that encompasses the relational dynamics of the many, on the move. It is this geography that some of the city planners and academics in Amsterdam in the 2000s were striving to articulate and that some of their contemporaries in Milan had in the back of their minds.

Implications for the Enterprise of Spatial Strategy-Making for Urban Regions

Some policy-makers and analysts, concerned to maintain a focus on distributive justice and environmental well-being in the face of strong pressures to prioritise economic competitiveness, are suspicious of a relational geography. A physicalist geography provides a clear way to identify whether space has been allocated for the provision of basic services and to assess the impacts of a potentially harmful development on its surrounding area. But, as urban planners are only-too-well aware, allocating sites does not ensure the provision of facilities and their use as intended. And those analysing the impacts of human activity on the natural environment are very conscious of the complexity of the connection between a specific development and the space–time reach of its impacts. If the ambition is to intervene strategically in evolving urban dynamics to keep considerations of distributive justice and environmental well-being 'in play' alongside the strong contemporary thrust to promote particular forms of economic vitality, and if the objective is to improve the conditions of urban life for the many and not just the few, then the insights of a relational geography provide a more sophisticated understanding of socio-spatial dynamics than a physicalist geography.

What kind of strategic enterprise does spatial planning then become in a relational geography of complexity? If urban areas are to be understood in the perspective of a relational geography, then the work of strategy formation becomes an effort to create a nodal force in the ongoing flow of relational complexity. This force is drawn forward through the effort of 'summoning up' conceptions of an urban area, in ways that selectively lock together some transecting relations, opening up connectivities to encourage new synergies to emerge, creating a strategy with persuasive and seductive power, which can become itself an 'actor' in the ongoing flow of relational dynamics and have effects on materialities and identities. This implies that planning efforts have to abandon the idea that there exist some pre-given spatial ordering principles that can provide a legitimate basis for interventions in the emergent realities of urban areas. It is the social process of the production of such principles that gives them legitimacy, a politics likely to include resistances (Natter and Jones 1997) and challenges (Massey 2004a), to express agonistic dramas (Amin and Thrift 2002; Hillier 2002). There are all kinds of spatial ordering principles that are manifest in the ongoing flow of particular systems of relational interactions in an urban region. These interact, conflict, dominate, generate creative synergy in all kinds of contingent ways, as they evolve. The result may destroy past socio-spatial patterns and orders, and may generate new ones, but the complexity of the multiple relations and their inventive dynamics is too great to allow prediction, except in the most stable of situations. The future is emergent, in a process of continual invention, not pre-designed.

The focus of attention of a strategy formation that focuses on the qualities of an urban area as such will still derive its legitimacy as an enterprise from its focus of attention to the qualities of places, as experienced as recognised juxtapositions in the flow of relational life in which the past and the future intermingle. There are, however, many dimensions to this experiencing, a range that is not readily captured in the tools of abstract analysis from an 'external observer' position. This means that strategic planning efforts need to find ways to link conceptions of the complex evolutions of urban dynamics to the experiential knowledge of people situated in many different relational positions and to the creative work of expressing that experience in cultural conceptions. It cannot be assumed that the 'places' of an urban area and the existence of an urban area itself are objectively 'knowable'. They are created through processes of 'recognition' – in the flow of daily life, in the labelling of imaginative production and assertion, and in analytical conceptions. Because there are many relational interactions, because processes of recognition are in continual formation, there are many experiences, imaginations and conceptions around, many 'cities' to be summoned up.

Strategic actors, seeking to shape the material and imaginative realities of urban life in some way, are in the midst of this evolving, relational multiplicity. Those who can read the early signs of emergent potentialities are likely to be 'streetwise' on many of the 'multiple trajectories' through whose encounters the future is being made. Their power derives from a capacity to 'read' emergent potentialities and to create arenas where multiple 'readings' encounter each other. The power of spatial strategies for urban areas, once articulated, arises from the way they develop new connectivities, give attention to emergent 'places' and call up new meanings of place qualities. In this way, spatial strategies and the processes of their articulation create an additional relational layer to add to the evolving mix. As such, they possess a particular dynamic of nodal force and extending threads, weaving into and across other relational dynamics, creating, if able to accumulate sufficient power, a kind of rhizomic force to influence how other actors-in-relations think and act.[22] They generate intellectual and relational resources that flow into and potentially enrich the 'public realm' of discussion about urban conditions. Such strategy-making efforts link conceptual power, Lefebvre's 'conceived space', to material resources and regulatory power. They may also generate or reassert meanings for places-as-juxtapositions that add to the store of imaginative resources available for the formation of identities by those attached in some way to a place. Through this conceptual power, 'seeing like a state', the geographies mobilised can have significant, if often unexpected, effects on emergent potentialities.

A relational geography encourages those producing spatial strategies for urban regions to focus attention on what, in a particular urban area and according to particular values, are critical juxtapositions and connectivities, and where these are located in a governance landscape. This implies a very focused selectivity. The tradition of holistic, comprehensively integrated strategies and plans for urban areas becomes, in this perspective, both intellectually impossible, as too hard and unpredictable to grasp, and politically dangerous, as likely to fix too much too narrowly (Parr 2005). But this does not dispense with the requirement for richness in breadth and in recognising and understanding the range and diversity of the relational mix to be found transecting an urban area. It is not the selective focus of the Cambridge Sub-Region strategy as such that is open to criticism, but the narrow imaginative perspective on urban dynamics through which that selection has been pursued. A relational geography can be a powerful management tool. It can be used to focus on a narrow set of relations, for example those that position a city in international economic relations, excluding from attention the many other relations that co-exist in an area with these specific relational webs.[23]

But a relational geography is also a valuable resource for demanding attention for wider, richer and more inclusive perspectives on urban dynamics. It has more potential to reveal the multiplicity of webs of relations, the way they intersect and the consequences of this for the distribution of access to opportunities and quality-of-life experience in a city. It encourages a broad imaginative perception of urban dynamics and the qualities of 'citiness'. It provides an intellectual structuring tool for those seeking to promote openness to new linkages and potentialities in urban experiences, to encourage mixity and 'hybridity' in the city. It helps in recognising the creative synergy of opportunities for encounter and conflict over meanings and access to material resources, as well as the need for some zones of comfort and safety, for traditions and memories, in the dynamic unfolding of the future. The significance of a relational geography lies not only in choices about what qualities of place and what connectivities become the subject of selective attention, but also in the way that the 'impacts' of interventions are imagined and calculated. Thus a relational geography offers a perspective on the spatiality of urban areas that is likely to increase the effectiveness of any governance effort in urban management. It also has more potential to make visible the way a strategy may affect the situated, lived experiences of the justice, environmental implications and economic consequences of particular strategic interventions in complex urban dynamics than a simple geography of physical proximity.

A relational geography emphasises dynamics and fluidities. But, in Chapter 6, I argued that, in some situations, there could be so much fluidity and instability that a strategy to promote better urban daily life conditions while safeguarding environmental well-being and economic vitality might seek to stabilise key parameters. In other situations, the emphasis might be on unsettling long-established rigidities (see Figure 6.2). The Holford strategy for Cambridge can be seen as a classic example of a strategy to hold on to a 'pre-modern' past. But, in the end, this sought through physical fixity to contain social and economic dynamics that were whirling ever wider in their connectivities. Its imagination was based on a geography that assumed a socio-spatial cohesion and equilibrium that could be achieved through the integration of key connectivities within the 'container space' of an urban 'region'. Such a geography, as still asserted in romanticised conceptions of place-based 'community', is likely, in a context of relational complexity, to reduce inventiveness, the richness of diversity and openness to new opportunities. As in the classic 'NIMBY' response to development threats, it promotes a univocal defensive resistance, rather than a multivocal assertion of mixity. In Figure 7.2, I link this axis of stability and change to a relational axis of proximity and connectivity to create a matrix through which to consider critical choices to be made about the focus of an urban 'region' strategy. The axis of 'energy' or force suggests that, in some situations, strategic effort should concentrate on consolidation, in others on opening up new opportunities. The axis of relational flow suggests that, in some situations, strategic effort should focus on enhancing qualities of proximity, creating new nodal opportunities while enhancing the qualities of existing and emergent ones, leaving the connectivities to evolve. In others, the effort may be better spent on focusing on the quality of connectivity in the urban context, for example, Amsterdam's 'accessibility' agenda, leaving place qualities to evolve.

In presenting a contrast between two geographical traditions through which to grasp the spatiality of urban dynamics and in encouraging spatial strategy-makers to think carefully about the geographies they are mobilising in various parts of their work, I have also underlined that any geography mobilised in a governance context carries power. It has the potential to change how we live in and experience urban life. The use of spatial concepts and metaphors in strategy-making is thus no innocent gloss on a bundle of policy issues that spatial strategy-makers feel is important. It is a politically potent and highly charged way of representing complex issues about the qualities and

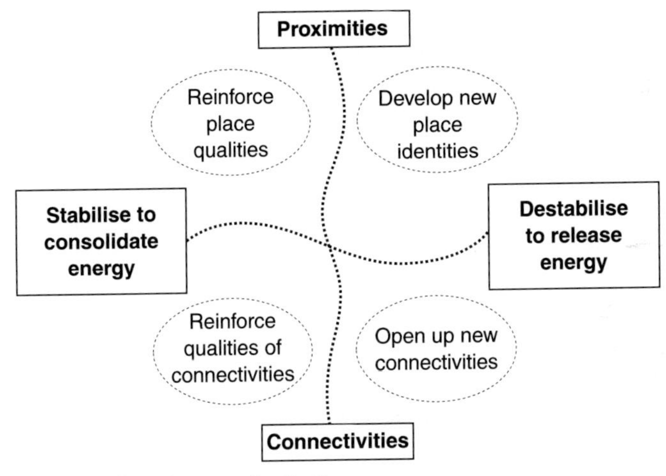

Figure 7.2 Focusing strategic attention

connectivities of places. Spatial concepts and metaphors have considerable 'travelling' and 'fixing' power. They are often appealingly seductive. For this reason, they are politically tricky. When deployed during strategy-formation processes, they may on the one hand mask the actual play of interest politics that underpins what the concepts are applied to. On the other hand, they may make it so clear who and what will be affected by a proposed change that they generate intense conflict.

Selecting critical juxtapositions and connectivities, and imagining effects through time thus become critical intellectual and ethical challenges for those involved in strategy-formation efforts. In this geography of relational complexity, strategic interventions are not about 'creating a future'. Rather, they involve throwing into the flow of relational interactions some ingredients that may have the effect of sustaining some relations, shifting others and generating new potentialities. They are 'risky bets' about what may make a difference. They are intentional contributions to 'future-forming', rather than deliberate attempts to impose a 'map' on the future. In the next chapter, I probe further into the intellectual and ethical challenges of such an approach by examining the processes of invention, discovery and knowledge accumulation deployed in strategy formation.

NOTES

1 See Amin and Thrift 2002; Lefebvre 1991; Massey 1984, 1994; Thrift 1996.
2 See Beauregard 1995; Fischler 1995; Healey 2002; Liggett and Perry 1995.
3 See Dematteis 1994; Gregory 1994; Massey 1984, 2005.
4 See Faludi and Van der Valk 1994; van Duinen 2004; Zonneveld 2005a.
5 See Boyer 1983; Gold 1997; Hall 1988.
6 See Dühr 2005; Friedmann 1993; Graham and Healey 1999. The term 'Euclidean' refers to the geometry of the ancient Greek mathematician, Euclid. The term 'cartesian' refers to the work of the seventeenth-century philosopher, Descartes. Sociologists have long been critical of such a perspective.
7 See Faludi 2002; Jensen and Richardson 2004; Richardson 2006; Zonneveld 2000, 2005b.
8 See Amin and Thrift 2002; Graham and Marvin 2001; Thrift 1996.
9 This is illustrated well in De Roo 2003, who discusses the way the Dutch approach to setting standards in environmental policy embodied a geography that emphasised proximate impacts over more distant ones.
10 See Brenner 1999, 2000; MacLeod 1999; Macleod and Goodwin 1999. These debates are strongly grounded in a regulation-theory approach, and, in the case of Brenner, in an analysis of German experiences.
11 See Dühr 2005; van Duinen 2004; Zonneveld 2005a.
12 See Amin 2002; Amin and Thrift 2002; Amin *et al.* 2000; Ingold 2005; Massey 2000; Thrift 1996.
13 This conception of development in terms of a sequence of patterns, comparative statics, is well-expressed in Foley 1964.
14 See the critical discussion in Amin 2002; Massey 2004a, b.

15 Amin 2002; Amin and Thrift 2002; Hillier 2007.

16 See, for example, the retail study undertaken for the Cambridge area in the early 1970s (DoE 1974) (see Chapter 5, note 19).

17 I draw here on Allen *et al.* 1998; Amin 2002; Castells 1996; Harvey 1985; Massey 1984, 2005; Thrift 1996.

18 Amin takes this concept from Thrift 1999.

19 As the insights of actor-network ideas in the work on socio-technical systems explains, see Latour 1987; Law 2004; Murdoch 1995.

20 Such as Clifford Geertz (1988) and Pierre Bourdieu (1990), and those examining socio-technical systems (Callon 1986; Latour 1987; Law 2004).

21 See Bourdieu 1977; Healey 1997; Ingold 2000; Wenger 1998.

22 See Kunzmann (2001) for an analysis of initiatives in such 'rhizomic' governance in the Ruhr, Germany.

23 Massey critiques the London Development Strategy for its failure to attend to the different trajectories that constitute the London economy, and for the lack of attention to the way the dominant financial nexus creates exclusions and inequalities (Massey 2005).

CHAPTER 8

GETTING TO KNOW AN URBAN 'REGION'

> Reading the landscape of knowledge production is no less important than knowing who might have particular information (Forester 1993: 97).

> To describe [the methods of inquiry of ordinary people and functionaries] calls for an emphasis on the varied, open, never-ending and inconclusive character of their investigations of a social world in motion (Lindblom 1990: 34).

KNOWLEDGE, MEANING AND URBAN 'REGION' STRATEGIES

In the previous chapter, I emphasised how spatial strategy-making for urban areas involves 'summoning up' an idea of an urban 'region'. In Chapter 6, I introduced key dimensions of such strategy-formation processes – filtering, framing and generating mobilising force through which such an idea or ideas and their policy implications are shaped. In these processes, a wide range of bits of information, of concepts, of ideas about problems and issues, of causes and effects, and of the qualities of places are drawn into strategy-making processes. These buttress strategy formation with information, models and concepts that give meaning to a strategy. Such meanings and their resonances help to legitimate and sustain a strategy, by making it more persuasive and seductive.

The strategic planners of the mid-twentieth-century planning movement were well aware of the need for 'knowledge' about the areas they were concerned with. They collected data about all kinds of phenomena and incorporated these in their reports. In Amsterdam, Milan and Cambridge, fired with a new social-scientific understanding of urban relations, planning teams in the 1970s undertook major surveys and analyses of urban conditions. In Amsterdam in the 2000s, the importance of a strong in-house research function in the planning department was stressed yet again. But the discussion of strategy-formation processes in Chapter 6 and of the multiple spatialities of urban areas in Chapter 7 raises significant questions about the nature of the knowledge being accumulated in this way. What kind of 'epistemologies' (that is, ways of knowing) are reflected in how 'knowledge' is accumulated and used in spatial strategy-making processes? How are encounters between different understandings and meanings negotiated? What kinds of learning go on in these processes of accumulation, encounter and negotiation, and how far do these encourage the kind of creative discovery processes through which new policy frames become recognised and new meanings of an urban region are 'summoned up'? Such questions raise issues about the relations between knowing, imagining, discovering and acting; about the linkages between formal 'research' and systematised 'knowledge', as in scientific papers or technical practice guidance, and knowledge which is implicit in techniques and procedures, in the day-to-day routines of policy practices. They focus attention on the relation between information and the frames of reference through which 'bits of information' are ordered into meanings and understandings. They raise issues about whose knowledge counts, what 'sources' of knowledge are drawn upon and about the relations between 'experts' in a particular area of knowledge and 'the rest of us'. Although spatial strategies focused on urban areas may often involve little more than minor revisions to well-established conceptions, all episodes of explicit strategy-making involve some kind of creative assemblage of what is 'known' and what is 'imagined', synthesising from myriad sources an idea of an urban area as a place, as it may be imagined now and as it could be in the future (Fischer 2000).

Spatial strategy-making episodes are not just social processes of constructing a new framing conception among a set of actors, through which a 'meaning' of the place of an urban 'region' and its consequences for their activities can be imagined, as discussed in Chapter 6. They are social 'construction sites', arenas in which multiple ways of knowing about what is significant, and about what could happen, are explored, conceptualised and symbolised, tested and, in instances where powerful new frames are formed, re-embedded into the ongoing flow of the various transecting relations, in the form of a new (or reinvigorated) idea of 'place' and the priorities that arise from this.

The activity of spatial strategy-making typically moves through several arenas – the planning office, the council chamber, special consultation forums, formal inquiry processes, informal meetings among professionals, etc. The 'management offices' of 'construction sites' also take different forms and are located in different arenas within any governance landscape. It may be a Mayor's office, a city-planning team, some kind of Strategic Partnership, the office of a consultancy firm, the meeting room of a lobby group, a section of a national government department, or several of these in some combination at different times in the formation of a strategic frame.

Thus, in Amsterdam, the key 'construction sites' for spatial strategy-making were the formal and informal arenas where planning officials from different levels and sectors of government met to work out approaches and compromises. In Milan in the 1970s, the 'construction sites' were the planning office and the informal party networks connecting planners and politicians with activists in neighbourhood organisations and in the arenas of the Province government. In the 1990s, in contrast, the 'construction sites' were drawn back into the City Council administration – the planning office, the formal and informal arenas where officials, consultants and politicians met. In the Cambridge area in the 1990s, in contrast, the strategic 'construction site' was a network association outside the formal government framework.

Wherever situated, those involved in spatial strategy-making activities, aware of the intellectual and political challenges of their enterprise, are continually seeking out ways of making their understandings and proposals more robust. In the case accounts, those centrally involved in strategy formation acquired information on phenomena such as demographics, traffic flows, the location of public services and green spaces, retail spending in different locations. They used models about the relations between traffic flows and traffic networks to predict the effects of a new road or rail link. They asked consultants to model the distributions of retail spending under different spatial scenarios, and sought to assess the relation between particular policies and projects and their environmental impacts. They worried about the social, economic, environmental and political consequences of strategies, policies and projects, and about how their strategies might relate to those of other major players. They sought to assess the 'capacity' of particular areas to accommodate more development and to evaluate the engineering and chemical challenges of the remediation of polluted land and the appropriate standards for 'decking over' road and rail routes. They puzzled over how to translate the 'feeling' and potentials of an ambience, such as accessibility and urbanity, into particular qualities that could be deliberately cultivated by public intervention. They thought about the relation between the strategies they were working on and the dynamics of the governance landscape. Thus strategic planning work mobilised many different areas of knowledge – about existing relations and dynamics, about the qualities of places, about who was investing what and where, about environmental impacts, and about what mattered to different stakeholders. Those involved searched for knowledge to help to build up the substantive content of strategies – the what, where, who and how much of particular interventions. They sought knowledge that would help them think about the justification, acceptability and operationalisation of strategies – the how and why of particular interventions.

The case accounts show that knowledge was sought out in different ways. 'Construction sites' in spatial strategy-making used both 'on-site' and 'off-site' methods. In Amsterdam, Milan and Cambridgeshire in the mid-twentieth century, the whole task of gathering knowledge was largely 'subcontracted' to highly regarded planning experts. In the 1970s, in contrast, most of the 'research and intelligence' activity was brought 'in-house' or 'on-site'. In both cases, the designing and researching were conducted alongside each other. By the end of the century, however, different sourcing approaches had developed in the three cases. In Cambridgeshire, strategic planning teams in the agencies responsible for producing strategies had only limited in-house research capacity. The knowledge-accumulation function was therefore in part subcontracted to consultants, and in part undertaken through a range of consultation processes and arenas where various groups of stakeholders met to discuss particular issues. In Milan, too, the internal knowledge-gathering function was very limited compared to the 1970s. It was being provided in part by a search process undertaken by staff on temporary secondments from the *Politecnico di Milano*, in part by the knowledge of experts drawn into the Evaluation Panel for particular projects and as expert advisers, and in part through consultations with stakeholders in negotiations about the various activities needed to realise particular projects. In Amsterdam, all these sources were well-established, but in addition there was a strong in-house research team which, in turn, made use of academic research to enlarge their conceptual and imaginative horizons.

Knowledge 'gathering' was thus an important activity in spatial strategy-making. But what did this involve and what happened when different forms and fields of knowledge were brought together, assembled, and probed in relation to the task of creating a strategy? The processes of 'knowledge assembly' certainly resulted in the accumulation of data, bits of information. This becomes visible in the technical reports that accompany formal strategies. Typically these cover assessments of existing conditions, forecasts and projections of various kinds and assessments of different types of impact.[1] Some of this material may be used to test strategies and evaluate whether particular criteria are being met.

But the use of bits of information already requires some prior concepts that focus their relevance and meaning. The consultancy studies undertaken in the Cambridge case in the 1990s/2000s collected bits of data to fit into technical routines designed to test alternatives, against an approach and criteria derived from national government policy guidance. In Milan, in the same period, technical evaluations were used to ensure conformity with legislation (for example, with respect to environmental impacts). But they were also undertaken to help the Evaluation Panel and 'round-tables' where project negotiations took place to probe the need and scope for negotiating 'public benefits'. The Cambridge approach fitted information into existing categories and the politics that framed them. The Milan approach sought to encourage learning processes through which clearer strategies might eventually be framed. The first approach used information to confirm and justify positions and choices; the second to help probe situations, to develop knowledge about the relations that produced impacts and shaped patterns of costs and benefits.

The production of 'knowledge' is thus not about the accumulation of information but about the development of understanding through the creation of meaning. Knowledge 'assembly' in the construction sites of spatial strategy-making typically involves the gathering up of all kinds of different understandings of issues that are juxtaposed in some kind of encounter. The case accounts suggest that the knowledge that 'reframes', that leads to new ideas about an urban area, is the product not of formalised, expert or scientific knowledge as such, but of social processes of debate, encounter and challenge where diverse perspectives conflict. The more introverted a knowledge-assembly process, the more limited the probing and challenging is likely to be. Reframing is encouraged when knowledge 'from outside' challenges and tests established conceptions. This raises important questions about where, in the diverse and dynamic multiplicity of relations that transect an urban area, what counts as 'knowledge' is produced and used, and how much of this 'knowledge' is accessed in the making, legitimising and diffusion of spatial strategies for urban 'regions'. Such questions, in turn, unsettle assumptions that 'experts' should be the primary source of the knowledge needed for spatial strategy-making.

In this chapter, I explore what a social process of creatively 'discovering' an urban 'region' in an open, dynamic and relational way can mean for developing spatial strategies. In doing so, I draw on recent work in interpretive policy analysis and on the insights of the production of 'scientific' knowledge developed in the study of 'socio-technical systems'. The next two sections develop an interpretive understanding of the nature of knowing, learning and discovering introduced in Chapter 1. The first contrasts an interpretive approach in the context of well-established approaches in the planning field. The second develops such an interpretive, post-positivist perspective on knowledge and learning. The core of the chapter draws out the implications of such an approach for spatial strategy-making as a social process of creative discovery, which draws in and on the multiplicity of experiences and ways of knowing about urban conditions and dynamics. The chapter concludes by considering the qualities needed in the institutional sites of spatial strategy-making if such processes are to promote richly aware, inclusive and open-minded encounters and arguments about urban potentialities.

IMAGINING FUTURES IN SPATIAL STRATEGY-MAKING: TWO MODELS

As discussed in Chapter 6, the work of strategy-making involves processes of filtering ideas and information, focusing and framing notions of potentialities and trajectories, and mobilising support for the resultant strategic ideas. Quantities of pieces of information, notions of causes and effects, and understandings of what is happening and what should be valued are stirred up in these processes. But these sorting processes are in no way neutral. They involve not only selecting from among the potential abundance of knowledge that which relates to dominant or emerging framing ideas of what is strategically significant. They also involve selection in terms of what is considered 'acceptable' or 'valid' knowledge.[2]

This becomes evident in the contrasts often made between the knowledge of 'experts' and that of 'citizens'. The dominant tradition of Western science emphasises the search for laws governing the relations between phenomena, that can provide an understanding of objectively determined cause–effect relations upon which 'policy theories' and techniques such as impact assessment can be based.[3] This epistemology, often called 'positivist', stresses a search for 'correspondence' between objective, material 'reality' 'out there' and scientific representation of it. Expert scientists are skilled in the production of this knowledge and their knowledge is therefore taken in society as more 'valid' than 'lay' knowledge.

But this conception of expert knowledge has been increasingly challenged by studies of the social production of scientific knowledge[4] and by interpretive policy analysts (Fischer 2003; Forester 1993). This puts forward an alternative epistemology in which the distinction between 'expert' and 'lay' knowledge is much less clear. This epistemology emphasises the interpretive work needed to give meanings to the observations and findings made in scientific inquiry, and the social processes which come to legitimate certain interpretations over others. This epistemology stresses that all knowledge is partial, structured by the purposes and perspectives of the inquirer, which are in turn situated in a particular historical and geographical context. It also values the knowledge acquired in the practical work of day-to-day engagement in the world and in all kinds of modes of reflection on the experience of 'living in the world'. In this epistemology, the scientific laboratory and the professional office are just special kinds of 'construction site' for knowledge production and use, not completely different, for example, from what building workers come to know on an actual construction site, or local residents come to notice as they experience the effects of a local waste-treatment facility. Knowledge in this epistemology is validated by the 'coherence' of the story in which it is located and the resonance of the story with the hearer's own experience.[5] In these stories, what is valued and what is experienced are wrapped together. As a result, a piece of data or a statement about a cause–effect relation carries with it, tacitly or explicitly, an interpretive framework of the social context within which the significance and meaning of the data or statement is established. Spatial strategy-making processes that draw in 'knowledge' from the different relational webs transecting urban areas are thus likely to bring into encounter not just multiple perceptions of urban experience and its significance. They will also bring into conjunction multiple rationalities and logics.[6] These two epistemologies are contrasted in Box 8.1.

A positivist epistemology underpins the 'Euclidean' geography of the mid-twentieth-century planners (see Chapter 7). Although, in presenting their work, they emphasised how 'survey' produced 'analysis' which then led to 'plan'; the plans they produced were less a product of this linear process than an imaginative bound to an idea of what the future city should be like. Van Eesteren in Amsterdam wanted to know about traffic

Model	Epistemology	For example . . .
The urban as a material social system, 'out there'	Rationalistic inquiry, positivist science	The studies undertaken in Amsterdam, Milan and Cambridge in the 1970s; processes of impact assessment in the 1990s/2000s
The urban as a socially situated representation, called into 'presence' through social processes	Interpretive, constructivist, post-positivist	The search, through debates and studies, for new understandings of the city in Amsterdam in the 1990s and 2000s

Box 8.1 Two epistemologies for understanding urban areas

flows, land conditions and population levels. But, in the end, his conception of the city of Amsterdam was shaped more by the discourses of modernism and its notions of the ideal city, moderated by an appreciation of the Dutch urban tradition. Holford and Wright in Cambridge in 1950 also made use of an ideal notion of Cambridge, an imagined 'essence' which could be maintained against the threats of twentieth-century development. They needed survey work, not to form an idea of this 'essence', but to address the politically charged issue of road alignments.

In effect, planners in this tradition saw themselves as key agents in 'summoning up' an imagined 'essence' of the city on behalf of the society. The 'construction site' was the planning 'studio', where filtering, focusing and framing came together in a design synthesis. This was then codified into categories and policy principles. In this deeply physicalist approach, the judgements of planners, through their analytical practices and the way they interpreted the desirability of physical forms and urban relations, locked together knowledge and value, generating plans and maps that were then imposed on urban areas through investment and regulatory powers. The legitimacy of this epistemology rested on society's respect for expert judgement.

This 'locking together' of knowledge, value and power in the production of a plan was deeply criticised by a subsequent generation of planners in the middle of the century because it was not 'scientific' enough. Their critiques have already been encountered in Chapters 6 and 7. They sought to prise apart the values shaping both knowledge acquisition and the production of strategies from the work of analysis and generation of alternative possibilities. They thought the way forward was to split the work of value setting and evaluating from analysis and developing alternatives. The former was, in a democratic context, they argued, the proper sphere of politicians. Planners should take on the role as technical analysts, providing advice on possibilities to politicians. The 'imaginative leaps' of planning experts were to be squeezed out by the logics of cause–effect relations 'discovered' by research inquiry structured by scientific procedures. The result, infused by concepts from management science, and particularly the work of Herbert Simon in the USA, was the 'rational planning model' discussed in Chapter 6, which has had such a pervasive influence on the practices of policy-making in Western governments. In this so-called 'rational policy' model, majoritarian political decisions select particular settings for values (as in the setting of environmental standards). Policy experts in the laboratory of the technical-planning office then draw on their 'knowledge' and skills in analysis and evaluation to develop these value orientations and goals into further analyses and administrative tools, and so arrive at a range of evaluated choices. These are then returned to the council chamber, for the key moment of 'choice'. The resultant strategy derives its authority and legitimacy from a combination of electoral mandate and respect for the formalised knowledge derived from scientific inquiry.

This scientifically grounded, goal-oriented approach to policy-making proffered gains in efficiency. Cause–effect relations were better understood and more democratic. Through the separation of fact and value, and through the explicit procedures of scientific inquiry, the 'leap' from values, analyses, policies and actual interventions could become more transparent. 'Discovery' and 'dreams' could be tied down by analytical routines to verifiable facts. This approach influenced the planning teams in all three cases recounted in the present book, and the strong emphasis during the 1990s on technical knowledge in local government in both Milan and Italy generally has echoes of this ambition (see Chapter 4). The model purports to generate a technical legitimation of public-policy choices for politicians nervous about making choices entirely based on their own political judgements. Its practices have become deeply engrained in much public policy-making practice across Western Europe.

Initial critiques of this approach focused on the difficulty of separating facts from values in knowledge production, on the lack of deductive linearity in the practice of plan-making and on the assumptions about the division of labour between politicians and professionals.[7] Critics argued that analysts are continually required to make judgements about what and how to analyse particular phenomena, which involve making value judgements about what is important. Politicians may be quite unable or unwilling to specify what they value in a way that can provide sufficient guidance to analysts. Either the result is a much more interactive process of knowledge manipulation than the model implies, or a substantial part of the politics of policy-making shifts from the institutional arenas in the command of politicians to those in the command of policy experts, as many claim has actually happened (Fischer 2003; Hajer 2003).

In the 1980s and 1990s, these critiques were reinforced by the recognition that citizens have a rich experiential knowledge of urban conditions. The social movements of the 1970s, the community development and neighbourhood improvement work that followed, and the promotion of community involvement in developing local environmentally sustainable practices, brought awareness that citizens are both knowledgeable and skilled about conditions that matter to them.[8] They know more about experiencing the city from their positions and perspectives than any outside expert. This knowledge is validated by observation and by sharing experiences with others. While mainstream policy-makers may have their own expert concepts and language, many on the outside, and often on the margins of governance processes, may already be 'multilingual', combining their own local knowledge with the language of science or public policy.[9] The 'experts', with their procedures for organising experience into 'scientifically' valid 'knowledge', may find communication between citizens and themselves difficult. But this may be because their conceptions, their 'conceived world', to use Lefebvre's term, is strange and alien to those who are not situated in their particular 'community of practice'.

This recognition leads towards the adoption of a different epistemology, one that recognises the multiple worlds in which 'knowledge' is produced. If, as discussed in Chapter 7, planning teams are themselves 'inside' worlds of multiple co-existing and co-evolving trajectories, each embedded in their own histories and geographies but, at the same time, transecting with and constrained by others, then they too are embedded in particular concerns and ways of thinking. Planning teams are often not just embedded in particular policy communities, but they are also tied together both by particular traditions that provide ways of thinking about issues and priorities, and by particular practices of manipulating knowledge. Their ways of thinking are shaped within the 'epistemic communities' formed through their social worlds of practice (Haas 1992; Knorr-Cetina 1999). Just as the residents of a particular place develop an experientially acquired 'local knowledge' of specific conditions, so expert groups, whether scientific groups or policy-making teams, have their own 'local knowledge'.[10]

This knowledge production becomes a social process of making meanings, shaped by the situations, trajectories, activities and values of particular social groupings. Urban areas are thus not just 'known' through the observations and analyses of experts. They are full of arenas in which knowledge is being used, formalised and accumulated, all through different positions and perspectives. In the flow of life, we develop what we come to know through observation, experience, discussion, practical activity and reflection, in which our emotions entwine with our reasoning, and our identities and values co-evolve with our experiences. If spatial strategies are to have sufficient 'resonance' with multiple urban experiences to gather legitimacy and mobilising force (see Chapter 6), then relying only on the traditional 'scientific' epistemology has serious limitations. Grasping the complexity of urban conditions and potentialities exceeds the capacities of the practices of science alone. Spatial strategy-makers need a broader epistemology, to encompass the existence of other ways of knowing and imagining urban conditions and potentialities. They need to develop an 'epistemological consciousness'.

The grounding for such a consciousness has been developing in the management, planning and policy literatures in recent decades.[11] It involves moving from a positivist to a social constructivist perspective (Fischer 2003). What counts as knowledge is 'mediated, situated, provisional, pragmatic and contested' (Blackler 1995, page 1040). Being situated, it arises from many positions. In the management literature, these positions are linked to the different perspectives of shop-floor workers, office staff, managers and executives. In the economic literature, positions are related to different firms in a multinational corporation, or a value-added chain. In firms, the accepted notion used to be that companies should have special teams, or use consultancies, to provide an 'R&D' capacity, producing knowledge that fed into strategy formation by senior executives. These days, there is much more emphasis on the 'learning organisation' and on mobilising the 'distributed intelligence' of organisational members in all kinds of situations in a firm, on developing the capacities of 'learning organisations'. This means accessing knowledge acquired through practical task performance as well as through review of the data sets acquired by means of some kind of recording routine, or through formalised manipulation of models of various kinds.[12] The challenge for spatial strategy-making is that there are very many different 'positions' through which the 'urban' may be experienced (see Chapter 7). Thus spatial strategy-makers who aim, when framing strategies, to involve 'stakeholders' from outside their own communities of practice and citizens from across an urban area cannot avoid encountering multiple frames of reference, multiple logics and multiple values. Nor can they avoid the politically charged process of moving from this multiplicity to foregrounding some meanings and backgrounding others (see Chapter 7).

By the 1990s, in all three cases in this book, those involved in spatial strategy-making were searching for ways of relating to other social groups and communities of practice. They were doing more 'talking' and less formal analysis. In this interactive work, they were partly searching for an alternative way of validating and legitimating their strategic ideas. But they were also exploring and probing what could be the content of strategies. Accumulating knowledge through 'talk', through social encounters, through discussion, debate, exchange of ideas, taps into the 'distributed intelligence' of urban areas (Innes and Booher 2001). It recognises the contribution of experiential knowledge as well as systematised knowledge. In the next section, I develop the epistemological and practical dimensions of this alternative model of situated knowing and creative discovering in an urban context.

Knowledge Formation and Use in Situated Practices

An interpretive epistemology

This model is given several labels in the literature. Policy analysts sometimes use the term 'interpretive', to refer to the work involved in creating meanings from disparate data (Hajer and Wagenaar 2003). Or they may use 'post-positivist', in contrast to the 'positivist' idea that knowledge can be developed through objective analysis, uncluttered by the social processes of its production (Fischer 2003). Others who emphasise these social processes use the term 'socially constructed' (Law 2004), as I have done earlier in this chapter. Clearly, such an epistemology infuses a relational geography and a sociological institutionalist account of planning and policy processes. Some advocates of this perspective present it in terms of a struggle to escape the limits of a dominating 'positivist' view of the world (Fischer 2003). But this is too dualistic a presentation. Anthropologists teach us that there are potentially many epistemologies, each with their own ways of 'making sense' of observations and experiences, resulting in particular logics and 'rationalities' through which arguments and claims for collective action may be structured and legitimated (Geertz 1983; Scott 1998). The 'interpretive' and 'constructivist' perspective allows such multiplicity to be brought into view, while also recognising the way different groups of scientists actually do their work. It is thus more encompassing than the positivist model, which sets up dichotomies between 'scientific' knowledge and 'lay' knowledge, 'rational' argument versus 'irrational' commentary. The work of scientific groups, and of other expert 'epistemic communities' is, in an interpretive approach, understood as a particular form of knowledge-production process.[13]

In summary, the assumptions of an interpretive epistemology are that our understandings of the world are inherently limited by our natural capacities and technologies.

In part, our knowledge of the world exists as a 'store', accumulated in our cultures and communities of practice. But it is also in constant production as we relate observation, experience, intuition to what we 'know' from this store. We learn through the flow of life, through practical engagement as well as through formal processes of 'studying' and structured reflection (Giddens 1984). Getting to know, 'knowing', is thus a process, a continuing activity (Blackler 1995), developing through our practical engagements, our living 'in the world'.[14] In this way, the process of knowledge development and its substance are co-produced (Forester 1999), 'validated' by practical relevance, by emotional resonance and by socio-cultural acceptance. What we know exists in many forms (see Figure 8.1), from systematised accounts and analyses, and practical manuals, to stories exchanged in the flow of social life, and skills exercised in doing practical work.[15] 'Knowledge' is embodied as well as expressed, implicitly present ('tacit knowledge') as well as explicitly articulated. When articulated, what we know may take shape in multiple registers, from the measured presentations of technical analyses to deeply felt, emotively expressed cares and concerns.

This conception of knowledge hugely expands the range of knowing that could be gathered, generated and mobilised in episodes of spatial strategy-making. The knowledge of the physical, natural and social sciences has an important contribution to make, as well as the technical knowledge of professionals engaged in practical tasks, but other knowledges are also important. For example, if, as in the Cambridgeshire discussion of development corridors and 'high-quality public transport', the ambition is to persuade people to use trains and buses rather than cars, transport planners will use data sets about how people both use and switch between different transport modes. But they also need to understand how different people living in different parts of an area experience journeys, how the activity and mode of travelling fits into their flow of life and

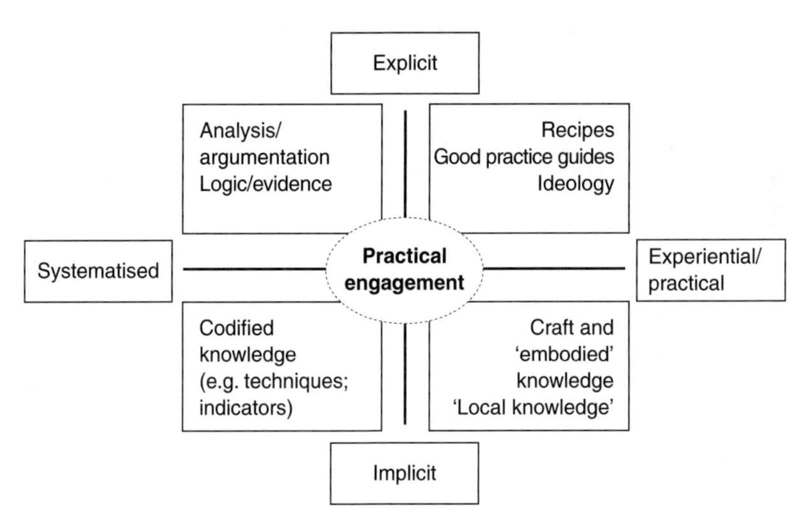

Figure 8.1 Forms of knowledge

their identities. It means getting to know the specifics of what it is like doing particular journeys – crossing streets to get to bus-stops, waiting for buses, standing on lonely, windy and wet stations at night, trying to combine shopping, commuting, picking up the kids and going to night classes all in one day. Drawing in such experiential knowledge not only helps to improve the understanding that informs policy. It may also help to make the arguments of strategy-makers more robust, through resonances between what the 'experts' say and the experiential knowledge of those with a stake in an urban area. But accessing such experiential knowledge is not easily done through traditional quantitative methods. It demands institutional space and time in which people can identify and articulate their experiences.[16]

An interpretive epistemology does not just expand the range of the knowledge needed to make a spatial strategy seem well-grounded in local realities. It emphasises the socially-situated nature of the way knowledge is produced and used. This means that the 'construction sites' for the development of a strategy are not just arenas where knowledge from around and about is accumulated and synthesised. They are also themselves an institutional site, an arena, of practical engagement, where those involved are struggling to get to know what is significant and to synthesise meanings out of the morass of knowledge streams flowing into the site. In these processes, knowledge, discovery and imagining are not separated. Instead, 'knowing' is understood as a complex process of empirical discovery, identification of new phenomena, empirical 'testing', logical deduction, recognition of what is valued, imagining consequences and impacts, creating meanings, discovering potentialities. These 'learning processes' proceed in non-linear ways, through interaction between people as they exchange information and ideas, through interaction between people and the tasks they are engaged in, through interaction between people and the processes of the natural world. Feelings, sensibilities, intuitions, imaginative bounds, careful experimentation, testing, assessing, story-telling about experiences – all mill around in this conception of what it is to know, to imagine, to dream about the future. As we are told by those who study how scientists work (Knorr-Cetina 1999; Latour 1987), scientific practices also have these qualities, though, in the explicit language of 'science' and in the presentation of 'validated' 'scientific findings', many of the dimensions of the production of their 'knowledge' are smoothed away.

This 'smoothing' can be seen in the work of strategic planning in the cases, when well-developed strategies are 'rolled forward', or ideas about future development are presented in a consolidated form as 'options' among which 'consultees' are asked to choose. In these cases, planning teams work in largely closed 'laboratories' and 'studios'. But this means that the 'knowing' available in and around an urban area arrives to the 'construction site' either in an already heavily processed form (for example, as demographic statistics, or the arguments of pressure groups, or the experience of key stakeholders such as bus operators or developers), or it is filtered by the experience, concepts and values of the dominant players in the key arenas. In Amsterdam, Milan and Cambridgeshire, the main players were a mix of members of planning policy communities, local politicians and a few other local-elite members. They accumulated knowledge, filtered it and made the synthesis into sets of choices about trajectories. This approach to knowledge management simplified their task, but heavily limited the challenging, the probing of established viewpoints, the unsettling of taken-for-granted positions of 'learning to know' about the urban 'region' they were summoning up. This, in turn, restricted the potential for innovation, for discovering different ways of seeing an urban area. It limited the creativity of the encounters between the diverse ways of thinking about urban areas, their qualities, threats, problems and potentialities. Data was collected about some dimensions of where people lived and worked, about how they moved around an area, and where services were located. But, except in the debates initiated in Amsterdam in the 1990s, little attempt was made to tap into the 'distributed intelligence' of urban life.

These processes of synthesising and smoothing of the multiplicity of knowledge resources available about urban dynamics have major consequences for the formation of strategic frames. On the one hand, policy-makers limit their recognition and understanding of the emergent potentialities in an urban area. As a result, they may persist in making all kinds of assumptions about cause-and-effect relations that have little grounding in what is actually going on, while significant relations are missed. The routines and technical practices of 'impact assessment' are little substitute for a rich awareness of the multiple ways project impacts may be experienced (Owens *et al.* 2004; Richardson 2004). In other words, strategy-makers are missing out on a key resource. On the other hand, strategy-making in this mode is likely to be challenged from all kinds of directions by those who have different experiences and understandings. A strategy may fail to persuade. If it does persuade the powerful, it risks having adverse consequences for those groups who are less articulate and less well-connected to the relevant governance arenas.

The value of an interpretive epistemology through which to approach spatial strategy-making focused on urban areas is that it provides a way to recognise the presence of multiple rationalities and allows for encounters between them to be valued. It reduces the tendency to smooth away complexity, diversity, multiplicity and conflict in the processes of knowledge gathering, filtering and synthesising (Lindblom 1990). It thus provides a grounding for a more democratic, and potentially more just, focus for the governance of urban areas (Corburn 2005). Combined with a recognition of the multiplicity of relational webs and forms of living in an urban area, an interpretive epistemology encourages strategy-makers to enlarge the social 'construction sites' for strategy production, to encourage learning processes situated among this multiplicity through which key potentialities and challenges for the shared experience of urban living can be 'recognised' and alternative possibilities of the 'urban' summoned up. In Amsterdam, the opening up of debate about urban futures in the mid-1990s sought to widen out their 'construction site' to encourage diverse experiences and perspectives to find expression and engage in debate as part of the early filtering, framing and focusing processes of strategy formation. Politicians in Milan, in contrast, soon pulled back from such an idea. In the Cambridge area, although the arena for strategy-making moved to an informal partnership in the 1990s, this rapidly shifted from framing to mobilising, with consultancy studies used in part to help shape options and in part to 'test' them, using techniques convincing to national government. Although, in all cases, those involved were 'learning' a good deal about the changing dynamics of the areas they were focusing on, only in Amsterdam was there any real attempt at some kind of 'collective learning' about urban potentialities across a governance landscape.

I now explore in more detail what such an interpretive epistemology implies for the practices of spatial strategy-making. What does it mean to put such a practice as the practical engagement at the centre of Figure 8.1? I focus on three dimensions. The first concerns how meanings taken as 'valid' are created. The second concerns the social processes through which knowledge is produced and validated. The third concerns the recognition of diversity and multiplicity in the 'communities of practice' which have a stake in emergent urban 'region' strategies. In this elaboration, I do not seek to eliminate the positivist model outlined in the previous section. I seek instead to situate it among the array of ways to 'read' and 'learn about' what is going on. Rather than a privileged rationality, I treat it as one of many. The challenge is to explore ways of promoting generative rather than destructive encounters between multiple rationalities.[17]

LEARNING THROUGH STORIES

Spatial strategy-makers call up some kind of idea of an urban 'region', its problems and opportunities. They also make use of assumptions about the causes of 'problems' and the consequences of possible interventions. They make assumptions, implicitly and explicitly, about critical connectivities. Much of the tradition of urban analysis that developed models of urban-system relations, for example in relationship to shopping behaviour, transport flows, housing-market dynamics, environmental impacts and accessibility to services, operated by extracting a few key variables from the complexity of relations, and then manipulating these to explore the effect of changing particular variables. Such analyses will always have some value, to prompt thinking and to criticise the myths that often circulate about cause–effect relations. But, as I have stressed in Chapter 7, the characteristic of a policy focus on 'place' qualities and dynamics involves a recognition of the multiple forces intertwining to create what becomes 'visible' to relevant stakeholders as a 'place'. Different economic nexuses interrelate with different socio-cultural worlds and with the dynamics of geological, meteorological and biological processes in this transecting and intertwining. The focus of spatial strategy formation is thus on processes that are shaped by multiple forces. The qualities that are recognised today, which are seen as inherited from the past and which might emerge in the future, are the consequence of multiple causes, contingently produced in a specific situation. Despite advances in multivariate analysis, multicausality makes any form of prediction both difficult, and, in a public policy context, potentially misleading and dangerous. Such analyses, and modelling efforts based on them, may have a useful function in exploring possible interactions and connections. They help to reduce specific technical uncertainties. But understanding urban dynamics is full of ambiguities that cannot be 'smoothed away' (Forester 1993). Urban modelling assumes a conception of a city, rather than the creation of a conception. It operates by fixing assumptions about cause–effect relations prior to analysis. But too great a reliance on models – of urban economies, of traffic flows or retail spending patterns or housing markets – limits and reduces the imaginative scope through which urban qualities and dynamics are perceived.

An interpretive approach suggests that spatial strategy-making processes would be enriched by treating the knowledge that comes forward as an array of 'myths' and 'stories', rather than focusing on a single model that is then refined and filled out by the gathering in of new knowledge (de Neufville and Barton 1987). If all our knowledge of the world is limited in some way, and if precise predictions are only possible over very short horizons in very stable situations, all knowledge, however produced, has some of the qualities of a 'myth', a story told to help explain, justify, focus attention and to mobilise action. A story brings to life a way of seeing the world, and reveals the significance of complex relations and action dilemmas. A story is an imaginative work, which situates what is referred to, and gives coherence through its interpretive work. The filtering processes involved in strategy-making may thus be understood in part as 'encounters' between multiple stories. The process of strategy-making then becomes the formation of a new 'story', its richness and resonance reflecting its relations to all these other stories.

But these encounters between stories are unlikely to proceed in a simple linear way, as emphasised in Chapter 6. New 'ways of seeing' urban dynamics may enter the arena at many different stages in the development of a strategy, perhaps from some new academic study, or some change in the mix of values that are being promoted, or the appearance of some new lobby group with a strong agenda. Or equally, new ways of seeing and hence the potential for a new strategic story to emerge are 'discovered' through the activity of strategy-formation and exploration itself.

For example, in the Cambridge case, studies were undertaken by stakeholders in the 1980s to highlight the effect of growth restraint on the emergent Cambridge economic 'cluster'. These fed into political struggles about the future of Cambridge, which led to a proliferation of studies about the nature of the 'cluster', about traffic concerns, about growth options, etc. Each study generated another agenda that needed exploration. Meanwhile, the agendas for the studies were continually shaped by actual and anticipated local protests, by changing national government policy and changing ways of understanding phenomena such as 'economic clusters' and 'transport systems'. Similarly, in Amsterdam in the 1980s and 1990s, the city-planning team was continually seeking to understand newly emergent issues and priorities, generated by the changing national and local policy landscape and by new ideas about urban dynamics emerging in academic arenas, generating a rich and continually changing array of claims and counterclaims about issues and policies.

This practical experience of knowledge production for policy purposes suggests that knowledge generation, far from being a linear sequence, proceeds in interactive, continually contested and often 'jerky' ways, across a range of institutional arenas. New information, or information newly seen as important, challenges established frames, while new framing ideas re-cast models and information. If those involved in the construction sites of strategy formation maintain a capacity for continual 'strategic reflection', then the work of filtering and framing goes on even as a strategy moves into consolidation phases. For example, those promoting a reframing of spatial strategy in the Cambridge area in the 1990s were continually pushing ahead with rethinking the content of development plans, as debates on regional strategies in formal inquiry arenas produced new knowledge about local conditions and new judgements about the balance of national policy directions. They were vigorously constructing and revising the discursive storyline of the 'relational layer' they sought to insert across the governance landscape relevant to the Cambridge area, moving among institutional sites as they did so. In Amsterdam, the city planners gave much more attention to listening to the many stories bubbling up and reverberating in the 'public realm' of discourse about the area, its challenges and qualities. The whole governance landscape was in part their knowledge-production laboratory, though they tended to keep the filtering and synthesising work to themselves. The non-linearity of strategy-formation processes is one reason why Mazza, in his work for the *Documento di Inquadrimento* in Milan, continually insisted that the *Documento* was not a strategy. A strategy might in the end emerge from the processes set in train by the innovations introduced alongside the *Documento*, but, as discussed in Chapter 6, the role of the initial move was to challenge and 'unsettle' previous assumptions and procedures.

Those involved in strategy-making processes that operate with an interpretive epistemology will have an awareness that a key challenge is to 'capture' a possibly emerging strategic storyline and to 'recognise' its potentialities as it floats past them, but without 'fixing' it so quickly that its possible limitations are ignored. As it emerges, other stories about causes and effects, potentialities, problems and challenges need to become attached to it and explored, to draw out the capacities of an emerging strategic story in focusing attention – its assumptions, inclusions and omissions, and its resonances with what is already known – so that its credibility and legitimacy can be assessed. Such strategy-formation processes develop knowledge and meanings through practices of probing different 'stories' to draw out their arguments and logics, their values and significances. In these encounters, what may start as a struggle between different social groups and perceptions become re-cast into arguments about values, potentialities and ambiguities.

In summary, if knowledge is understood as a process of always incomplete knowing about conditions and potentialities affecting the relational lives of those in and around urban areas, then those involved in spatial strategy-making need to attend to the dimensions of how they 'get to know' about the urban set out in Box 8.2.

Dimension	Implications of an interpretive epistemology	Suggestions
Understanding cause–effect relations	Outcomes have many causes They are contingent on situational specificities	Avoid too great a reliance on simple cause–effect models Keep different ideas about cause and effect in play
What can be known	Knowledge of 'what goes on' is always partial It involves the production of 'meaning' from diverse inputs, in which values, facts and meanings are wrapped together	Listen for stories and use stories as well as analyses and models to probe ideas about present problems and future potentials
Knowledge-development processes	'Getting to know' is an interactive, non-linear process of 'sense-making'	Avoid compartmentalising knowledge production and use into stages Probe through argumentation using multiple 'logics'

Box 8.2 Learning through stories

THE SOCIAL PRODUCTION OF KNOWLEDGE

The challenges of non-linearity and ultimate unknowability reinforce the recognition that the production of knowledge through scientific practices, and in other ways, is a social construction through which 'sense' and 'meaning' is generated. 'Science', in other words, is not a neutral, objective tool with which to legitimise policy-making and public action. What is called scientific knowledge is produced and transmitted by complex social practices within research institutes and universities. The same applies to the knowledge produced and used in policy contexts. In both cases, knowledge is produced through processes of negotiating about the focus of attention, through social processes of experimentation and discovery, through testing, validating and legitimating 'results' and policy ideas. In strategy-making in governance contexts, the key motivation is to address some perceived 'problems', or to provide direction to a disparate set of activities, or to create an orienting identity through which key relations and values can be advanced more effectively, or maybe merely to meet some procedural requirement to produce a strategy. Knowledge and ideas are sought, not as valuable in themselves, but to perform political and policy-development work. For strategy-makers, a 'result' is a strategy that convinces key audiences – politicians, citizens, business groups, pressure groups, professional peers – and in turn refocuses attention on new action priorities. It is validated not just by the 'authority' of peers, but by its resonances with diverse 'epistemic communities', and their expectations of the effects it may have.

Strategy-makers in governance contexts thus face complex and diffuse contexts in which their efforts are validated and legitimated. They need to pay attention to the 'sense' a strategy makes not just to politicians and to professional peers, but also to the variety of those who are affected by a strategy and may challenge it, and to the arenas that make up the 'construction site' for strategy formation. They can hardly avoid recognising the social processes through which the knowledge they use and develop is generated, filtered, validated and legitimated. In the past, strategic planners could proceed more like traditional scientists, insulated by their laboratories and the social worlds of their peers. Van Eesteren in Amsterdam could look to his peers in the CIAM movement to validate his approach. Holford in Cambridge similarly drew on his standing among his peers, his knowledge of local politics and among those shaping the emergent national planning system in the 1940s. Today, in Amsterdam, Milan and Cambridge, the arenas within which ideas about strategic frames are produced and tested extends to the 'public realms' of urban governance, and the resultant ideas are validated and legitimated in many different ways (see Chapter 6). This raises critical questions about who gets to be involved in strategy-formation processes and in what arenas. A strategy-making process that aims for awareness of the multiplicity of intersecting trajectories needs to find ways to link to diverse participants and to draw multiple perspectives and understandings into conjuncture in public realm arenas, to enable discursive encounter to occur.

Despite differences in the task focus and contexts of spatial strategy-making and scientific research inquiries, those interested in policy-making processes can learn from studies of how scientists work. In her studies of multinational scientific research teams, Knorr-Cetina (1999) shows how the social organisation of the team varied with the context, the research task and the traditions of the particular scientific community. Research teams involved in high-energy particle physics operated very differently from those in molecular biology. In other words, processes of knowledge development and 'discovery' are shaped by the content as well as the context of what people are seeking to 'know'. In Amsterdam, Milan and Cambridge, all the strategies sought some kind of knowledge of the 'urban'. But emphases varied according to the critical tasks that a strategy was to address. A focus on expanding cities to provide more and better housing and living conditions will emphasise different sources, forms and content of knowledge development than a focus on reducing the environmental impacts of urban living, or coordinating development investment more effectively with infrastructure provision.

In Amsterdam, Milan and Cambridge by the 1990s, those centrally involved in strategy formation were immersed in ongoing learning experiences in which the nature of the task they were involved in continually shifted. In Amsterdam, the city-planning strategy team sought to identify the special qualities of Amsterdam as a city through concepts of 'openness' and 'urbanity', in order to arrive at a specification of qualities that could be used both in arguing for the need for special investment attention from national government and the province, and in focusing negotiations with private-sector developers. In Milan, the central strategic planning task was to provide some kind of framework to constrain and focus the initiatives of developers, politicians and various public agencies concerned about impacts and about negotiating contributions to public services. In Cambridge, the task, as defined by those who lobbied for a change in strategy, was to get acceptance of substantial growth and to capture the investment required to provide the necessary infrastructure. Knowledge was sought out in each case through the lens of the initial task focus. The cases varied, however, in the openness of thinking about this focus. In Amsterdam, drawing on many sources of inspiration and challenge, the core strategic planning team was continually rethinking its understandings and reformulating its task focus as it drew out the knowledge it acquired. In contrast, in Cambridge, knowledge was primarily drawn upon to buttress and fill out an established position. In Amsterdam, the emphasis was on learning about conditions and potentials; in Cambridge knowledge was used to validate and legitimate an already-formed policy shift.

But the issue of the relation of knowledge to tasks is more than just a question of the way the purposes of strategy-makers shape what they set out to know. In the complex multirelational context of the urban, all kinds of knowledges are produced through the way people in different relational contexts engage in their various activities. Some of this is known explicitly and is readily conveyed into data or opinions or stories about experiences. But people also come to know through practical engagement, through their work, maybe driving buses around a city, or collecting a city's rubbish, or running a day nursery, or just walking around of an evening. This kind of knowledge, acquired through bodily engagement as well as mental formulation, is difficult to access in the 'public realm' of debate about what an urban 'region' is and could be like, and what qualities are important and to be valued. Yet it is critical to how stakeholders come to respond to strategic ideas and opportunities for debate. It is also critical to developing some kind of understanding about how a policy may potentially unfold in specific situations in a particular urban area. This demands that key actors in spatial strategy-formation processes need to find ways to become 'streetwise' on the many streets of an urban area.

This implies that strategy-formation processes that seek to reach out to the multiplicity of conceptions and experiences of urban conditions need to develop both an outward orientation and a capacity to interact with diverse communities of practice. This puts a premium on communicative capacities. This is not just a question of the language used in presenting documents for public consultation and information. It is also a question of the capacity to interact with others in multiple ways and to hear what is being conveyed in interchanges with diverse groups and individuals. Planners have been much criticised for assuming that everyone speaks their own language of practice, and for requiring those who engage in consultation practices to adopt this technical language.[18] If those involved in spatial strategy-making are to become 'streetwise' on many streets, they need not just to 'walk' the streets, but to learn how to 'hear' what people are

Dimension	Implications of an interpretive epistemology	Suggestions
The arenas of strategy formation	Arenas are multiple The selection of arenas and of participants is a critical choice	Open up arenas Involve many participants
The task focus of strategy formation	The task focus for strategy formation is not fixed but evolves	Emphasise learning processes and critical probing Allow for shifts in frames and focus as understandings and meanings evolve Allow for creative discovery
The languages of strategy formation	Access the languages and epistemologies of different communities of practice	Develop 'streetwise' skills Recognise the many dimensions of communicative practices

Box 8.3 The social production of urban knowledge

showing and telling them, and to 'speak' in multiple registers, in order to exchange information, concepts, values and meanings. As emphasised in the literature on communicative practices, discursive competence is much more than a capacity to articulate appropriate words. It involves ways of expressing meanings through gesture, emphasis, facial expression. It may involve some degree of practical engagement, as in participatory research, or focus groups, or design theatres.[19] Thus, those centrally involved in strategic planning need not only allow themselves to look around and about an urban area for sources of knowledge and inspiration. They also need to develop awareness of the peculiarity of their own 'policy languages' and to become 'multilingual', able to recognise other languages and communicative practices through which knowledge and ideas might be conveyed or summoned up into consciousness, so as to enlarge and enrich debates about urban futures. Box 8.3 summarises these dimensions of the social production of knowing for strategy formation.

THE MULTIPLE ONTOLOGIES AND EPISTEMOLOGIES OF URBAN LIFE

Strategy formation involves selectivity and synthesis (see Chapter 6). But, as the urban modellers of the 1960s discovered, the range of meanings, facts and 'valuings' that can be known about urban areas defies the capacities of human intelligence; and there is much that cannot be 'known'. A relational geography and an interpretive policy analysis add to this recognition an awareness that, in our multiple identities and relations with others, we not only come to perceive and evaluate the conditions of life in urban areas in different ways. Our perceptions and 'valuings' are themselves co-produced as we develop identities (ontologies) and modes of understanding and reasoning (epistemologies). As a consequence, in an urban area, there are potentially very many ways of experiencing and knowing the place qualities of the urban. The challenge for spatial strategy-making initiatives that seek to promote a rich and inclusive meaning of an urban 'region' is how to acknowledge this multiplicity and its diverse forms and locations, while at the same time developing a focus of attention from which ideas about the 'urban' can be summoned up, critical potentialities imagined and discovered, and desirable interventions identified.

It is sometimes assumed that different forms of knowledge can be associated with different social groups. Planning teams could then turn to scientists for analysis and systematic evidence about phenomena, to consultancies and government advice for 'good practice' recipes, to their own professional store of techniques for codified knowledge and to 'local communities' (directly, or via politicians) for their 'local knowledge', their experiential knowledge. But the studies of communities of practice, whether of policy groups, scientific teams or local neighbourhoods, suggest that all the forms of knowledge in Figure 8.1 may be drawn upon in the flow of any practice. What differs between communities of practice is partly the 'content mix', how systematised knowledge, of analysis, of good practice manuals, of craft knowledge and localised experience, are combined. This relates to both the nature of the task of the practice, but also to its context and history. Beyond this, differences between communities of practice lie in the processes through which what counts as valid knowledge and legitimate inference is established. That is, the differences lie in the approach to argumentation, in the logics and rationalities used. An interpretive epistemology acknowledges a meaning of 'scientific rationality' not as a singular logic, but as a socially situated mode of reasoning and drawing inferences, one of many possible logics.

This means that spatial strategy-makers seeking to acknowledge the diversity of ways of sensing the 'placeness' of forms of living in an urban area need to be prepared to seek out and bring into conjunction not just the knowledge held by stakeholders in all kinds of positions in a governance landscape. They also need to recognise the multiplicity of logics that may be manifest as people in different positions and with different trajectories of experience puzzle about what an urban area, a place or a connectivity means to them. If episodes in spatial strategy-making seek to promote richly inclusive conceptions of an urban 'region' and keep in conjunction the potentially conflicting values of distributive justice, environmental well-being and economic vitality, then those involved need to find ways to bring forward this multiplicity, to encourage what is hidden and 'tacit', to confront and engage with what is explicit, systematised and codified. It involves searching for different angles of vision and experience, and articulating these in ways that can become at least partially visible to each other. An effort in urban spatial strategy-making that seeks to ground the selectivity and synthesis of a strategy in a rich awareness of urban complexity needs to finds ways of bringing this diversity into some kind of encounter. This may produce challenge and conflict, but is itself a probing process, through which the politics in any strategy can become more visible.

How was this being done in Amsterdam, Cambridge and Milan? In all the cases, strategy teams sought out the knowledge of other government departments. They also checked back with politicians. They drew on their own experience and memories of encounters with all kinds of groups and situations, particularly in Amsterdam and Cambridgeshire, where key players in strategy-making in the late 1990s and early 2000s had been around from the 1970s. In the 1970s in Amsterdam and Cambridge, and in Amsterdam ever since, there were strong in-house research teams to 'accumulate' and interpret information. In Cambridge from the 1980s, strategy teams looked to consultancies to provide 'intelligence' about aspects of urban dynamics, while in Milan and Amsterdam, more use was made of university expertise. From the 1970s, in Amsterdam and Cambridgeshire, consultation with citizens and all kinds of other stakeholders had become normal practice, though often undertaken in a highly structured and episodic way.

By the 1990s, the importance of accessing the multiple experiences of urban areas in strategy-formation processes was acknowledged in the practices of planning teams. Strategy-makers mixed formal studies, syntheses of consultation exercises, the outcomes of discussions with those they recognised as key stakeholders and their own experiential knowledge. In this way, formalised knowledge interacted with these less-formalised and less-visible influences on how planning teams selected and synthesised what became key strategic concepts. However, in all the cases, previous planning concepts and 'waves' of ideas within the planning policy community framed the 'summoning up' of ideas about urban dynamics and about the content of strategy. Other voices might be welcomed to test and inform strategy ideas, but were often ignored because connections between what people were saying and their identities and modes of reasoning were not recognised. The voices of those outside the nexus of established policy-making had to rely instead on the capacities and motivations of those within the policy community to widen their own horizons and capacity to access, 'learn about' and absorb into their own 'knowledge' what others feel, experience and know about the evolving qualities of places and connectivities.

The politically aware 'outsiders' mobilised into protest groups and special-interest lobbies to make sure that they were 'heard' in ways that actually changed strategy-making practices, as happened in Amsterdam in response to the urban social movements of the 1970s. But, in the 1990s and 2000s, the voices of the more marginalised were less articulate in accepted political languages and more difficult to hear. The elderly and poor immigrants in Milan's increasingly segregated neighbourhoods were being 'discovered' by academic analysis and the work on the *Piano do Servizi*, rather than as an active 'voice' in Milan's urban governance landscape. The multiple epistemologies and ontologies of urban experience thus continued to encounter each other primarily in the minds of strategic planners, in their debates and discussions in the main arenas through which strategies were produced. They were manifest again in arenas for consultation, formal objection and inquiry into a strategy. The planners filtered the multiplicity, which then reappeared in formally structured arenas where those with sufficient interest and knowledge could challenge, test and legitimate the selectivities and syntheses arrived at by the planners.

It is not the selectivity and filtering work in itself that can be criticised from the perspective of an interpretive epistemology. This is inherent in the creation and use of strategic frames. A strategy-making process cannot avoid moments when only a few people undertake the work of synthesis, whether as a recognition of an emergent strategic idea, or the production of a new 'way of seeing' an urban area and its strategic priorities. What makes the difference between a narrowly grounded and a richly aware consciousness of the diversity of experiences and 'valuings' of urban conditions is the experience and capacity for insight of those few people. As they develop, imagine and frame strategic ideas, or re-express and reassess what has gone before, whose are the experiences and values they draw upon? If they are 'on stage', in the foreground of the encounters between understandings, 'valuings' and rationalities, who do they acknowledge as in the background? What non-present others are drawn into 'presence' as strategies are discussed? Here the contrast between recent strategy-formation processes in Amsterdam and Cambridgeshire emerges clearly. In Amsterdam, those involved in strategic planning were themselves 'streetwise' in at least some of the streets of the city. They organised and drew on debates about the city and its conditions. They commissioned studies to help them learn more. They anticipated all kinds of challenges from active voices in the governance landscape, and were faced with claims that they still did not know enough and were approaching their work too narrowly. All kinds of non-present others were drawn into the strategy-making 'construction site' as work progressed. In Cambridgeshire, in contrast, most of the encounters were limited to those politicians, planners and university/business interests who were arguing for more space for economic development and more attention to the transport dimensions of environmental sustainability; that is, to highly professionalised views of the experience of the area. And the critical non-present 'other', whose presence loured over the various arenas of strategy formation, was the national government department in charge of the 'planning system', as guardian of the key power of land-use regulation and as the source of public investment funds.

Accessing the multiple ontologies and epistemologies of urban life is thus no easy route to producing a spatial strategy that will be widely accepted as relevant, valid, socially-just and legitimate. Opening up strategy-making processes, widening the arenas of direct encounter between multiple communities of practice and multiple rationalities, involves complex social and political dimensions, full of danger of manipulation. If one actor openly reveals a practice or interest, the resultant visibility may lead another to destroy or capture it. Or one group's mode of reasoning may 'crowd out', or be 'converted' into another's mode. Or one actor's perspective and sensibility may be manipulated, distorted and used to legitimate another's power play, as illustrated in Flyvbjerg's famous case of Aalborg, Denmark (Flyvbjerg 1998). Those involved in strategic spatial planning thus face complex managerial and ethical judgements as they consider the

Dimension	Implications of an interpretive epistemology	Suggestions
Multiple logics and rationalities (epistemologies)	Recognise the diversity of logics and 'rationalities'	Respect different ways of seeing, and of modes of expression and reasoning
Multiple identities, positions and trajectories (ontologies)	Recognise that what is sensed, valued and understood varies with identities, positions and trajectories	Search out experiences from multiple positions
Multiple sites of encounter between ontologies and epistemologies	Promote arenas where different ontologies and epistemologies can encounter each other in creative, rather than destructive, ways	Encourage generative encounters between identities and rationalities that aid challenge and discovery

Box 8.4 The multiple ontologies and epistemologies of urban experience

arenas within which conceptions of an urban area may be summoned up, 'discovered' and clothed with meanings and implications for policy interventions. In making these judgements, they create a particular ambience in which diverse ontologies and epistemologies encounter each other. Box 8.4 suggests probing questions through which to explore the qualities of that ambience of encounter.

ENCOUNTER, ARGUMENTS AND DELIBERATIVE PRACTICE

Strategy-making focused on urban futures involves invoking a particular 'sense' about the potentialities and trajectories of urban dynamics, within which specific connectivities and place qualities rise into focus as critical considerations in shaping action now in order to encourage or ward off perceived potentialities. This chapter has sought to emphasise that this selective invoking and focusing needs to draw not just on systematic research studies, technical assessments and the 'store' of knowledge within a professional group. It needs to recognise that what is 'known' is the outcome of a process of making meanings in social contexts. A clear separation cannot be made epistemologically between expert 'planners', guardians of systematic technical knowledge, and lay knowledge, channelled into strategy-formation processes via politicians. An interpretive understanding of the way knowing is achieved emphasises that meanings and valuing are wrapped together and co-evolve (Fischer 2000). What is of strategic significance is more an emergent creative discovery than a logical deduction from established evidence. This production of a strategy, in turn, generates yet another set of meanings and 'valuings' that flow around and among the diffused 'knowings' of an urban area. A spatial strategy-making that seeks to create a strategic frame centred on the 'summoning up' of an idea of an urban 'region' in a rich and open way needs to gather in the 'knowings' and imaginings generated in diverse relational worlds into some kind of conjunction, some kind of encounter. In creating a relational layer around the strategic development of an urban 'region' (see Chapter 7), spatial strategy-making thus also contributes to the governance culture prevailing in a place, to the qualities of the public realm of urban life.

Many episodes of spatial strategy-making are conducted in arenas created through procedures specified in formal planning systems. Many such systems are designed in a way that privileges professional knowledge and a linear logic. One result is that many planners think of 'encounters' as the often-uncomfortable experiences of public consultations with apparently disinterested citizens, or highly charged public fights, or their experience of formal inquiry arenas where they have to 'demonstrate the validity' of their arguments. Other stakeholders, including citizens who seek to engage actively in governance processes, often report similar uncomfortable or frustrating experiences when they enter the arenas set up by government actors for what is variously called consultation, participation, engagement and empowerment. An interpretive, non-linear perspective on knowledge emphasises that 'encounter' among diversity is not confined to such formal arenas. It happens in the flow of interaction throughout strategy-formation processes and is likely to be part of the ongoing 'flow of life' of all those who 'encounter' urban governance processes in some way. Those designing spatial strategy-making processes informed by the considerations discussed in this chapter might seek to locate the sites of encounter through which 'knowings' about an urban area are gathered up and co-probed nearer to the flow of life going on in the multiple webs of relations transecting and interweaving in an urban area. Such encounters might have more of the character of assemblages (Fischer 2000), of bricolage (Innes and Booher 1999b; Melucci 1989), than systematic analysis and appraisal:

> Exploration requires *bricolage*, the gathering and piecing together of clues, the following of tracks that lead back to the starting point, the recognition of signs that are instantly recognisable, and the discovery of other signs that were missed the first time round (Melucci 1989: 13).

These terms imply that meanings, values and information from multiple directions are thrown up, looked at, allowed to float around and combine in all kinds of ways, in potentially creative processes of exploring and discovery. They are 'probed' (Lindblom 1990) and argued about. Through probing and arguing, lots of 'bits' of ideas and facts as they tumble about are not only made visible, brought to the foreground, but are reflected upon – where do they come from, what meanings have generated them, what assumptions are embedded in them? In this way, the bricolage shifts into an array of potentially important place qualities and connectivities, and an assemblage of arguments about what is significant, for whom and why. Through such processes, new meanings are generated and new patterns perceived, through which to focus on strategic priorities and appropriate interventions as necessary, if any come into view. The formation of strategies in urban governance contexts in this way lies at the heart of contemporary discussion of 'deliberative democracy'. In these discussions, a model of democratic governance practice is celebrated that emphasises the creation of public spaces for encounter and argumentation which are open and accessible to multiple languages and logics and to many modes of expression and forms of argument.[20] A critical quality of open and accessible forms of argumentation is that the emphasis is less on 'closure' and more on 'discovery'. The emphasis is less on reducing uncertainties (will this or that happen? what is the actual cause of this problem?), and more about revealing potentialities and opportunities, about opening up ambiguities, tensions and difficult choices.[21]

This implies that spatial strategy-formation processes that aim for an open-minded perspective on place qualities and connectivities, richly aware of the multiplicity of trajectories interweaving through what an urban area could be, need to pay careful attention to the nature of the institutional sites through which encounters among different perspectives, imaginations and knowledges take place. These are inherently 'performances', with front stages, back stages, choruses and audiences (see Chapter 7). The nature of these performances reveals much not just about how an urban 'region' is imagined, but who a strategy is for and what work it is meant to do. If the front-stage performance is conducted in professional jargon with specialised references to key policy texts and government actors, then it is unlikely to attract much of an audience from the wider public. If the play is conducted in ways that bring to the front stage in recognisable ways the many lives of those in, around and beyond an urban area, and raise the dilemmas and responsibilities of the shared experiences of urban life, then it may capture the attention of more people. It may also reduce the extent to which the less vocal and less well-networked are ignored or marginalised in the construction of strategic frames. A key challenge for strategy-makers seeking to develop more deliberative ways of doing governance work is to open out the 'construction site' of strategy-making to the front stage of a public performance, to open up to the possibility of 'civic discovery' (Fischer and Forester 1993: 7).

But it is possible to get carried away by a performance, by the rhetorics of a play. It is easy for some characters and some ideas to command the stage too much. This may produce strategies that fail against 'practical realities'. Or strategic ideas may end up too narrowly attached to the interests of one group of players. In this case, a strategy and the meanings that sustain it could cut off, oppress or render invisible the concerns about urban conditions of many others. The result could be an oppressive strategic frame, limiting innovation and encouraging the hegemony of a single perspective – of the business corporation, or a pressure group focused on just transport, or the conservation of historic buildings, or of a local elite seeking to manage a local area to exclude those unlike themselves.

It therefore matters what meanings and what values are mobilised as the work of strategy-making moves from the 'bricolage' of filtering an 'assemblage' of all kinds of knowledges, and settles into a strategic synthesis. This implies that those involved in creating, probing, testing and challenging strategic ideas as they arise need to maintain not only a continual critical attitude. They need to reflect on the ontological and epistemological assumptions being used to evaluate, justify and buttress a strategic idea. This critical reflection needs to focus not just on what the effects of a strategy might be and how these may get distributed, among groups and through time. It needs to disentangle the kinds of assumptions being made in arguments about effects and impacts. Such critical reflection needs an 'epistemic consciousness'. The value of such a critical consciousness is that alternatives are kept in play; not just alternative information, or even alternative cause–effect models, but the recognition that there are alternative ways in which situations can be framed, understood and valued (Massey 2005). Such a 'strategic performance' leaves spaces for a politics of contestation and challenge, a politics of open evaluation and probing of policy frames (Lindblom 1990). It allows 'voices from the margins' to find expression and, from the very differences of views from marginalised positions, reveal more clearly the biases and omissions of the primary actors (Sandercock 2003a).

But strategy-making for multivocal, multilingual and multi-epistemological urban areas is not quite like the two analogies I have used in this chapter – a construction site or a staged performance; for the 'building' is never finished and the play is never over. The work of strategy formation emerges not as a final, enduring 'product' but, instead, it is a continual 'work in progress', continually unfinished and evolving (see Chapter 6). In this always evolving, always uncertain, always contested institutional space of multiple encounters, what emerges as a strategy is a framing conception (or conceptions), a focus of attention and some kind of agreement about what, as a consequence of the framing and fixing, needs to be done now. This fixing is always provisional, always contestable, always at risk. The 'fixes' could be of many kinds. It could be a material investment (the location of a new settlement, or a high-speed train station) or a location for special attention (Milan's *T-rovesciata*, Amsterdam's *Zuidas*). Or it could be a key value translated into a policy criterion, such as the promotion of more low-cost housing or accordance with critical environmental principles. Or it could be the formation of new arenas of encounter in which strategic orientations can emerge (Milan's Evaluation Panel, the seminars organised by the Cambridgeshire Horizons agency). The 'discovery' processes involved in strategic spatial planning are thus not just about developing an idea of an urban 'region', but about working out appropriate fixes for policy attention and policy action in a continually emergent, continually contested urban reality.

The implication is that a spatial strategy-making that seeks to be open-minded and richly-aware of diverse trajectories and the emergent potentialities through which the 'place' qualities of an urban area are created is deeply contingent on specific conditions, situated in the particularities of a collection of trajectories and their encounters. A strategy that has the capacity to 'travel' and to 'endure', to continue to do useful work in developing urban qualities and potentialities, needs both to be grounded in multiple ways of knowing and experiencing an area, and to contribute to creating that grounding (see Chapter 6). It needs to contribute to cultivating a richly aware urban intelligence. It is very unlikely that this can come about through following some standard prescription or good-practice 'guidance' on how to prepare a spatial strategy. Nor can strategy-makers expect to rely solely on one kind of knowledge. Nor is it easy to sub-contract the task of strategy-making to some outside agency, such as a consultancy or a university team, although these agencies may have a contribution to make. Strategy-making that endures does so by becoming embedded in governance processes, part of the governance landscape, as both the Amsterdam and Cambridge cases show so clearly. It is difficult to sub-contract to outsiders, such as consultants, the complex political and intellectual work of gathering in multiple forms of knowledge from multiple positions and perspectives, and framing strategic ideas through encounters between multiple conceptions and experiences. Attention has to focus instead on creating some kind of 'public realm' through which multiple perspectives may encounter each other. In such 'public realm' encounters, it may be possible to explore each other's ways of knowing and acting, their images and values, their projects and 'pains' (Forester 1999), both in episodes of strategy-making and in the unfolding of governance discourses and practices through which strategies emerge to be seen and to perform institutional work. This has a wider implication than the sphere of urban planning and management. It suggests what the qualities of a creative, inclusive and open-minded 'knowledge society' and inclusive learning culture of governance might actually be (Amin and Cohendet 2004). This moves the discussion on to the issue of governance capacity. In the final chapter, I draw together the insights presented in the previous three chapters to address the implications for urban governance of a relational and interpretive approach.

NOTES

1 Many such documents were produced in the various episodes in Amsterdam, Milan and Cambridge.
2 See Fischer 1989, 2000; Flyvbjerg 1998.
3 See Fay 1996; Fischer 2003; Law 2004; Lindblom 1990.
4 See Knorr-Cetina 1999; Latour 1987; Law 2004.
5 See Fischer 2003 for the distinction between a correspondence and a coherence way of establishing 'truth'. Fischer provides a helpful synthesis of a multidisciplinary wave of intellectual development on the nature of knowledge, perception, learning and discovery. See Barnes 1982; Blackler 1995; Forester 1993; Hajer 1995; Ingold 2000; Innes 1990; Law 2004; Nonaka *et al.* 2001; Wynne 1991; and Fischer's own extensive earlier work.
6 See Fay 1996; Fischer 2003; Lindblom 1990; Salzer-Morling 1998.
7 See Breheny and Hooper 1985; Healey *et al.* 1982; Paris 1982.

8 See Corburn 2005; Holston 1998; Macnagthen and Urry 1998; Scott 1998; Taylor 2003.
9 See Corburn 2005; Hillier 2002; Sandercock 2003a.
10 This perspective, which casts scientific knowledge as a distinctive form of local knowledge, is developed in both cultural anthropology (see Geertz 1983), and Science and Technology Studies (see especially Knorr-Cetina 1999; Latour 1987; Law 2004).
11 See Blackler *et al.* (1999) and Nonaka *et al.* (2001) on organisational learning, Amin and Cohendet (2004), Lagendijk and Cornford (2000) and Morgan, K. (1997) on knowledge as a resource for firms, and Fischer (2003), Forester (1993) and Hajer (1995) on knowledge and learning in policy contexts.
12 See Amin and Cohendet 2004; Blackler 1995; Lagendijk and Cornford 2000; Morgan and Murdoch 2000; Nonaka *et al.* 2001; Takeuchi 2001.
13 This is emphasised strongly among scholars in the field of 'Socio-Technical Studies' (see Knorr-Cetina 1999; Law 2004; Latour 1987).
14 For the implications of 'living in the world', see Kitching (2003) for a discussion of Wittgenstein's understanding of this, and see also Ingold (2000). Such a perspective is also found in the work of planning writers drawing on pragmatist inspiration (see Forester 1993; Verma 1998).
15 Figure 8.1 has been created from multiple sources, particularly Blackler 1995; Fischer and Forester 1993; Forester 1999; Innes 1990; Melucci 1989; Nonaka *et al.* 2001; Thrift 1996, chapter 6.
16 There has been an escalation of qualitative research methods in the social sciences in recent years to help policy-makers access such experiential knowledge. For valuable contributions in the planning field, see Corburn 2005; Eckstein and Throgmorton 2003; Flyvbjerg 2001; Roy 2003 (appendix).
17 The importance of combining forms of knowledge in developing policy debate is stressed by many authors, see especially the work of Owens and Cowell 2002; Owens *et al.* 2004.
18 See Hillier (2002), Corburn (2005), Davoudi (2006), and many examples from studies of urban regeneration experiences involving community participation.
19 There is a substantial literature on techniques for such practices (see Burns and Taylor 2000; Plummer 2000; Susskind *et al.* 1999) and an increasing number of studies evaluating how different techniques work in different contexts (see Burby 2003; Gunton *et al.* 2003; Innes and Booher 1999b; Taylor 2003).
20 See Amin and Thrift 2002; Dryzek 2000; Fung and Wright 2001; Healey 1997; Massey 2005; Melucci 1989; Sandercock 2003a; Schlosberg 1999.
21 See Forester (1993: 104) and Fischer (2000: 70) on the distinction between uncertainty and ambiguity.

CHAPTER 9

RELATIONAL COMPLEXITY AND URBAN GOVERNANCE

> Spatial planning is best viewed as a set of interdependent processes involving multiple actors that seek to create more liveable, life-enhancing cities and regions (Friedmann 2005: 213).

> The modern city is ... so full of unexpected interactions and so continuously in movement that all kinds of small and large spatialities continue to provide resources for political invention as they generate new improvisations and force new forms of ingenuity ... the city is brimful of different kinds of political space (Amin and Thrift 2002: 157).

> In the present transition ... we stand especially in need of planners who will assist with the recovery of an active political community (Friedmann 1987: 417).

A 'Relational View' of the Planning Project

This book has explored what is involved in focusing governance attention on the place qualities of urban areas and the space–time dynamics of the relations and interactions that take place in such areas. In particular, it has examined what it means to intervene in shaping place qualities through conscious attention, through some kind of strategy, which embodies and expresses a conception of the place of an urban area, whether this is what may be commonly called a city, a city with its surrounding landscapes or a collection of urban and rural settlements. I have been especially interested in how far such strategically focused governance attention has the capacity to contribute to a double transformation: in the material trajectories of urban dynamics and the potentialities they afford to the multiple social and environmental relations that inhabit and pass through urban areas; and in the forms of governance directed at shaping the qualities of places experienced as urban. I have also emphasised the complex interplay between explicit attention to place qualities and connectivities, and the wider context in which such attention is situated, both in terms of evolving governance processes and wider social, environmental and economic dynamics and governance cultures. To understand the interplay of proximities and connectivities in urban areas, and the interactions that form governance processes, I have used the intellectual lens of an interpretive policy analysis and a relational geography. These focus attention not just on individual agents or on structural driving forces but on interactions, on how meanings are made, on how relations are understood, and how action is shaped in social contexts. I have focused this lens on the politically charged processes through which collective action is imagined, mobilised, organised and practised to 'make a difference' to urban conditions.

This chapter draws the discussion together around the issue of governance processes in urban areas. The first section consolidates the approach developed throughout the book, to situate the planning project as an activity centred on the governance of place. The second section builds on the approach to governance introduced in Chapter 2 to elaborate more clearly a relational approach to urban governance. The third section then makes one final 'visit' to the case studies to assess how far they have realised the potential of a planning project for the governance of place in ways that enrich rather than reduce the multiple experiences of urban daily life, and promote policy attention conjointly to distributive justice, environmental well-being and economic vitality. The final section offers suggestions for those in practical contexts seeking to pursue such a spatial strategy-making project.

The experiences presented in Chapters 3, 4 and 5 provide accounts of evolving practices of place-focused governance attention to urban qualities and dynamics. They show recurrent efforts to set specific practices of place-management and project development in the context of some conception of the qualities and dynamics of a wider urban 'region'. At the core of such efforts is the ambition to maintain an awareness of wider horizons in time and space, while engaging in the ongoing flow of allocating resources to projects and programmes and regulating development activity. I have shown how the concepts mobilised in the search for appropriate understandings and management tools have shifted substantially over time. In the confident mid-twentieth century, an urban area was understood as a coherent entity, with a concrete physical pattern expressing a simple relation between economic and social dynamics on an environmental surface. Such an entity could be managed by strong spatial plans backed by powers of public ownership of land and engagement in development activity. This conception has given way to the contemporary recognition of the complexity of the relations that co-exist and transect in urban areas and the range of governance processes which affect how these relations evolve. This complexity finds expression in the burgeoning academic literature in the sciences and social sciences on 'complexity theory'.[1] But it is also appreciated through a practical awareness of the multiplicity of relations that demand and need attention, of the indeterminacy and unpredictability of what happens in urban areas, and the effort involved in seeking ways of linking up, coordinating and jointly focusing the activities of an array of governance involvements that contribute to how the proximities and connectivities of an urban area evolve and are recognised.[2]

The discussions presented in Chapters 6, 7 and 8 have used an interpretive and relational intellectual lens to develop a way of thinking about this complexity. My aim has been partly to provide a more systematic academic treatment so that practices can be better understood, developed and changed. I have specifically sought to provide a critical perspective through which to grasp what is involved in making a spatial strategy for a 'conceived' urban 'region' and to help make judgements about whether such an effort is desirable or achievable in particular times and places. I have sought to detach the notion of a strategy both from its use as a cosmetic rhetorical invocation required to meet some legal or funding requirement, and from a conception of a rigid plan, whether in the form of a comprehensive spatial pattern or a coordinated, sequential programme of action. Instead, I have emphasised two dimensions of strategy: as an orientation, a frame of reference that gives direction across a diffused governance landscape; and as justification for specific interventions that arise from the perceptions, meanings and values embodied in the frame (see Chapter 6). Strategies exist as revisable, fluid conceptions continually interacting with unfolding experiences and understandings, but yet holding in attention some orienting sensibility. Such a notion of evolving strategy, continually in formulation, is a necessary complement to the recognition of the relational multiplicity of the lived experiences of contemporary urban worlds.

This double shift, to a more complex appreciation of the multiplicity of relational dynamics to be found in urban areas, and to a perception of strategies as fluid, revisable frames of reference, unsettles many of the established conceptions of the planning policy community, and of wider governance cultures. It is no longer possible to understand material places and social nodes as 'the local', the sub-regional, the regional, etc., positioned in a one-dimensional hierarchy of scales (see Chapter 7). Instead, 'placings' are produced through the dynamics of relational webs that connect the place of one node with others, near and far, and locate it in times of memory as well as future potentialities. There is no objectively coherent functional entity that can be called a 'city' or an urban 'region', or Amsterdam, Milan or Cambridge. Instead, the concept of an urban 'region' has to be 'called to mind', 'summoned up' into imagination. It has both to 'make sense' and resonate with existing understandings and experiences to gain persuasive and seductive force, and to 'create sense', by adding meaning and identity to an array of ongoing struggles, perceived problems and opportunities. An explicit conceptualising of an urban 'region' attracts governance attention to the extent that it adds value and gives value to aspects of the ongoing stream of collective action. I have emphasised that the imaginative resources for the creation of an urban 'region' strategy, which has the potential to perform such institutional work, cannot be formed through a narrow conception of the relational dynamics of an urban area. However selective and focused the identification of key qualities and the interventions proposed for key strategic attention, an urban 'region' strategy that has the potential to add value and give meaning to relationships in continual formation and interrelation involves a rich knowledge of the multiple webs of relations and communities of practice of urban life. This implies a capacity among those involved in strategy formation and revision to interact with and tap into multiple sources of knowledge about, and ways of giving meaning to, urban life and an ability to relate different forms of knowledge to each other.

Throughout, I have emphasised the value of recognising multiplicity, diversity and heterogeneity. This is partly because this is what 'tapping into' the experiences of urban life tells us. Multiplicity is a significant quality of contemporary material and imaginative urban existence.[3] Any governance strategy that ignores this will encounter problems of effectiveness. But I have also stressed this because a recognition of multiplicity and diversity is an essential optic through which to keep in the focus of attention the complex ways in which distributive justice, environmental well-being and economic vitality can be generated and experienced. If the ambition in initiatives in urban governance is to keep all these values in conjunction in the foreground of policy attention, then it is important to resist tendencies in spatial strategy-making towards reductionist, monotopic, hegemonic viewpoints, and to recognise the messy and multiple daily experiences of urban conditions. For governance actors brought up in the welfare states of Western Europe, this is a challenging political and intellectual project. It demands a viewpoint different from that of the delivery of individual services organised through vertically structured policy communities revolving around the nation state, or large regions in federal states. Instead, it requires perceiving an urban area from the viewpoint of the lived experiences of citizens and businesses, in daily, weekly, yearly and intergenerational timespans. At the same time, it emphasises awareness of the multiple connectivities of people and firms in urban areas with all kinds of other people, in other places, now, in the past and in the future. It demands attention to the 'liveability' of urban areas (Friedmann 2005), as experienced through a 'daily life' perspective (Amin and Thrift 2002; Healey 1997), whilst also being fully aware of a potentially global reach of relational associations and responsibilities (Massey 2005).

In such a perspective, the planning project has the potential to become much more than just an attempt to insert a spatial perspective into public policy, or to encourage better coordination or more effective ways of tracing the impacts of a proposed development project. It offers a different way of thinking about what governance attention could achieve and how governance interventions might be designed and operationalised. It becomes a way of thinking about the dynamically evolving relations between the multiple proximities and connectivities of places that are called up as a focus of governance attention. It involves a way of imagining how such proximities might develop, and how this may affect different relational webs and their intersections. It sustains a practice of mobilising policy attention to how values of distributive justice, environmental well-being and economic vitality may be compromised by the neglect of attention to qualities of place. It demands a capacity for strategic judgement about which actions might carry the capacity to make significant contributions both to material improvement and to the values and sensibilities that diverse groups have about the 'place' of the urban.

The project of urban planning thus becomes an activity explicitly concerned with the governance of place. But such a planning does not necessarily work through an explicitly formalised spatial strategy or legally required development plan. Its contribution, its value within the array of governance processes in an urban area, lies in maintaining focus on an awareness of place effects and relations among multiple governance arenas and practices. To this end, those involved in planning have a responsibility to maintain a strategic consciousness of place dynamics. But how is such a 'planning project' positioned in relation to wider landscapes of governance involved in shaping urban futures?

THE RELATIONAL COMPLEXITIES OF URBAN GOVERNANCE

In Chapter 2, I introduced a conception of governance as the mobilisation of collective action for public-interest purposes across the spheres of the state, the economy and civil society. This mobilisation operates through all kinds of webs of relations, that connect the organisations and procedures of formal government with informal governance arenas and networks, and the wider society. I argued that particular actors, tied into one or more networks, and operating in different institutional sites or arenas, are embedded in the discourses and practices of governance processes, even as they resist, challenge and strive to create alternatives. Governance processes, although sometimes appearing stable and immovable, are in continual tension, as actors struggle for dominance over arenas, for control over discourses and practices, making use of and being undermined by 'reform' movements inside formal government as well as mobilisations from outside. The patterning of power dynamics and routines of practice are held in place but are also unsettled by movements in the wider society, the spheres of civil society and the economy. These movements create expectations and demands, providing a shifting judgemental ground that shapes the perceived legitimacy of governance activity. The power dynamics of governance are thus played out at multiple levels of practical consciousness, as expressed in Table 2.1, and make use of different forms of power. Spatial strategy-making initiatives are located within these governance dynamics. The case accounts show how the institutional sites of their articulation have been affected by wider shifts in governance organisation. In the mid-twentieth century, such initiatives strained against the grain of the hierarchical, sectoral organisation of welfare states. At the end of the century, they promoted a place-focused perspective as an alternative to the perceived fragmentation of governance activity occurring in urban areas. All the cases show the significance of the authoritative power of the formal state, backed by the expertise of professional groups. They also show a growing recognition that, in a more diffused governance context, more persuasive and seductive forms of power become significant.

This approach to governance challenges concepts of government and urban governance as either a unity, or a single actor (Stone 2005). Governments do not act. Actors in positions in government act, drawing on all kinds of resources and social practices which shape their perceptions, ethics, remits and responsibilities. Materials produced by government activity, such as laws and statements, 'act' through the authoritative power embodied within them. In the processes described in the cases, 'governance' emerges as a complex array or assemblage of relations and rationalities – of formal politics interacting with policy communities across diverse arenas; of the logic of formal law interacting with the logics of immediate interests and the logics embedded in evolved practices and discourses; of the logic of formal authority interacting with an array of networks and challenges that shift through time. It may sometimes appear as if a consolidated unity, a regime, exists. Amsterdam City Council, with strong formal powers supported by a widespread perception among citizens and business interests of its importance and legitimacy, has at times had the quality of a regime (Harding 1997). Yet it has shifted discourses and practices over the years, moving from a social-democratic programme for delivering liveable neighbourhoods to ordinary residents, to a more recent emphasis on a cosmopolitan perception of the qualities of urbanity and the promotion of 'world-class' business environments. A closer look at Amsterdam's governance processes reveals complex tensions between different discourses and communities of practice, linked to different constituencies across the city, and changing at different speeds (see Chapter 3).

Rather than a unified regime, it is perhaps better to consider urban governance landscapes in terms of their degree of consolidation. In some places, an enduring hegemonic regime may develop, sometimes centred on formal government, or within a nexus between a particular group and the state. In countries such as the UK and the Netherlands, policy communities around particular sectors of government activity have tended to exercise a strong stabilising force, a contrast with the USA, where business groups have commonly exercised this role (Stone 2005). It is these policy communities that are being challenged by the government reform movements in both countries. In Milan, as in Italy generally, political party networks played a similar role in the post-war period until the 1990s (Vicari and Molotch 1990). But however firm the consolidation of an urban governance landscape, the forces excluded from it will continue to assert their challenges, unsettling the hegemony. In the 1970s, Offe (1977) argued that any enduring stability was unlikely. Instead, he claimed that governance processes embodied a continual 'restless search' for a resolution to unresolvable contradictions among social groups (see Chapter 6). This suggests that, in exploring governance landscapes, rather than searching for the existence of stable regimes, it may be more productive to attend to the degree of 'restlessness' and the trajectories of searching, to the complex ways in which what is fixed at one time and place becomes fluid and unsettled in another, and vice versa.[4]

In the present period, in Western Europe at least, formal governments at all levels are struggling with challenges to the 'fixes' inherited from the mid-twentieth-century welfare settlements. The critiques of the 'welfare state' mode of policy formation and delivery are of long-standing. Business interests argue that government procedures and practices place too much emphasis on bureaucratic rule-compliance rather than releasing entrepreneurial energy. Those concerned with social welfare complain of the paternalist attitudes embedded in the practices of welfare-delivery agencies and government departments. Environmental lobbies stress the neglect of environmental considerations in programmes targeted at social welfare and business support. But these critiques rarely support an argument for weaker formal government. Instead, the emphasis has been on transforming governance discourses and practices, and on changing organisational arrangements and procedures (Pierre and Peters 2000). Functions formally undertaken by government agencies have been privatised, devolved, shifted to special agencies and partnerships, with diverse arrangements for establishing their legitimacy. The result in urban areas is a much more diffused governance landscape than that dominant in the post-war period, in which the relations and interactions between the spheres of the state, civil society and the economy are more complex and less well-understood than in the past. The big struggles between key social groups (the conflicting interests of labour and capital, for example) have been displaced by a multiplicity of struggles over diverse issues, unsettling past fixities, but not necessarily succeeding in generating new ones (Lascoumes and Le Galès 2003). Moments of opportunity to unsettle fixities have opened up all over governance landscapes, creating opportunities for innovation, but also for exploitation of one group, or interest, by another. Governance landscapes, as evident in all three cases, seem much more unstable, and the task of reading their emergent trajectories much harder. In such conditions, it is wise to avoid treating urban governance as a single unit or assuming that the 'urban level' of government will respond in a similar way as before to exogenous forces and policy directives from higher tiers of government.

In this context, initiatives in spatial strategy-making, calling on the promise of the planning project conceived as a consciousness of the qualities of the proximities and connectivities emerging in an urban area, offer a way of bringing the diffusion of initiatives and arenas, of discourses and practices, into some kind of encounter. A strategic framing of urban qualities and possibilities has the potential to stabilise diffused governance initiatives around some shared direction, to unsettle embedded trajectories from the past, and to challenge or encourage emerging trajectories. By bringing an awareness of the place of the urban from the background of attention into the foreground, a spatial strategy-making project has the potential to contribute just by raising awareness of linkages and tensions, and pointing out the discriminations created by current governance activities when played out across the places and accessibilities of an urban area. Such a project may be able to go further; first to create a strategic frame that resonates with the concerns of many, that mobilises actors in diverse positions across a governance landscape, that accumulates mobilising power to reach arenas where significant authoritative power is exercised; and, second, to travel to many arenas of governance across an urban area, so that the frame is frequently 'called up' to shape and legitimate diverse actions. The activity 'works' by creating a discourse and a nexus of relations that interpenetrates with other flows of activities-in-relations, shaping attention and potentially sedimenting into governance cultures. In such a context, as argued in Chapter 7 (see Figure 7.2), the activity of spatial strategy-making for a diffused urban governance landscape may contribute to a temporary stability in an imagined urban trajectory, a temporary 'fixing'.

To create such an effect, an initiative in spatial strategy-making needs to reach key institutional sites where authoritative power over investments and regulations are exercised, and to filter into the wider 'public realm' of discourse about urban conditions. Thus, in Amsterdam, notions of a multicentric urban 'region' and corridors of development were continually pulled back into conceptions of compact development and firm physical distinctions between town and country. In Milan, the notion of the urban area remained firmly riveted to a conception of the '*cuore*', or 'urban heart', as the essence of the area's urban qualities. In Cambridgeshire, any proposals for growth had to be shaped into the longstanding conception of a city region with a clear settlement hierarchy. In many instances of spatial strategy-making, dominant discourses from other governance arenas may limit what is imagined about the place qualities of an urban area. The emphasis on 'economic competitiveness' promoted by EU, national and local government actors and strategies exercised a strong influence on the development of spatial strategies in Amsterdam and the Cambridge Sub-Region by the 1990s, and was being picked up in Milan in the mid-2000s. This discourse, promoted through public-funding priorities and legislation, was difficult to challenge at local levels of government, without strong governance cultures prepared to mobilise to keep other priorities in play. This capacity existed in both Amsterdam, around concerns for neighbourhood quality and urban liveability, and in the Cambridge area, around concerns for environmental quality.

The economic 'competitiveness' discourse, where it becomes hegemonic, reduces spatial strategy-making to a monotopic and univocal expression of the urban. It 'sees' and privileges only a small selection of the relational webs that transect an urban area. It loses the promise of an inclusive awareness of the multiplicity of relational webs that inhabit and transect an urban area, and the complexity of the interactions, synergies and conflicts between them. In the new English Planning and Compulsory Purchase Act 2004, an attempt has been made to limit such an exclusive preoccupation with economic relations through defining the purpose of the planning system, and of spatial strategies within it, as the pursuit of 'sustainable development' (ODPM 2005a). This philosophy is presented as an ambition to encourage strategies that are place-focused, integrative and inclusive. But this seems to suggest that there is a comprehensive, robust 'integration' to be found between conflicting potentialities and priorities. Strategy-making involves selectivity and simplification (see Chapter 6). Integration around a new focus means disintegration from a previous focus. It is never clear in the rhetoric of 'sustainable development' what is to be foregrounded and what slips out of policy attention.[5] Instead, the concept of 'sustainable development' belongs to a rhetoric that suggests that a harmonious 'fix' between the competing concerns of economic competitiveness, environmental sustainability and social cohesion can readily be found (CSD 1999; ODPM 2005a). Yet this is challenged even within the governments which promote it.[6] In the wider society, mobilisation around specific issues, such as particular business interests or associations promoting the welfare of particular non-human species (such as birds), or habitats, or landscapes, or groups concerned with access to low-cost housing, or with the redress of particular inequalities, struggle with each other to get policy attention. This leaves the field apparently open to well-organised lobby groups and established policy communities. These continue to define policy agendas and discourses, and group issues into sectors for policy attention, while at the same time making calls for more holistic or 'joined-up' approaches (6 *et al.* 2002; Wilkinson and Appelbee 1999).

However, the idea of integrating policy attention around place qualities is not just a product of a 'sustainable development' movement that has developed into a policy community. There are other tendencies in civil society and the economy that press for an awareness of the way issues interrelate in places – the places of the neighbourhood and of urban areas in particular.[7] The inspiration for a different awareness comes from the experience of daily life conditions in urban areas, whether of individuals, or firms, or other species and physical processes. It derives from changes in social perceptions, habits and aspirations. It is driven by the experience of the 'throwntogetherness' of being in urban areas (Massey 2005), of overlapping and overlayered co-existence in shared spaces (Healey 1997). This creates a political ground for the articulation of policy momentum around the everyday experience of living in urban areas. This impulse motivated the planners of the mid-twentieth century, but they assumed a socially and spatially integrated urban order. Such an impulse was revived in the urban social movements of the 1970s, but informed by a more critical view of the power relations between different groups in the city. Now such an impulse continues to inspire citizen protests and complaints about urban conditions. It also lies behind many of the demands from business interests for help in smoothing out the conflicts in finding and keeping a labour force, or moving goods around. It is being articulated in a revival of interest by political groups in the 'liveability' of urban environments. It is this momentum that widens and counteracts the narrow emphasis on 'economic competitiveness' and that enriches the weak umbrella of the idea of 'sustainable development'. But, compared with earlier movements, this impulse is more diverse and fragmented, reflecting the changes in the relations of urban dynamics overall and in the relations of urban governance in particular.

A revival of interest in the place qualities of urban areas is thus more than just another turn in policy fashion or a narrow pre-occupation with developing the economic asset base of cities. The urban comes into focus because it 'makes sense' as the locale of the complex intersections of much daily life experience. It is more than a neighbourhood, where the patina of daily life around the home, the workplace, the school, the leisure centre, the entertainment district, the park, etc., creates a physical awareness of place qualities and encounters. It is the locale that people live within and move across, the place of alternative opportunities and potentialities, of multiple nodes and connectivities, of diverse private and public places, of encounters with all kinds of 'others' (Amin and Thrift 2002; Sandercock 2003a). Such a recognition in society generally could power initiatives in spatial strategy-making for urban areas, if they were developed in ways that resonated with such an understanding. But the urban 'in mind' in this appreciation of daily life has no fixed boundaries. The relations that weave within and across urban areas extend to all kinds of other places and scales. Determining the qualities to express within a strategic frame of reference and the priorities to focus on when developing a strategic programme therefore requires careful, situated, working out, in conjunction with many parties across an urban area.

In all the three cases, those involved in spatial strategy-making were attempting to reach beyond inherited intervention mechanisms and to re-cast them in recognition of the contemporary realities of the 'daily life' of households, businesses, property developers, visitors, etc. Amsterdam planners sought to expand the 'area of search' for the focus of the urban by widening the institutional arena for considering spatial strategy and by enriching their long-established emphasis on the qualities of urban life. In Milan, those involved in spatial strategy-making struggled to consider the impacts of projects in new ways and to make richer connections between real-estate development processes and the negotiation of public benefits. In the Cambridge Sub-Region, those involved in strategy-making sought to co-align the needs of a particular sector of business with the daily life of households.

In each case, there was an undercurrent of concern about the tensions between the urban area, as experienced in this daily life way, and the administrative jurisdictions of the formal government system. But in none of the cases had a formal organisation emerged at a level beyond the municipality (or county, in the case of the Cambridge Sub-Region) with judicial and political powers, even though it was widely recognised that the intersecting relations of the economy, state, civil society and environment were better understood from such an institutional site rather than in an arena focused on a neighbourhood or a particular development site, or the arenas of a nation state or a large province or region. Attempts to create formal metropolitan-area agencies had failed in Amsterdam. In Milan, the political relations between the province and the city inhibited such an arrangement and, in any case, the area of the province was too small to capture the continually expanding urban area. In Cambridgeshire, the formal strategic planning powers of the county were weakened in favour of a much larger region, for political reasons that had little to do with the development challenges faced in the area. As Salet *et al.* (2003) have argued, the creation of formal region agencies above the level of municipalities has proved problematic in Europe.[8]

Many continue to argue, however, that new formal structures and procedures are necessary to carry forward the momentum behind spatial strategy-making focused on urban areas. This is often presented as the only way to give sufficient authoritative power and legitimacy to the exercise. But formalising strategy-making into a particular arena and set of procedures is not necessarily conducive to releasing the creative and persuasive understandings and imaginations with which to infuse a strategy with mobilising force. Formalisation could stifle mobilisation capacity. Networking between existing formal and informal arenas may be just as effective a way of raising strategic consciousness about the qualities and connectivities of an urban area. Although any strategic idea needs to attach itself to authoritative power at some stage if it is to deploy public resources and powers legitimately, it is just as important that a mobilising force is developed to carry strategic ideas forward. This requires more attention to persuasive and seductive power. The alternative is to develop ways of keeping a view of the urban 'in play', as discussed in the previous chapter, across the array of governance institutions in an urban area. Informal mobilising around issues of specific concern to those with a stake in an area is more likely to generate a real sense of collective concern for the future of an urban area than any formal reconfiguration of government agencies and responsibilities.

In any case, explicit attention to the place qualities of an urban area may not always be a high priority and there may be no momentum locally to carry such an idea forward. Rather than proposing general organisational arrangements or general formulae requiring the production of strategies, those seeking a more integrated and inclusive approach to urban governance might examine specific local conditions by assessing the local momentum behind attention to place qualities and the degree of awareness of the multiple dynamics shaping urban futures. Figure 9.1 suggests such a scoping tool. Amsterdam could perhaps be positioned in the top-left quadrant, Milan on a trajectory from bottom-right towards the left and the Cambridge Sub-Region positioned uncertainly in the middle.

Formalisation may also generate tendencies to create spatial strategies in reductionist ways. Legal requirements or professional fashions that demand a 'strategic vision' or 'core strategy' may be met by borrowing statements about an urban area from examples from elsewhere, or from 'good practice guidance' (Gaffikin and Sterrett 2006; Shipley 2002). But an urban area is not a singular object with a governance arena and intervention strategy to match. I have stressed throughout this book that urban areas

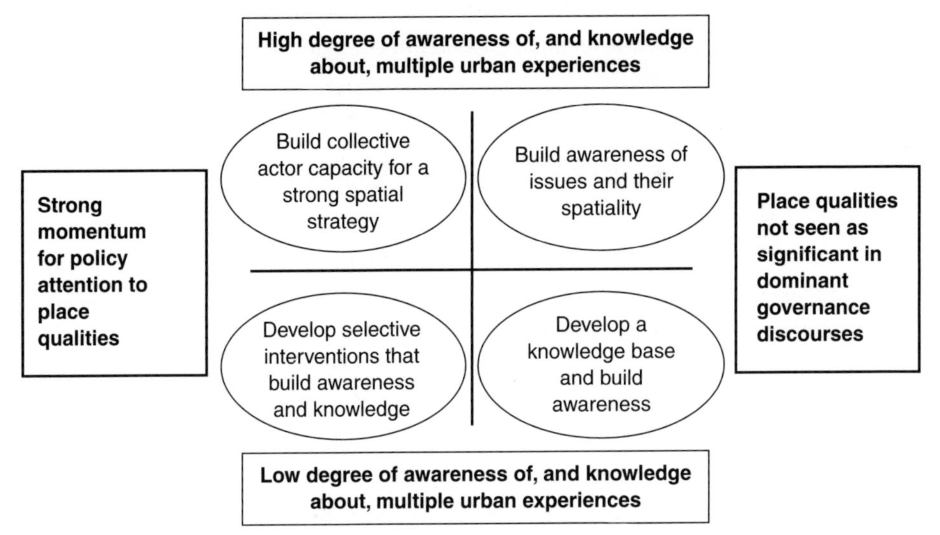

Figure 9.1 Scoping the potential moment of opportunity for explicit spatial strategy-making

need to be understood as a complex array of intersecting relations. Its proximities and connectivities are specific to its particular histories and geographies, and are in continual evolution. So governance attention needs, in turn, to find ways both to 'foreground' attention to the daily life of urban experience in specific places and to avoid a narrow fixing of understanding and of strategy on abstract dimensions that do not resonate with locally lived experiences.

The art of generating a momentum for strategic attention to the place qualities of urban areas thus lies in the ability to grasp and enlarge local and external moments of opportunity and to position strategic ideas in ways that relate to the evolving concerns and experiences of stakeholders in such place qualities. The responsibility of those generating such a momentum is to attend to the relevance and legitimacy of their initiatives. This means more than just capturing control of the exercise of authoritative power to distribute investment resources and to structure how land-use regulation powers are used. It means an ethic of critical consideration as to why and for whom attention to the 'place' of an urban area has value and meaning, as to how such attention is developed, and as to the multiple ways in which the legitimacy of such a policy programme may be established.

In summary, spatial strategy-making initiatives in urban areas in Western Europe at the turn of the century are governance interventions designed to challenge the sectoral divisions of welfare-state organisation and promote a more locally sensitive understanding of the daily life experience of co-existence in shared localities. Although sometimes confined to narrow arenas of influence, or produced in reductionist ways, or sometimes too fragile to accumulate sufficient power to capture governance attention and access to authoritative power over resources and regulatory systems, they may also succeed in accumulating force to spread across a governance landscape. In such circumstances their power can be considerable. This means that such strategies need to be under continual critical review to avoid too hard a stabilising effect on future evolutions. In the next section, I suggest key questions for such critical review.

THE STRATEGIC PLANNING PROJECT IN PRACTICE

I now make a final 'visit' to the three cases used in this book, focusing on the most recent initiatives. As the overall case accounts emphasise, the consequences of initiatives undertaken at the turn of the century will take time to play out through the complexities of urban relations and may not be experienced or perceived until much later. So any assessment now can only be of their possible potentialities. How effective, how transformative and how beneficial may these turn out to be? How far do they promote and achieve the normative concerns raised at the end of Chapter 1 and elaborated throughout this book? I centre the assessment around three sets of questions.

First, how far have these been informed by a daily life perspective and an attempt to hold the combined values of distributive justice, environmental well-being and economic vitality in critical conjunction? Have they been too narrowly focused on economic 'competitiveness'? Have they been too boxed in to sectoral arenas to have significant effects? Have they come to over-stabilise the emergent potentialities of their urban areas? Second, what has been their impact in transforming the urban governance landscape? How far have initiatives been able to shift embedded governance discourses and practices and change debates in the 'public realm' of the wider governance landscape? Finally, what impacts have they had on material conditions and on place identities? Whose lives and relations have been or may be changed by the potentialities opened up by a strategic initiative? How different are such changes from what has gone before?

A MULTI-VALENT DAILY LIFE PERSPECTIVE

Economic considerations permeated the recent experience in all three cases. Exogenous forces encouraged more attention to financial considerations as a result of national shifts in the financing of public projects and agencies. Reductions in general funding for local government encouraged greater attention to ways of combining private investment with public purposes. This approach to development was well-established in England, generating a 'planning' practice long dominated by the regulation of development. The challenge for spatial strategy-makers in the Cambridge Sub-Region was to mobilise formal government to enable a move to a more proactive approach to development management. But national government wanted to squeeze as much funding from private developers as possible. The discourse of 'economic competitiveness' was very strong in this case, but held in continual check locally by a civil society that demanded careful attention, both to environmental qualities and to affordable housing for 'key' workers. The concerns of more marginal social groups and the demands of areas outside the 'sub-region' tended to slip out of attention in this context.

In Milan, the discourse of global economic competitiveness was only weakly present until the mid-2000s. The focus of planning-practice innovation was on relations with real-estate developers. The key ambition was to shift attention from designing projects to actually achieving them, through changing formal bureaucratic procedures and encouraging more collaborative working among key stakeholders. An abstract sensibility to the need to balance distributive justice with making development projects economically viable led to the effort to negotiate public benefits from private development, but knowledge about needs and demands was initially very limited. Environmental issues were addressed primarily through the technical machinery of formal environmental assessment requirements. The new arenas thus opened up small possibilities for some kind of conjunction between the social, environmental and economic dimensions of projects, but exactly how specific relations negotiated around a project were connected to each other and to other governance relations remained uncertainly identified.

The Amsterdam case illustrates the most energetic attempt to build on a traditional focus on the 'liveability' of the city and to update this to reflect contemporary perceptions of diversity and multiplicity. This was informed by a pragmatic sensibility in which a practice of doing projects was continually shaped by wider strategic frames about urban development. But a combination of national devolution pressures demanding more attention to the wider urban area, along with the developing power of the city's districts and the new kinds of public–private partnerships emerging around major projects, was diffusing the consolidated control over building the city that once centred in the City Council planning department. In this situation, there was an emerging tendency to construct different groupings of interest. Large corporate companies and major infrastructure investors clustered in 'platforms' related to major development projects, while residents discussed issues of 'liveability', related to established neighbourhoods in arenas related to the district councils. In this context, concerns for economic vitality and distributive justice could drift apart.

TRANSFORMING URBAN GOVERNANCE

How far, and in what way, have the initiatives shifted previously dominant urban governance discourses and practices, and provided new dimensions to the wider realm of public discourse about urban conditions and their management? In Amsterdam, planning officials have long been accustomed to a powerful role in shaping both the materialities of urban development and governance attention to the development needs and qualities of the city. But the discourses and practices developed within the arenas of the planning policy community have brought into being practices of continuous interaction with both other policy communities and the diverse arenas of public discussion about urban conditions in the city. If anything, both the wider discussion and the planners' response have speeded up in recent years. Yet there is still criticism that spatial strategy-making remains too contained within the social world of the planners. The ongoing struggle has been to maintain a kind of mutual transformation process that weakens the old boundaries of a sectoralised, public-sector-dominated governance structure, to allow different ways of practising and different ideas about issues such as accessibility, urbanity and liveability. There remains a danger that the strategic force holding in conjunction concern for distributive justice, environmental well-being and economic vitality, centred in the arena of the City Council planning department, will weaken or focus too strongly on major projects, rather than on the fine-grained interventions that help to sustain the complex daily life co-existence of many webs of relations across the city.

In the Cambridge Sub-Region, the effort for strategic mobilisation has had considerable success. It has shifted discourses away from a conservation agenda towards a development focus. But now this has to be translated into new practices of proactive development promotion. A much bigger agenda of projects has to be managed, each involving complex partnerships with major public- and private-sector actors, with major difficulties in coordinating public-sector agencies responsible for different fields of investment, and under the watchful gaze of a critical and articulate local citizenry. Once, the County Council provided a critical institutional site for coordinating investment in transport and the regulatory processes of allocating development sites. By the 2000s, its formal position in this respect had been removed, with new sites consolidating in the much-more-distant arena of the East of England region, and in the more specifically focused development-management agency, Cambridgeshire Horizons. It is now (2006) a major challenge to find an appropriate long-term agency form for this task, and to combine the practices of proactive development promotion with those of regulating the way private interests are constrained in the development process. The first has to look to persuasive power and political legitimation; the second to authoritative power and legal legitimation. The Cambridge Sub-Region strategy is held in place operationally by the presence of particular people who have shifted formal arenas but nevertheless carry their networks and their strategic orientation with them. The danger for the strategy promoters now is that a narrow focus on delivering housing numbers and finding ways to push the funding of infrastructure onto private developers may drive out consideration of the quality and liveability of the development produced, let alone the fine-grained management of neighbourhood change elsewhere in the area. In Milan, the wider impacts of shifts in the practices of development negotiation are still too limited to assess their longer-term potentialities in transforming the urban governance landscape.

IMPACTS ON MATERIAL CONDITIONS AND PLACE IDENTITIES

What impacts are the recent strategic initiatives likely to have? This question was continually addressed, both politically and technically, in all the cases. The consideration of impacts raises issues such as: whose lives are lived better because a new piece of city has been created? Whose accessibilities have been enhanced, to what and where? In what ways have new development and new ways of managing existing development reduced the environmental harm generated by urban development and increased the distributional justice of opportunity in an urban area? What contribution has been made to economic vitality? How have the connectivities within and beyond an urban area been enhanced or reduced? Such questions have no easy answers in the short term and require a different kind of research to that undertaken here. So I can merely comment in an impressionistic way.

In Amsterdam, many of these issues have long been held in the forefront of policy attention, both across the City Council and in the wider governance discourse, but would be compromised if the relations with different programmes and different stakeholders became separated, as may already be happening. The most important issue, therefore, is how far the concerns about urbanity, accessibility and liveability will be kept 'in play' as the new 'pieces of city' in the major projects are developed. In Milan, the potential for separating considerations into different arenas is also a key challenge, as well as anticipating where the next burst of private-sector development investment is likely to focus. In Cambridgeshire, the debates on amounts and locations of development have largely been settled, with a major impact on the built environment of future decades. What remains to be struggled for is the quality of new development. Who will live and work in the new projects and how will the opportunities created in the new 'pieces of city' impact on and connect to the opportunities for others? This is especially significant given the overlaying of linkages centred on and around Cambridge with those in the wider sub-region and the London metropolitan region as a whole. Where will the new residents of Northstowe and the other major development allocations actually work, shop, get services and enjoy their leisure time?

What is striking is that, among the three cases, only in Amsterdam was there a vigorous debate about the qualities of the urban area and the proximities and connectivities weaving through it, within which the impacts of the particular developments on the qualities and living experiences of others in the area could be considered. It is in this respect that the qualities and capacities of governance cultures come into play. In Amsterdam, a governance culture has emerged with an active public realm, with multiple arenas, multiple opportunities for discourse and some connectivity between them. This allowed for continual critical contestation and exploration about the qualities of urban life. This, in turn, demanded an ongoing strategic awareness by key governance actors of the importance of respect for the liveability of the city, and its diverse locales and connectivities, understood in terms of the diversity of the city's social groups. In both Milan and the Cambridge Sub-Region, there are strong traditional conceptions of place, but recent strategic episodes have paid only limited attention to the connectivities between the emergent place qualities of the new development being promoted, and the qualities of places outside the strategic focus. Nor are local governance cultures demanding such attention.

THREE CONCLUSIONS

The recent experiences of the cases suggest three important conclusions. First, the political ideologies and social movements that supported attention to the qualities of the daily life experience of ordinary citizens in the mid-twentieth century, and reasserted again in the 'radical movements' of the 1970s, are no longer so actively present to sustain such attention. In the context of a more diffuse landscape of urban governance, it is much harder for any single governance node to keep the different governance initiatives in creative encounter. To the extent that corporate economic interests have the resources to drive partnerships to achieve major urban development initiatives, and to the extent that politicians and technical policy communities buy into their agendas, then concerns for environmental well-being and distributive justice may be displaced into the background of policy attention. Such a shift is facilitated where citizen involvement in policy development is separated off into different relations within governance processes from those involving major economic actors. In effect, governance activity, once organised into sectoral policy communities with their relational webs, is tending to regroup around new institutional sites and relational layers, each with a different conception of the 'urban'. This 'fragmentation' is convenient for powerful corporate interests and strong lobby groups seeking to escape the restraints inherent in paying attention to agendas other than their own, but may also frustrate economic interests seeking a coordinated approach to, for example, development-site allocation and infrastructure provision. Spatial strategy-making efforts that seek to resist a segmented and narrowly focused approach to urban development may well encounter tough struggles. Their leverage may depend not just on the commitment of politicians and officials, but also on the traditions and capacity for mobilisation around daily life conditions among urban residents and interest groups.

Second, while the dominance of technical expertise, and especially that of a particular professionalised planning policy community, has been a valuable resource for maintaining attention to the place qualities of urban areas, the arenas of planning systems and the expertise of planners are not necessarily the most appropriate starting points for initiatives in spatial strategy-making. The technical experts in the cases, mostly planners linked by the culture of their professional community to a particular experience of discussion and practice, were increasingly aware of this, seeking in one way or another to convince those outside their own arenas to pay attention to place qualities and connectivities. This was most clear in the Amsterdam case. But this means not just that technical experts within formal government arenas should attempt to build rich relational links across the many 'streets' of the daily life of urban citizens and businesses. It also means accepting that a momentum for a strategic initiative may arise quite outside the arenas of formal government. It means accepting the value of mobilisation effort generated within civil society, however uncomfortable this may seem to be. It means appreciating that others in many arenas and networks in an urban area may have a better capacity than technical experts to 'summon up' an idea of an urban 'region' that has widespread resonance and mobilisation force within a particular governance context. In such a context, technical experts experienced in dealing with place-focused policies and with complex development impacts need to perform a delicate balancing act between valuing and facilitating the energy of civil society initiatives, just as they now do to initiatives from the world of business, while at the same time assisting in mediating between conflicting perspectives and agendas. Politicians, through their concern with their 'constituents', and expert planners, through their knowledge of the diversity of social groups and connectivities in urban areas, are always faced with the ethical challenge of paying attention to the many, while welcoming and fostering diverse initiatives 'outside the state'.

Third, there are difficult contemporary tensions within the government reform agendas pursued in EU and national arenas between encouraging a more strategic approach by local governments, and the pressures to reduce and target funding as a way to ensure that public actions are legitimate and accountable. Within planning systems in particular, the pressures to develop a more proactive, strategic and inclusively collaborative planning process still have to co-exist with regulatory procedures that allocate rights to use and develop land and property, which are legitimated through legal processes. Where development finance is involved, local place governance agendas that focus on place qualities may be driven by the search to capture grants from other tiers of government or from outside agencies. They may then become enmeshed in audit requirements that restrict how funds are spent. Demonstrating the legitimacy and accountability of public-sector actions is a necessary element of an active and transparent democratic polity. But where the primary driver of strategy preparation is to meet requirements demanded by external funders, such strategies are likely to end up narrowly based and inattentive to their multiple implications. The problem here relates to the degree of autonomy available to particular arenas and agencies of government. If the arena where a strategy is produced and operationalised is too weak in its ability to access public investment funds and regulatory power, then it will be difficult to escape the logics and practices of accountability and legitimacy pursued by those arenas and agencies of government that control the necessary powers and resources. In this way, the power to release additional energy from civil society and the economic sphere through the persuasive seduction of a spatial strategy for an urban 'region' may be undermined by the continued urge by other levels of government to exercise too commanding an authoritative power over local initiatives.

Acting Strategically for Urban Futures

In this book, I have argued for a view of the 'planning project' as concerned with the governance of place. I have also argued that a strategic orientation to place dynamics, and especially to the dynamics that sustain, create and change qualities of urban places, adds value to specific projects and initiatives by providing an orienting frame in which the impacts and interrelations of projects and the ongoing flow of urban dynamics can be explored. But throughout the book, I have stressed the difficulty and challenge of developing and sustaining a strategic focus in the governance of place. All too often, strategic efforts reach no deeper into governance processes than legitimising rhetorics. Or they are captured by a narrow conception of urban qualities and dynamics. Such narrow conceptions, such as a singular emphasis on 'urban competitiveness', do little to escape the functional sectoralism of welfare-state organisations that split urban governance into separate policy communities around sectoral fiefdoms. Or they promote a 'single issue' politics. They fail to grasp the multiple dimensions of the experience of daily life as lived in multiple relations in space–time. They ignore the significance of how these relations intersect to produce the place qualities and relations that affect people's quality of life, the qualities of environmental relations and the vitality of all kinds of economic activity. I have emphasised, through a relational geography, that it is in the flow of daily life that governance interventions have their impacts. The case for a governance of place focused on 'summoning up' a conception of an urban area rests on the ability to reach not only beyond any rhetoric, but on the capacity to work across sectoral divisions and single-issue politics. It centres on the demand for, and capacity to, keep in play and in conjunction the values of distributive justice, environmental well-being and economic vitality, not as abstract principles, but as manifest in the co-existence in time and space of daily life experience in many different relational webs.

I now conclude this chapter, as I have done in previous chapters, with some suggestions about the practical consequences of this way of thinking about spatial strategy for those involved in some way or other in urban governance. In making these suggestions, my particular concern is to give some encouragement to those struggling to shape urban futures in ways that will enhance the well-being of the diverse others whose lives and enterprises are attached to specific urban areas. I insist, however, that my suggestions are not to be treated as recipes or guides to be rolled out in specific instances. Instead, I intend them as probes for thinking (Lindblom 1990), particularly for those seeking for ways of improving governance processes in their urban areas to enable multiple voices and relations to find expression and to promote an encounter and productive co-existence with the potentially conflicting demands of distributive justice, environmental well-being and economic vitality, as experienced in daily urban lives.

I present these suggestions in six boxes. They reflect no particular order or weighting. The attention given to them in specific situations will depend on local particularities. The six issues are:

- Imagining the urban.
- Creating arenas for strategy formation and review.
- Creating frames of reference and specific strategies.
- Generating mobilising force.
- Nourishing strategic understanding.
- Nourishing a vigorous public realm.

These suggestions centre on how urban dynamics are understood. I have argued in Chapter 7 for a relational understanding of the multiplicity of webs of relations that intersect and overlap in urban areas. I have also emphasised the complexity and multiplicity of emergent trajectories arising in urban arenas. If this understanding is accepted, then an ability to take the pulse of these dynamics and to 'read' emerging potentialities and conflicts is a critical capability for any initiative in spatial strategy-making which seeks some kind of effectiveness. If the ambition is to keep distributive justice, environmental well-being and economic vitality in creative rather than destructive conjunction, then the

- Appreciate the daily life experiences of urban relations, their proximities and connectivities.
- Recognise the diverse ways urban life is lived and valued.
- Look for potentialities and restrictions, synergies, barriers and conflicts.
- Identify what is a 'problem', for whom, when and where.
- Beware narrow simplifications and imported conceptualisations.

Box 9.1 Imagining the urban

- Consider arenas as sites of encounter between many dimensions, interests, perceptions and values.
- Look for arenas appropriate to those whose attention is needed – to commit and to legitimate the commitment of resources and regulatory power.
- Find ways to tap into multiple sources and forms of knowledge.
- Connect arenas and stakeholders to formal powers as well as generating informal mobilising force.

Box 9.2 Creating arenas for strategy formation and review

ability to 'read' emergent trajectories, not from the heights of some privileged position but from multiple positions and through multiple perspectives in and around urban areas, becomes a vital resource (see Box 9.1).

Spatial strategy-making takes place in particular institutional sites in a governance landscape. Those initiating episodes of strategy-making need to think carefully about what actors and networks can, should and will come to be drawn into active engagement with such work. They need to consider which institutional sites are likely to provide suitable arenas within which framing ideas might be generated and explored, who gets involved in these and where different kinds of conflict and tension may be resolved. This involves some capacity to map governance arenas and relational webs, and to highlight where new governance relations and arenas may need to be built. Considering such issues leads to asking questions about the different discourses and practices to be found across an urban governance landscape and how far and how they may be brought into some kind of encounter. It requires an appreciation of the qualities and contribution of the relational webs that the effort of spatial strategy-making itself may generate (see Box 9.2).

The critical issue here is what kind of frame of reference and what specific strategies are 'called up' in the work of strategy formation. Strategy-making involves linking an understanding of urban dynamics to a valuing of particular qualities and potentialities, in a way that creates a frame of reference for specific choices that will affect future trajectories in significant ways. As discussed in Chapter 6, this work is not easily conducted in a

- Recognise that a strategic frame is about focusing attention, not just a technical and management exercise.
- Distinguish between a rich understanding as a frame of reference and the key selective choices that make an actual strategy.
- Consider whose actions need to be influenced for a strategy to have effects.
- Actively involve stakeholders whose actions and judgements will affect the power of a strategy, in encounter with each other as far as possible.
- Nourish vibrant, challenging debate inside and outside arenas of explicit strategy formation and learn from this.
- Avoid recipes for making strategies.

Box 9.3 Creating strategic frames for specific choices

- Do not assume a strategy focused on the place of the urban is necessarily desirable.
- Relate arguments for a strategic effort and a specific strategy to local conditions.
- Accept that strategic concepts will be shifted as they accumulate mobilising force and that failure in the short term may generate long-term energy.
- Maintain a critical attitude to frames and strategies as they develop momentum.
- Recognise that those who grant legitimacy to a strategy are as significant as those whose actions will realise a strategy.

Box 9.4 Generating mobilising force

systematic, linear way. A strategy-making process that seeks to maintain a rich and multiple understanding of urban dynamics while promoting a broad conception of what makes an urban area 'liveable' for many different groups cannot proceed as a technical exercise controlled by a single policy-community. It has to reach out to the multiplicity of experiences and values in all its phases, and accumulate attention and legitimacy through a range of forms of deliberation and challenge (see Chapter 8). If it develops momentum, a strategy-making process cannot avoid contributing to the 'public realm' of debate about the qualities of daily life in urban environments, how to improve these and, where conflicts break out, whose concerns should take priority. It therefore is both shaped by, and helps to shape, the governance processes and cultures of its context (see Box 9.3).

Spatial strategy-making is inherently political. It involves calling up some proximities and connectivities as strategically important and leaving others in the background. It involves building constituencies of support for a strategy. Its power grows as a strategic idea accumulates mobilising force to push a strategic frame around the governance arenas that need to take the ideas on board to allow a strategy to have material effects. Wide-ranging discussion around how to understand urban dynamics and significant choices may help to promote acceptance of a strategic frame, but a frame also has to resonate with the values and concerns of those whose actions will affect future choices – that is, citizens, different policy communities, politicians at different levels of government, key investment agencies in government and the private sector, companies and campaigners of many kinds. A critical and difficult moment in strategy-making is the shift

- Maintain a capability to develop an understanding of evolving urban dynamics, to read emergent patternings.
- Draw in and on multiple forms of knowledge.
- Allow for new insights and new phenomena to arise.
- Keep choices and framing concepts under continuous review.

Box 9.5 Nourishing strategic understanding

from creating a frame to promoting it. The art of a spatial strategy-making that helps to shape futures without becoming restrictively hegemonic is to maintain a critical attitude while taking a strategy 'forward', allowing its sense to evolve, in continual interaction with multiple constituencies (see Box 9.4).

Spatial strategy-making may take place in episodes when explicit attention is given to 'summoning up' a conception of the urban through processes that generate rich discussion in a variety of arenas. But the relational dynamics flowed on before, and will flow on after, such exercises. If a strategic frame develops the power to penetrate into the discourses and practices of many communities and relational webs across an urban governance landscape, and if its orientation emphasises its contribution to paying attention to the daily life conditions of the 'many, not the few' (Amin *et al.* 2000), then attention will need to be given to maintaining multiple ways of 'reading' emergent trajectories from multiple positions. This means much more than measuring progress against various indicators and asking different groups about their experiences. It means a capability to relate all kinds of different bits of information, and different bits of 'sense' through which people are experiencing and studying urban relations, and to draw these together to probe for signs of new relations and new patternings through which urban trajectories are emerging. Such a capability needs imaginative engagement as well as systematic analysis and, above all, needs time to listen, observe, test and to think (see Chapter 8) (see Box 9.5).

Thinking is not only conducted in 'think tanks'. It is also a key quality of public debate, as experiences are exchanged, issues contested, actions challenged and possibilities explored. A spatial strategy-making process that maintains continual critical attention to the qualities of the daily life experience of urban conditions, and which endures beyond specific episodes of concentrated attention, feeds a particular dimension into these debates and itself has the potential to enrich the range and quality of debate. It creates a relational layer as it feeds others (see Chapter 7). All kinds of issues are likely to crop up in the flow of public debate, but a key contribution of a relational approach is to keep in focus the multiple connectivities, both within an urban area and with other forces and other areas. An ethical, responsible urban polity will consider how what is pursued within its area may impact on others elsewhere, while a politically shrewd polity will keep a watch on the potentials and dangers of evolving exogenous forces. Overall, then, spatial strategy-makers are wise to consider how their activities will help to enrich the public realm of

- Recognise exogenous forces and exploit their potentialities, but avoid domination by their power to define ideas and practices.
- Build internal and external connectivities to increase understanding of urban conditions, and increase relational richness around strategic agendas.
- Build awareness of connectivities and responsibilities within and beyond a particular urban area.
- Use strategy-making work to enrich public debate and argument.

Box 9.6 Nourishing a vigorous public realm

understanding of urban life and its governance, as experienced in particular urban localities and in all the places to which those who find themselves there connect to other people in other places. In other words, they need to consider how their work contributes to sustaining and transforming the governance cultures in which they find themselves (see Box 9.6).

CODA

This book has presented a way of thinking about what it means to think and act strategically with respect to the governance of place. It has emphasised the complexity and the intellectual and political challenges of such an enterprise and stressed the importance of appreciating the diversity of situations and the varying opportunities in which it may be undertaken. It has argued for an evolutionary perspective on how new initiatives interact with situated emergent trajectories, in which new approaches to the governance of place may as often result from the slow working out of complex contradictions internal to governance processes as from major crises and forceful exogenous pressures. It has underlined that the activity of producing spatial strategies is full of risk and uncertainty. This arises not just because strategies may not achieve what their producers intended. In the complexity of urban dynamics, the unexpected and unintended may sometimes open up new, unforeseen possibilities. The most risky element of spatial strategies arises if they become powerful shapers of future potentialities. Then, their inherent contradictions may close off emergent potentialities which later come to be seen as desirable. Thus those involved in formulating and carrying forward such strategies bear a heavy ethical burden of responsibility to society, demanding continual reflexivity and preparedness to review and revise understandings, 'valuings' and practices.

The artistry of spatial strategy-making lies in combining a continually evolving strategic imagination about urban dynamics and potentialities, fed from multiple perspectives and positions, with a capacity for selective focusing on critical relations and choices where action now will make a difference. It involves creative synthesis among competing possibilities and values in ways that have a broad resonance with those whose judgements and feelings give legitimacy to governance interventions, while also showing how different strands of argument and their proponents feed into and are affected by the synthesis arrived at. It is a skilled activity, performed through the contributions of many actors in many scenes of action, drawing on a complex backstage of experiences, analyses, conflicts and lobbying, and continually playing to a critical audience, of interested stakeholders, watchful observers of governance practice, and all kinds of people who happen to drop in on the performance from time to time.

Audiences judge a play, film or novel according to how it resonates with some aspect of human experience, although specialists may also be interested in the technical skill of its construction. A strategic spatial frame of reference for the governance of urban 'regions' as imagined places, however skilled its technical construction, will have little legitimacy unless it too resonates – with people's experience of daily life in urban places and their imagination of what being in a particular urban area means to them. Such a strategic 'play' proceeds not by analytically smoothing out the messy complexity of urban life, but by calling it to mind in all its wondrous, frightening, routine, unexpected, comic and tragic manifestations. The quality of the 'play' of spatial strategy-making lies in finding ways of expressing the wonder of the urban world, and the frightening responsibility of acting within it, in ways that can be appreciated as real, shared dilemmas by the audience. This art of strategic imagination and judgement (Vickers 1965) succeeds by capturing just that fleeting essence that expresses a multidimensional feeling for the unknowable, multidimensional, emergent 'placeness' of the urban.

The urban condition is already the subject or backcloth of many books and films, and is studied from a range of different analytical perspectives. But these plays and studies are performed for particular audiences who choose to be interested. Governance activity, however, touches us materially and imaginatively in ways we cannot avoid. Some kind of collective action is a quality of all societies. A governance of place speaks about and to audiences without exit strategies from co-existing in urban areas or from collective action through which many of the qualities of urban areas are produced. Instead, what matters are the qualities of governance. Both those producing spatial strategies and those who will be touched by them are 'thrown together' through the co-existence of their relational lives in the places and connectivities of urban areas. A strategic governance of place speaks to the conditions of that co-existence. The touchstone of judgements about the performance of a strategic play about place governance will be how well it resonates with the multiple, particular experiences of that co-existence.

NOTES

1 See Byrne 2003; Innes and Booher 1999a; Urry 2005.
2 For examples of accounts by practitioners of their experiences, see Albrechts 2001; Goodstadt and Buchan 2002.
3 See Amin and Thrift 2002; Fay 1996; Healey 1997; Sandercock 2003a.
4 Here I am challenging, as does Offe, equilibrium conceptions of social order, emphasising instead the inherent potential for conflict between social groups, in terms of values, access to resources, and ways of thinking about society and its governance.
5 See Healey 2006b; Owens and Cowell 2002; Tewdwr-Jones and Allmendinger 2006.
6 By 2005, the ODPM was also advising planners to give priority to 'reading' housing markets (ODPM 2005b).
7 See Moulaert *et al.* (2000, 2005) for a general discussion of these tendencies, around the theme of 'social innovation'. Moulaert uses the term 'holistic' to encompass an approach that integrates economic considerations into a frame centred on social life and human flourishing.
8 There are always exceptions, however, but they are often recognised for their exceptionality, as in the case of the Hanover area in Germany (see Albrechts *et al.* 2003).

APPENDIX

ON METHOD

I have been asked, especially by other researchers doing qualitative case study work, how I have undertaken the cases for this book and how the cases have been woven into the overall book project. In many ways, because I worked on this project as a 'lone researcher' for about four years, it has felt like doing a PhD again. But the experience has also been different, in part because, as a 'known academic' at the end of my career, I have had a different 'persona' with respect to those I have interacted with in the cases than would a PhD student at the start of a research career. But also, I have used the cases to help develop an argument about the nature of spatial strategy-making, understood in a particular way. So I have not set out with a carefully refined and researchable 'research question', developed and tested hypotheses, arrived at 'findings', and concluded about what further research would be a good idea. In this brief appendix, I outline how I got started on the project, developed a particular orientation, and selected themes and cases to explore. I then explain how I did the casework, and finally how I used the case material in the thematic chapters of the book.

SHAPING THE PROJECT

The project started from a growing curiosity about the practice of spatial strategy-making in urban contexts as it was being promoted in Europe in the 1990s. This built on my own work in the 1980s on the practice of 'development plan-making' and its impacts in the UK (Healey 1983; Healey *et al.* 1988). These were early efforts in what later became recognised as an 'institutionalist' way of understanding planning practices. But, as I discuss in the book, the pressure for strategic planning in the 1990s was being promoted by a particular conjuncture of ideas about the 'competitiveness' of urban economies, about the 'sustainability' of urban relations and trajectories, and about the capacity of urban governance arrangements to develop 'integrated' and 'strategic' responses to these perceived challenges. Studies of the practice emerging in the 1990s, to which I had contributed, stressed the complexity and difficulty of the enterprise of strategic spatial planning for urban areas.[1] But yet, it seemed to me, that the attempt to develop a strategic approach to the development trajectory of the places of the urban lay at the heart of the 'planning project'. In the early 2000s, as I was contemplating moving to a semi-retired relation with university activity, I was primarily interested in developing my 'institutionalist' work on urban governance and governance capacity.[2] I had thought that it would be useful to clarify my understanding of the nature of strategy-making and of what it means to focus on the spatiality of phenomena. But, as with many PhD projects, what starts out as an apparent sideline moves unexpectedly into centre stage, and this is what happened to me. I retired from university administration and teaching in autumn 2002, and with (some of) the time this gave me, I embarked on the present project.

In one sense, I was free to pursue the work in whatever way I liked! I did not 'have to be' a 'proper academic'. But in another sense, I had been talking with PhD students and other researchers about the craft of research, both in its empirical dimensions and in the relation between concepts and research inquiries. So I was acutely conscious of what I was doing as I went along. For several months, I proceeded in an open-minded and interactive way, exploring the 'literature', developing some knowledge about particular cases, visiting some cases and talking to academics and practitioners about them, drawing out what seemed to be key themes. I did not start off with 'a theory', or a hypothesis. I started off with a puzzle about a practice. What was its nature? How could it be understood? How could I make sense of practices that were considered by those involved as in some way efforts in spatial strategy-making? How could I relate them to normative ideas about what such a practice should be or could become? Of course, my conceptual mind could not be a complete blank, as it was full of ideas about 'institutionalist' ways of understanding governance practices, but I had no clear 'fixes' to guide me through the messy reality of spatial strategy-making practices.

After a while, just as when doing a PhD, I felt the need to discipline myself into a more structured orientation and focus. This became the material now to be found, after many iterations, in Chapters 1 and 2. I realised that I was combining a 'sociological institutionalist' perspective on urban governance with related work on interpretive policy analysis. But the challenge for me was to link this together with a relational geography to assist in understanding what 'spatiality' could mean. Even though I had been surrounded by colleagues at Newcastle working on these issues,[3] it has taken me some time to work out how to link them to the 'planning project'.[4] I consolidated my approach into the four themes that now structure the book. I initially called them: governance capacity, knowledge resources and systems of meaning, the treatment of place/space, and strategic focus and selectivity. Under each heading, I had derived a whole array of questions that I thought I needed to address, from the literature and from some of the comments about possible cases. These questions were partly linked to analytical issues about where my focus of attention should be if my perspective was an institutionalist one, concerned with how social realities were produced and patterned through complex relations and interactions. But, having a planning background, I also felt obliged to consider not just what effects the practices of spatial strategy-making might have. I also felt the obligation to assess whether these practices were, or could be, 'good', or 'progressive' in some way. I specified this as a normative concern with the extent to which spatial strategy-making practices could keep in critical conjuncture the potentially contradictory values of distributive justice, environmental well-being and economic vitality, related to the richness and diversity of daily life experiences and aspirations of those living in urban areas. By early 2004, I found I did actually have three research questions, which are to some extent addressed in the book![5]

THE CASES

I did not have to do the detailed cases that are now in the book. I could have merely reviewed the issues through generalised discussion, abstracted from the existing literature. Or I could have used accounts of particular cases that had already been written up in the literature. Or I could have attempted a survey of practices, selecting key informants from many places. But none of these research strategies would have reflected my sense of the deeply situated and contingent way in which governance practices evolve. It is this sensibility that had led me towards the 'sociological institutionalism' through which I had been exploring the fine-grain details of urban partnership practices (Healey 2006d). An institutionalist understanding of governance phenomena emphasises the importance of an evolutionary perspective, of locating practices as trajectories with pasts and futures, and stresses the complex interplay between agency and structuring dynamics. I therefore concluded that I needed to explore how practices of spatial strategy-making were accomplished in considerable depth, as case 'histories' or narratives. This meant that I could only undertake a few cases and that I had to treat the cases to be found in the literature very circumspectly, as their situated contingencies could not easily be assessed in the often short accounts of them.[6] Eventually, I settled on the three cases presented here.[7]

The selection of cases for in-depth, qualitative research is always more a practical question than the product of systematic choice criteria. I wanted cases in very different institutional contexts and used my knowledge of planning experiences in western Europe to find cases where there was some prospect that I could get an in-depth understanding. I had been talking with Italian planners and academics since the mid-1980s, had many planning contacts in Milan, and could read and speak Italian and knew of the critical discussion about the Milan *Documento di Inquadrimento*. I had excellent contacts in the Netherlands, and Dutch planners and academics write and speak well in English.[8] I took some advice and eventually settled on Amsterdam. I 'came across' the case of the Cambridge Sub-Region by chance when invited to speak at a seminar the day before the launch of the draft *Cambridgeshire and Peterborough Structure Plan* in early 2002. In preparation for this, I read both the plan and a good deal of other contextual material. It was clear to me from our earlier work in the UK (Vigar *et al.* 2000) that this was a very interesting case, exemplifying the challenge of strategic planning in affluent, economically dynamic southern England. In each case, I thought I would focus on the most recent 'episodes' of spatial strategy-making, which were around the late 1990s/ early 2000s.

By October 2003, I had constructed a research protocol for the case-study work. This had the form of a matrix, relating the four themes of my inquiry to the 'institutionalist' way I thought I would write the accounts. This had the following headings: context, process and practices, content and discourses, impacts and outcomes. In each box of the matrix, I listed questions I needed to ask to develop my accounts. I used these fairly systematically, as I searched different sources to create the material from which I could write the case accounts. As the research methods textbooks tell us to do in case-study research, I used a variety of sources. I read general literature about urban policy and planning in Italy and the Netherlands, and specific material about economic, social, cultural and political issues, especially where related to the case areas. I consulted academics doing research on, or working with actors in, the case areas, and I owe a big debt not merely to the senior academics, including many friends of my own generation, but to doctoral students and post-doctoral students who helped me greatly (see my acknowledgements at the start of the book). I owe an especially big debt to (nearly Dr) Stan Majoor and Dr Filomena Pomilio. In addition to other actors, I also had the privilege to talk with practitioners in all three cases about their experiences of struggling with the challenge of spatial strategy-making. Some of my best insights came as I talked with them about their work and their relations.[9] The following table summarises those I had such discussions with, sometimes several times, between autumn 2003 and autumn 2005 (with some emails and telephone discussions since).

	Amsterdam	*Milan*	*Cambridgeshire*
Practitioners	5	4	10
Academics as practitioners	3	5	1
Other academics	8	5	6

I refer to my meetings with these people as 'discussions', not 'interviews'. I did have an agenda for each meeting, derived from my research protocol. In social-science research, interviews are anticipated to be 'with strangers' (Weiss 1995). The interviewer has to balance the delicate social relation of establishing credibility, pursuing a specific agenda and letting the person interviewed talk in as free and open a way as possible. But I was usually not a stranger. I came with a baggage as a 'known' person, with expertise in what I wanted to have a discussion about. So our meetings proceeded interactively. From time to time I had to express what I actually thought, rather than holding back too much. I tried to do this through probes, through suggesting interpretations, through raising alternative viewpoints, to seek out agreements and disagreements. So, rather than interviews, or even just discussions, what happened were very informative and fascinating 'conversations' about the themes and experiences I was interested in. In this way, this book has been the product of my own interactions with others involved in the cases, as well as the interaction between my themes and the accounts of practices I have produced in the cases.

This interactive quality was reinforced as I came to write up the cases. I wanted the cases as they became chapters in the book to read as interesting narratives. So I broke away from the headings of my research protocol and instead presented the cases as historical narratives. By the time I came to write up the first draft of the first narrative, in spring 2004, I had discovered that I could not just tell the story of the most recent episodes of spatial strategy-making. I had to delve back in time. I learnt this when I came across an exhibition in autumn 2003, celebrating 75 years of Amsterdam's planning tradition. This alerted me to something I should have known from the start, that the past significantly shaped the present, and that the vocabulary, emphases and metaphors of present policy debates are difficult to understand without recognising their resonance with past discourses and values. So my narratives came to have a time span of over 50 years. I thought a great deal about how to write such narratives.[10] The structure of the narrative centres on episodes of spatial strategy-making, as these have developed through time in each case. The narrative focuses around the content of strategies and discourses of place and space, and around the processes through which strategies were produced and accumulated influence. In telling the stories, I have woven into the flow of the story the material on the wider context (economic, political, social) to show how context shapes practices and vice versa. I have tried to embed in the accounts the material I would refer to later in the theme chapters. Following my 'institutionalist' perspective, my focus was on the practices of spatial strategy-making, on the dynamics of the relations through which these practices were lived, and the discourses mobilised within these practices, particularly as regards concepts of place and space.[11] Because each case is different, each story has a different 'style' and different emphases.

As every qualitative researcher knows, I have had to leave out an enormous amount of what I learned about from my 'sources'. I could have written large monographs about each case. In editing the accounts, I continually had to tighten them to keep to the main story and not be deflected down fascinating byways. This meant that I had to keep my own focus in mind, but at the same time, I had to keep in my thoughts what I had learned. As I wrote, I could often hear the comments of those I had talked with, or see a piece of text, which challenged me to avoid an easy conclusion. Then I sent the accounts back to key 'discussants' in each case, including practitioners, who carefully read and commented on my accounts. This made me think even more, but also made me feel a bit reassured that my accounts had resonance with those more close to the cases than I could ever be. However, a book is not quite the same as an academic research thesis. In a research thesis, authors have to make their methodology explicit. This can often be irritating for the reader more interested in the flow of argument or in a case narrative. For this reason, I have left out all direct references to my 'conversational discussions', though I have earlier versions of the chapter texts where I can locate where I have used each source.

DEVELOPING THE 'THEMATIC' DISCUSSION

To an extent, Chapters 6 to 9 of the book are a kind of 'analysis', drawing out 'evidence' from the case stories, in relation to my focusing themes and questions. In part, they have the form of an 'evaluation' about how far the practices, in the past and in recent times, match up with a relational approach and a normative orientation of the kind I have emphasised. But, as some who have read individual chapters have remarked, the chapters do not proceed as a typical analysis or evaluation. They are developmental. My puzzle about spatial strategy-making for urban areas has not just been about 'I wonder what has been going on'. It has been about 'what is the potential for such a policy focus to make life "better" in urban areas, for the many and not the few?' Would that potential be realised more effectively if a more relational approach to strategy-making for urban places were developed? What actually would it mean to evolve such an approach?

So the chapters themselves proceed in an interactive way, exploring shifts in modes of thought as presented in academic literature, relating these to the shifts in discourses and practices as reflected in the cases, and drawing these together to make normative suggestions to help those involved in the 'craft' of practising spatial strategy-making. The cases do not become, in this treatment, exemplars of 'good practice', or archetypal cautionary tales. They become stories of situated experiences through which, I hope, readers can get a flavour of what it means to engage in such practices as part of the flow of governance activity in particular places and times. The normative suggestions are offered as probes to imaginative endeavour for those embroiled in spatial strategy-making, designing interventions to encourage it to happen or considering mobilising collective action to create an episode of spatial strategy-making.

And the book itself has been a voyage of discovery, as, through continual interaction between the chapters and the themes, I have had to revise earlier ideas, tidy up loose concepts (such as my use of 'scale' and 'urban region'), and make the whole edifice of the book hang together as best I could. In this voyage, I have been enormously helped by many people, who have read and commented on individual chapters and on the book as a whole. I hope I have remembered everyone in the acknowledgements. I cannot stress too much the value of an array of 'critical friends' in the development of my academic work, and particularly in the present book. I may be alone at my desk but I am never alone as I write, thinking of what the books, cases and critical friends might say if I cast a sentence one way or another, or draw a conclusion in one way or another.[12] This can be an alarming thought, making writing almost impossible. The antidote to this is to keep on trying to find a way through, while paying careful attention to the actual and anticipated critique. In the end, you have to say: well, that's it. I've tried as best I can. Its time to stop and let the project go, warts and all. So now, readers, as the 'examiners' of an elderly academic, it's down to you to see what you make of it!

NOTES

1 See Albrechts *et al.* 2001; Healey *et al.* 1997; Motte 1995; Vigar *et al.* 2000.
2 See Cars *et al.* 2002; Healey 2006a, d; Healey *et al.* 2003.
3 Especially, in the 1990s, Ash Amin, Kevin Robins, Ali Madanipour, Stephen Graham and the late Jon Murdoch. In the 2000s, it has been very helpful to work with Frank Moulaert and his trans-European research networks focused on the nature of 'social innovation' and a 'holistic' view of local development (Moulaert *et al.* 2000, 2005), and have the continual supportive, yet critical, comments of Jean Hillier, and other colleagues and PhD students in SAPL, University of Newcastle.
4 I had introduced a 'relational' approach in my work on *Collaborative Planning* (Healey 1997), but critical comment on this book focused instead on my ideas about governance practices, in planning theory debates that were then arguing heatedly about the relative merits of a Habermasian or a Foucauldian view of the 'planning project' and its potentials.
5 These were: (a) What leads to an enduring capacity for spatial strategy-making at the urban region scale? (Note that since then I have problematised the use of the term 'urban region' and avoided using the word scale as far as possible!); (b) when does place-focused strategy-making get accepted as effective/legitimate and also promote distributive justice, environmental care and economic health? (And when is a place-integrated strategic focus at the urban scale of value?); (c) what roles do planners and planning systems play in all this?
6 There are some fine, detailed accounts of the evolution of urban policy, politics and practices in particular places that have this narrative quality. I have been especially inspired by Stone's wonderful account of Atlanta (1989), Punter (2003) on Vancouver, and Abbott (2001) on Portland, Oregon. See also the studies of the politics of planning practices, notably Meyerson and Banfield (1955) on Chicago and Flyvbjerg (1998) on Aalborg, Denmark.
7 I had initially intended to look at a case of cross-border urban strategic making in the Newry/Dundalk area, but I realised that undertaking four cases was too ambitious and the three cases were rich enough.
8 Despite this, and the great help given to me by Dutch academics and practitioners, I am sure I have missed important nuances of the Amsterdam experience through not being able to read Dutch.
9 I especially appreciated discussions with planners Caroline Combè, Micheal Hargreaves, Allard Jolles, Malcolm Sharpe, Paolo Simonetti, Peter Stoddert, San Verschuuren and Mark Vigor.
10 There has been much recent writing on narrative accounts in planning and policy analysis (see Eckstein and Throgmorton 2003; Flyvbjerg 2001; Sandercock 2003b), but I have also been helped by participating in the Barter Books Book Group, Alnwick, Northumberland.
11 To the extent that I have 'units of analysis', these are concepts of discourses and relations in practices.
12 It is this 'holding in mind' of potential interlocuters, of course, that I suggest as a key quality of an 'inclusive' practice of spatial strategy-making.

REFERENCES

6, P., Leat, D., Seltzer, K. and Stoker, G. (2002) *Towards Holistic Governance: the New Reform Agenda*, Palgrave, Houndmills, Basingstoke.

Abbott, C. (2001) *Greater Portland: Urban Life and Landscape in the Pacific Northwest*, University of Pennsylvania Press, Philadelphia.

Albrechts, L. (2001) From traditional land use planning to strategic spatial planning: the case of Flanders. In *The Changing Institutional Landscape of Planning*, Albrechts, L., Alden, J. and da Rosa Pires, A. (eds), Ashgate, Aldershot, pp. 83–108.

Albrechts, L. (2004) Strategic (spatial) planning reexamined. *Environment and Planning B: Planning and Design*, 31, 743–758.

Albrechts, L. (2005) Creativity as a drive for change. *Planning Theory*, 4(3), 247–269.

Albrechts, L. and Mandelbaum, S. (eds) (2005) *The Network Society: a New Context for Planning?* Routledge, London.

Albrechts, L., Alden, J. and de Rosa Pires, A. (eds) (2001) *The Changing Institutional Landscape of Planning*, Ashgate, Aldershot.

Albrechts, L., Healey, P. and Kunzmann, K. (2003) Strategic spatial planning and regional governance in Europe. *Journal of the American Planning Association*, 69, 113–129.

Alexander, E.R. (2002) Metropolitan regional planning in Amsterdam: a case study. *Town Planning Review*, 73, 17–40.

Allen, J. (2003) *Lost Geographies of Power*, Blackwell Publishing, Oxford.

Allen, J. (2004) The whereabouts of power: politics, government and space. *Geografisker Annaler*, 86B, 19–32.

Allen, J., Massey, D. and Cochrane, A. (1998) *Rethinking the Region*, Routledge, London.

Amin, A. (ed.) (1994) *The Post-Fordist Reader*, Blackwell, Oxford.

Amin, A. (2002) Spatialities of globalisation. *Environment and Planning A*, 34, 385–399.

Amin, A. (2004) Regions unbound: towards a new politics of place. *Geografisker Annaler*, 86B, 33–44.

Amin, A. and Cohendet, P. (2004) *Architectures of Knowledge: Firms, Capabilities and Communities*, Oxford University Press, Oxford.

Amin, A. and Thrift, N. (eds) (1994) *Globalisation, Institutions and Regional Development in Europe*, Oxford University Press, Oxford.

Amin, A. and Thrift, N. (2002) *Cities: reimagining the Urban*, Polity/Blackwell, Oxford.

Amin, A., Massey, D. and Thrift, N. (2000) *Cities for the Many Not the Few*, The Policy Press, Bristol.

Ave, G. (1996) *Urban Land and Property Markets in Italy*, UCL Press, London.

Bagnasco, A. and Le Galès, P. (2000a) Introduction: European cities: local societies and collective actors. In *Cities in Contemporary Europe*, Bagnasco, A. and Le Galès, P. (eds), Cambridge University Press, Cambridge, pp. 1–32.

Bagnasco, A. and Le Galès, P. (eds) (2000b) *Cities in Contemporary Europe*, Cambridge University Press, Cambridge.

Bailey, J. (1975) *Social Theory for Planning*, Routledge and Kegan Paul, London.

Balducci, A. (1988) La vicende del Piano: una periodizzazione. *Urbanistica*, 90, 50–59.

Balducci, A. (2001a) New tasks and new forms of comprehensive planning in Italy. In *The Changing Institutional Landscape of Planning*, Albrechts, L., Alden, J. and de Rosa Pires, A. (eds), Ashgate, Aldershot, pp. 158–180.

Balducci, A. (2001b) Una riflessione sul rapporto tra politiche per i quartieri e politiche per la citta. *Territorio*, 7–24.

Balducci, A. (2004) Creative governance in dynamic city regions. *DISP*, 3, 21–25.

Balducci, A. (2005a) Una Visione per la Regione Urbana Milanese. In *Milano, nodo delle rete globale*, Bassetti, P. (ed.), Bruno Mondadori, Milan, pp. 231–264.

Balducci, A. (2005b) Strategic planning for city regions: the search for innovative approaches. Paper to AESOP Congress, Vienna, July.

Ball, M. (1983) *Housing Policy and Economic Power*, Methuen, London.

Barker, K. (2004) *Delivering Stability: Securing our Future Housing Needs*, The Stationery Office, London.

Barnes, B. (1982) *T.S. Kuhn and Social Science*, Macmillan, London.

Barrett, S. and Fudge, C. (1981) *Policy and Action*, Methuen, London.

Bassetti, P. *et al.* (2005) *Milano, Nodo della rete globale*, Bruno Mondadori, Milan.

Beauregard, B. (1995) If only the city could speak: the politics of representation. In *Spatial Practices*, Liggett, H. and Perry, D.C. (eds), Sage, Thousand Oaks, CA, pp. 59–80.

Bertolini, L. and Dijst, M. (2003) Mobility environments and network cities. *Journal of Urban Design*, 8, 27–43.

Bertolini, L. and le Clercq, F. (2003) Urban development without mobility by car? Lessons from Amsterdam, a multi-modal region. *Environment and Planning A*, 35, 575–589.

Bertolini, L. and Salet, W. (2003) Planning concepts for cities in transition: regionalisation of urbanity in the Amsterdam structure plan. *Planning Theory and Practice*, 4, 131–146.

Blackler, F. (1995) Knowledge, knowledge work and organizations: an overview and interpretation. *Organization Studies*, 16, 1021–1046.

Blackler, F., Crump, N. and McDonald, S. (1999) Organizational learning and organizational forgetting. In *Organizational Learning and the Learning Organization*, Easterby-Smith, M., Araujo, L. and Burgoyne, J. (eds), Sage, London, pp. 194–216.

Boeri, S., Lanzani, A. and Marini, E. (1993) *Il Territorio che Cambia: Ambiente, paesaggio e immagini della regione milanese*, Abitare Segesta Cataloghi, Milan.

Bolocan Goldstein, M. (2002) Governo locale e operazioni urbanistiche a Milano tra gli anni '80 e '90. *Urbanistica*, 119, 90–102.

Bonfanti, B. (2002) Urbanistica in Milano (special issue). *Urbanistica*, 119, 81–138.

Bonomi, A. (1996) *Il trianfo della moltitudine*, Bollati Boringhieri, Torino.

Booher, D. and Innes, J. (2002) Network power for collaborative planning. *Journal of Planning Education and Research*, 21, 221–236.

Boriani, M., Morandi, C. and Rossari, A. (1986) *Milano Contemporaneo*, Designers Riuniti Editore, Torino.

Bourdieu, P. (1977) *Outline of a Theory of Practice*, Cambridge University Press, Cambridge.

Bourdieu, P. (1990) *In Other Words: Essays Towards a Reflexive Sociology*, Polity Press, Oxford.

Bourne, L.S. (1975) *Urban Systems: Strategies for Regulation: a Comparison of Policies in Britain, Sweden, Australia and Canada*, Clarendon Press, Oxford.

Boyer, C. (1983) *Dreaming the Rational City*, MIT Press, Cambridge, MA.

Breheny, M. and Hooper, A.J. (eds) (1985) *Critical Essays on the Role of Rationality in Planning*, Pion, London.

Brenner, N. (1999) Globalisation as reterritorialisation: the re-scaling of urban governance in the European Union. *Urban Studies*, 36, 431–452.

Brenner, N. (2000) The urban question as a scale question: reflections on Henri Lefebvre, urban theory and the politics of scale. *International Journal of Urban and Regional Research*, 24, 361–378.

Bridge, G. and Watson, S. (eds) (2000) *A Companion to the City*, Blackwell, Oxford.

Brindley, T., Rydin, Y. and Stoker, G. (1989) *Remaking Planning: the Politics of Urban Change in the Thatcher Years*, Routledge, London.

Bryson, J. (1995) *Strategic Planning for Public and Nonprofit Organizations: a Guide to Strengthening and Sustaining Organizational Achievement*, Jossey Bass, San Francisco.

Bryson, J.M. (2003) Strategic planning and management. In *Handbook of Public Administration*, Peters, G.B. and Pierre, J. (eds), Sage, London, pp. 38–47.

Bryson, J. and Crosby, B. (1992) *Leadership in the Common Good: Tackling Public Problems in a Shared Power World*, Jossey Bass, San Francisco.

Buchanan, Colin and Partners (2001) *The Cambridge Sub-Region Study*, Colin Buchanan and Partners, London.

Buchanan, Colin and Partners and GVA Grimley (2004) *A Study of the Relationship Between Transport and Development in the London–Stansted–Cambridge–Peterborough Growth Area*, ODPM, London.

Burby, R.J. (2003) Making plans that matter: citizen involvement and government action. *Journal of the American Planning Association*, 69, 33–49.

Burns, D. and Taylor, M. (2000) *Auditing Community Participation: an Assessment Handbook*, Policy Press, Bristol.

Burtenshaw, D., Bateman, M. and Ashworth G.J. (1991) *The European City: a Western Perspective*, Halstead Press, New York.

Byrne, D. (2003) Complexity theory and planning theory: a necessary encounter. *Planning Theory*, 2, 171–178.

Calabrese, L.M. (2005) Notes on official urbanism and the field of action of the urban project. In *Working for the City*, Meyer, H. and van den Berg, L. (eds), TU Delft, Delft, pp. 80–85.

Callon, M. (1986) Elements pour une sociologie de la traduction. *L'Annee sociologique*, 36, 169–208.

Callon, M. and Law, J. (2004) Guest editorial – presence, circulation and encountering in complex space. *Environment and Planning D: Society and Space*, 22, 3–11.

Cambridge City Council (CCityC) (1968) *Report of the Working Group on the Future Size of Cambridge*, Cambridgeshire City Council, Cambridge.

Cambridge City Council (CCityC) (1996) *Cambridge Local Plan*, Cambridgeshire City Council, Cambridge.

Cambridge City Council (CCityC) (2004) *Cambridge Local Plan: Redeposit Draft*, Cambridgeshire City Council, Cambridge.

Cambridge Joint Regional Town Planning Committee (CJRTPC) (1934) *Cambridgeshire Regional Planning Report*, Cambridgeshire University Press, Cambridge.

Cambridgeshire County Council (CCC) (1961) *Report: The First Review of the Town Map for Cambridge*, Cambridgeshire County Council, Cambridge.

Cambridgeshire County Council (CCC) (1979) *Cambridgeshire Structure Plan: Report on Public Participation and Consultations*, Cambridgeshire County Council, Cambridge.

Cambridgeshire County Council (CCC) (1980) *Cambridgeshire Structure Plan: Approved Written Statement*, Cambridgeshire County Council, Cambridge.

Cambridgeshire County Council (CCC) (1989) *County Structure Plan – 1989 revisions*. Cambridgeshire County Council, Cambridge.

Cambridgeshire County Council and Peterborough City Council (CCC) (2003) *Cambridgeshire and Peterborough Joint Structure Plan Review: Planning for Success*, Cambridgeshire County Council, Cambridge.

Cambridgeshire County Planning Officer (CCPO) (1977) *Consultation Report on the Cambridge Sub-Area*, Cambridgeshire County Council, Cambridge.

Cars, G., Healey, P., Madanipour, A. and de Magalhães, C. (eds) (2002) *Urban Governance, Institutional Capacity and Social Milieux*, Ashgate, Aldershot.

Castells, M. (1977) *The Urban Question*, Edward Arnold, London.

Castells, M. (1996) *The Rise of the Network Society*, Blackwell, Oxford.

Ceccarelli, P. and Vittadini, M.R. (1978) Un piano per la crisi. *Urbanistica*, 68/69, 58–88.

Chapin, F.S. (1965) *Urban Land Use Planning*, University of Urbana-Champagne Press, Urbana, IL.

Cherry, G.E. and Penny, L. (1986) *Holford: a Study in Architecture, Planning and Civic Design*, Mansell Publishing Ltd, London.

Chesterton Planning and Consulting (1997) *Cambridge Capacity Study*, Chesterton plc, London.

Christensen, K.S. (1999) *Cities and Complexity: Making Intergovernmental Decisions*, Sage, Thousand Oaks, CA.

Coaffee, J. and Healey P. (2003) My voice, my place: tracking transformations in urban governance, *Urban Studies*, 40, 1979–1999.

Cockburn, C. (1977) *The Local State*, Pluto Press, London.

Cognetti, F. and Cottoni, P. (2004) Developers of a different city. *City*, 7, 227–235.

Collarini, S., Guerra, G. and Riganti, P. (2002) Programmi integrarti di intervento: un primo bilancio. *Urbanistica*, 119, 83–90.

Committee for Spatial Development (CSD) (1999) *The European Spatial Development Perspective*, European Commission, Luxembourg.

Comune di Milano (2000) *Ricostruire la Grande Milano: Documento di Inquadrimento delle politiche urbanistiche communali*, Milan, Edizione Il Sole 24 Ore.

Cooke, P. (2002) *Knowledge Economies: Clusters, Learning and Competitive Advantage*, Routledge, London.

Cooke, P. and Morgan, K. (1998) *The Associational Economy: Firms, Regions and Innovation*, Oxford University Press, Oxford.

Cooper, A.J. (2000) *Planners and Preservationists: the Cambridge Preservation Society and the City's Green Belt 1928–1985*, Cambridge Preservation Society, Cambridge.

Corburn, J. (2005) Street Science: *Community Knowledge and Environmental Health Justice*, MIT Press, Cambridge, MA.

Corry, D. and Stoker, G. (2002) *New Localism: Refashioning the Centre–Local Relationship*, New Local Government Network, London.

Cortie, C. (2003) The metropolitan population. In *Amsterdam Human Capital*, Musterd, S. and Salet, W. (eds), Amsterdam University Press, Amsterdam, pp. 199–216.

Cowling, T.M. and Steeley, G.C. (1973) *Sub-Regional Planning Studies: an Evaluation*, Pergamon, Oxford.

Crang, P. and Martin, R. (1991) Mrs Thatcher's vision of the 'new Britain' and the other sides of the Cambridge phenomenon. *Environment and Planning D: Society and Space*, 9, 91–116.

Cremaschi, M. (2002) Un ritratto di famiglia. *Urbanistica*, 119, 33–36.

Crouch, C. (2004) *Post-Democracy*, Polity Press, Cambridge.

Crouch, C., Le Galès, P., Trigilia, C. and Voelzkow, H. (2001) *Local Production Systems: Rise or Demise*, Oxford University Press, Oxford.

Cullingworth, J.B. (1972) *Town and Country Planning in Britain*, fourth edition, George, Allen and Unwin, London.

Cullingworth, J.B. (1975) *Environmental Planning: Volume 1: Reconstruction and Land Use Planning 1939–1947*, Her Majesty's Stationery Office, London.

Curti, F. (2002) I tre corni di un dilemma trattabile. *Urbanistica*, 119, 107–111.

Davidge, W.R. (1934) *Cambridgeshire Regional Planning Report*, Cambridge University Press, Cambridge.

Davies, H.W.E., Edwards, D., Hooper, A. and Punter, J. (1989) *Planning Control in Western Europe*, Her Majesty's Stationery Office, London.

Davoudi, S. (2006) The evidence – policy interface in strategic waste planning: the 'technical' and the 'social'. *Environment and Planning C: Government and Policy*, in press.

Dawe, P. and Martin, A. (2001) *In our Back Yard: a Vision for a Small City*, P. Dawe Consulting Ltd, Oakington, Cambridge.

de Jong, M. (2002) Rijkswaterstaat: a 1978 French transplant in the Netherlands. In *The Theory and Practice of Institutional Transplantation*, de Jong, M., Lalenis, K. and Mamadouh, V. (eds), Kluwer Academic Publishers, Dordrecht, pp. 55–70.

de Magalhães, C. (2001) International property consultants and the transformation of local property markets. *Journal of Property Research*, 18, 1–23.

de Neufville, J.I. and Barton, S.E. (1987) Myths and the definition of policy problems: an exploration of home ownership and public–private partnerships. *Policy Sciences*, 20, 181–206.

de Roo, G. (2003) *Environmental Planning in the Netherlands: Too Good to be True: From Command-and-Control to Shared Governance*, Ashgate, Aldershot.

de Vries, J. and Zonneveld, W. (2001) Transnational planning and the ambivalence of Dutch spatial planning. Paper to *World Planning Schools Congress*, Shanghai, China, p. 25.

Dematteis, G. (1994) Global and local geo-graphies. In *Limits of Representation*, Farinelli, F., Olsson, G. and Reichert, D. (eds), Accedo, Munich, pp. 199–214.

Dente, B., Bobbio, L. and Spada, A. (2005) Government or governance of urban innovation? *DISP*, 162, 41–52.

Department of the Environment (DoE) (1974) *Strategic Choice for East Anglia: Report of the East Anglia Regional Strategy Team*, Her Majesty's Stationery Office, London.

Department of the Environment (DoE) (1980) *Circular 9/80: Land for Private House Building*, Her Majesty's Stationery Office, London.

Department of the Environment (DoE) (1985) *Circular 14/85 Development and Employment*, Her Majesty's Stationery Office, London.

Department of the Environment (DoE) (1991) *Regional Planning Guidance for East Anglia*, Her Majesty's Stationery Office, London.

Department of Transport, Local Government and the Regions (DTLR) (2000) *Regional Planning Guidance for East Anglia (RPG6)*, Her Majesty's Stationery Office, London.

Department of Transport, Local Government and the Regions (DTLR) (2001) *Planning: Delivering a Fundamental Change*, DTLR, Wetherby.

Dieleman, F.M. and Musterd, S. (eds) (1992) *The* Randstad*: a Research and Policy Network*, Kluwer, Dordrecht.

Dienst Ruimtelijke Ordening, Gemeente Amsterdam (DRO) (1985) *Structuurplan: Amsterdam Stad centraal Amsterdam*, DRO, Gemeente Amsterdam.

Dienst Ruimtelijke Ordening, Gemeente Amsterdam (DRO) (1994) *A City in Progress: Physical Planning in Amsterdam*, DRO, Amsterdam.

Dienst Ruimtelijke Ordening, Gemeente Amsterdam (DRO) (1996) *Open Stad: Structuurplan 1996*, Gemeente Amsterdam, Amsterdam.

Dienst Ruimtelijke Ordening, Gemeente Amsterdam (DRO) (2003a) *Plan Amsterdam: Het structuurplan 2003*, Gemeente Amsterdam, Amsterdam.

Dienst Ruimtelijke Ordening, Gemeente Amsterdam (DRO) (2003b) *Het Structuurplan: summary Amsterdam*, DRO, Gemeente Amsterdam, Structure Plan Report no. 4/5.

Dijkink, G. (1995) Metropolitan government as a political pet? Realism and tradition in administrative reform in The Netherlands. *Political Geography*, 14, 329–341.

Dijkink, G. and Mamadouh, V. (2003) Identity and legitimacy in the Amsterdam region. In *Amsterdam Human Capital*, Musterd, S. and Salet, W. (eds), Amsterdam University Press, Amsterdam, pp. 331–355.

Drake, M., McLoughlin, B., Thompson, R. and Thornley, J. (1975) *Aspects of Structure Planning*, Centre for Environmental Studies, London.

Dryzek, J. (1990) *Discursive Democracy: Politics, Policy and Political Science*, Cambridge University Press, Cambridge.

Dryzek, J. (2000) *Deliberative Democracy and Beyond*, Oxford University Press, Oxford.

Dühr, S. (2005) The visualisation of network space in European spatial planning. Paper to AESOP Congress, Vienna, July.

Dyrberg, T.B. (1997) *The Circular Structure of Power*, Verso, London.

East Anglia Economic Planning Council (EAEPC) (1968) *East Anglia: a Study*, Her Majesty's Stationery Office, London.

East of England Regional Assembly (EERA) (2004) *East of England Plan: Draft Revision to the Regional Spatial Strategy (RSS) for the East of England*, EERA, Bury St Edmunds.

Eckstein, B. and Throgmorton, J. (2003) *Stories and Sustainability: Planning, Practice and Possibility for American Cities*, MIT Press, Cambridge, MA.

Elson, M.J. (1986) *Green Belts: Conflict Mediation in the Urban Fringe*, Heinemann, London.

Etzioni, A. (1973) Mixed-scanning: a 'third' approach to decision-making. In *A Reader in Planning Theory*, Faludi, A. (ed.), Pergamon, Oxford, pp. 217–229.

Fainstein, S. and Fainstein, N. (eds) (1986) *Restructuring the City: The Political Economy of Urban Redevelopment*, Longman, New York.

Faludi, A. (2000) The European spatial development perspective. *European Planning Studies*, 8, 237–250.

Faludi, A. (ed.) (2002) *European Spatial Planning*, Lincoln Institute of Land Policy, Cambridge, MA.

Faludi, A. and van der Valk, A. (1994) *Rule and Order in Dutch Planning Doctrine in the Twentieth Century*, Kluwer Academic Publishers, Dordrecht.

Faludi, A. and Waterhout, B. (eds) (2002) *The Making of the European Spatial Development Perspective*, Routledge, London.

Fay, B. (1996) *Contemporary Philosophy of Social Science: a Multicultural Approach*, Blackwell, Oxford.

Fedeli, V. and Gastaldi, F. (eds) (2004) *Pratiche strategiche di pianificazione: riflessione a partire da nuovi spazi urbani in costruzione*, Franco Angeli, Milan.

Fischer, F. (1989) *Technocracy and the Politics of Expertise*, Sage, Newbury Park, CA.

Fischer, F. (2000) *Citizens, Experts and the Environment: the Politics of Local Knowledge*, Duke University Press, Durham, NC and London.

Fischer, F. (2003) *Reframing Public Policy: Discursive Politics and Deliberative Practices*, Oxford University Press, Oxford.

Fischer, F. and Forester, J. (eds) (1993) *The Argumentative Turn in Policy Analysis and Planning*, UCL Press, London.

Fischler, R. (1995) Strategy and history in professional practice: planning as world-making. In *Spatial Practices*, Liggett, H. and Perry, D. (eds), Sage, Thousand Oaks, CA, pp. 13–58.

Flyvbjerg, B. (1998) *Rationality and Power*, University of Chicago Press, Chicago.

Flyvbjerg, B. (2001) *Making Social Science Matter: Why Social Inquiry Fails and How It Can Succeed Again*, Cambridge University Press, Cambridge.

Flyvbjerg, B. (2004) Phronetic planning research: theoretical and methodological reflections. *Planning Theory and Practice*, 5, 283–306.

Foley, D.L. (1964) An approach to metropolitan spatial structure. In *Explorations in Urban Structure*, Webber, M.M. (ed.), University of Pennsylvania Press, Philadelphia, pp. 21–78.

Foot, J. (2001) *Milan Since the Miracle: City, Culture and Identity*, Berg, Oxford.

Forester, J. (1993) *Critical Theory, Public Policy and Planning Practice*, State University of New York Press, Albany.

Forester, J. (1999) *The Deliberative Practitioner: Encouraging Participatory Planning Processes*, MIT Press, London.

Friedmann, J. (1987) *Planning in the Public Domain*, Princeton University Press, Princeton.

Friedmann, J. (1992) *Empowerment: the Politics of Alternative Development*, Blackwell, Oxford.

Friedmann, J. (1993) Towards a non-Euclidean mode of planning. *Journal of the American Planning Association*, 59, 482–484.

Friedmann, J. (2004) Strategic spatial planning and the longer range. *Planning Theory and Practice*, 5, 49–56.

Friedmann, J. (2005) Globalisation and the emerging culture of planning. *Progress in Planning*, 64, 183–234.

Friend, J. and Hickling, A. (1987) *Planning Under Pressure: the Strategic Choice Approach*, Pergamon, Oxford.

Friend, J., Power, J. and Yewlett, C. (1974) *Public Planning: the Intercorporate Dimension*, Tavistock Institute, London.

Fuerst, D. and Kneilung, J. (2002) *Regional Governance: New Modes of Self-Government in the European Community*, Hannover ARL, University of Hannover.

Fung, A. and Wright, E.O. (2001) Deepening democracy: innovations in empowered participatory governance. *Politics and Society*, 29, 5–41.

Gabellini, P. (1988) *Bologna e Milano: temi e Attori dell'Urbanistica*, Franco Angeli, Milan.

Gabellini, P. (2002) Guardare Milano e l'urbanistica italiana. *Urbanistica*, 119, 102–107.

Gaffikin, F. and Sterrett, K. (2006) New visions for old cities: the role of visioning in planning. *Planning Theory and Practice*, 7, pp. 159–178.

Gamble, A. (1988) *The Free Economy and the Strong State: the Politics of Thatcherism*, Macmillan, Houndmills.

Gario, G. (1995) Intergovernmental relations in Lombardy: provinces, regions and cities. *Political Geography*, 14, 419–428.

Garnsey, E. and Lawton Smith, H. (1998) Proximity and complexity in the emergence of high technology industry: The Oxbridge comparison. *Geoforum*, 29, 433–450.

Geddes, P. (1915/1968) *Cities in Evolution*, Ernest Benn Ltd, London.

Geertz, C. (1983) *Local Knowledge*, Basic Books, New York.

Geertz, C. (1988) *Works and Lives: the Anthropologist as Author*, Stanford University Press, Stanford, CA.

Giddens, A. (1984) *The Constitution of Society*, Polity Press, Cambridge.

Gieling, S. and de Laat, L. (2004) The inner city: sunny side up. In *Cultural Heritage and the Future of the City*, Deben, L., Salet, W. and van Thoor, M.-T. (eds), Aksant, Amsterdam, pp. 311–318.

Gieling, S. and van Loenen, H. (2001) *De lagen van de stad: opmaat tot een nieuw structuurplan (Plan Amsterdam 12)*, Amsterdam Dienst Ruimtelijke Ordening van de Gemmente, Amsterdam.

Gold, J. (1997) *The Experience of Modernism: Modern Architects and the Future City: 1928–1953*, E & FN Spon, London.

Gomart, E. and Hajer, M. (2003) Is that politics? For an inquiry into forms of contemporary politics. In *Looking Back, Ahead – The Yearbook of the Sociology of Sciences*, Joerges, B. and Nowotny, H. (eds), Kluwer, Dordrecht, pp. 33–61.

Gonzalez, S. and Healey, P. (2005) A sociological institutionalist approach to the study of innovation in governance capacity. *Urban Studies*, 42, 2055–2070.

Goodstadt, V. and Buchan, G. (2002) A statutory approach to community planning: repositioning the statutory development plan. In *Urban Governance, Institutional Capacity and Social Milieux*, Cars, G., Healey, P., Madanipour, A. and de Magalhães, C. (eds), Ashgate, Aldershot, pp. 168–190.

Graham, S. and Healey, P. (1999) Relational concepts in time and space: issues for planning theory and practice. *European Planning Studies*, 7, 623–646.

Graham, S. and Marvin, S. (2001) *Splintering Urbanism*, Routledge, London.

Granovetter, M. (1985) Economic action and social structure: the problem of embeddedness. *American Journal of Sociology*, 91, 481–510.

Gregory, D. (1994) *Geographical Imaginations*, Blackwell, Oxford.

Gualini, E. (2001) *Planning and the Intelligence of Institutions*, Ashgate, Aldershot.

Gualini, E. (2003) The region of Milan. In *Metropolitan Governance and Spatial Planning*, Salet, W., Thornley, A. and Kreukels, A. (eds), Spon Press, London, pp. 264–283.

Gualini, E. (2004a) Integration, diversity and plurality: territorial governance and the reconstruction of legitimacy. *Geopolitics*, 9, 542–563.

Gualini, E. (2004b) *Multi-level Governance and Institutional Change: the Europeanisation of Regional Policy in Italy*, Ashgate, Aldershot.

Gualini, E. (2004c) Regionalisation as 'experimental regionalism': the rescaling of territorial policy-making in Germany. *International Journal of Urban and Regional Research*, 28, 329–353.

Gualini, E. and Woltjer, J. (2004) The re-scaling of regional planning and governance in the Netherlands. Paper to *AESOP Congress*, Vienna, July.

Gunn, S. (2006) Mind the gap: an emerging hole in the way we think about the environment and plan for housing. *Town Planning Review*, in press.

Gunton, T.I., Day, J.C. and Williams, P.W. (2003) Evaluating collaborative planning: the British Columbia experience. *Environments*, 31, 1–11.

Haas, P.M. (1992) Introduction: epistemic communities and international policy coordination. *International Organization*, 46, 1–35.

Hajer, M. (1995) *The Politics of Environmental Discourse*, Oxford University Press, Oxford.

Hajer, M. (2001) The need to zoom out. In *The Governance of Place*, Madanipour, A., Hull, A. and Healey, P. (eds), Ashgate, Aldershot, pp. 178–202.

Hajer, M. (2003) Policy without polity? Policy analysis and the institutional void. *Policy Sciences*, 36, 175–195.

Hajer, M. (2005) Setting the stage: a dramaturgy of policy deliberation. *Administration and Society*, 36, 624–647.

Hajer, M. and Versteeg, W. (2005) Performing governance through networks. *European Political Science*, 4(3), 340–347.

Hajer, M. and Wagenaar, H. (eds) (2003) *Deliberative Policy Analysis: Understanding Governance in the Network Society*, Cambridge University Press, Cambridge.

Hajer, M. and Zonneveld, W. (2000) Spatial planning in the network society – rethinking the principles of planning in the Netherlands. *European Planning Studies*, 8, 337–355.

Hall, P. (1966) *World Cities*, Weidenfeld and Nicolson, London.

Hall, P. (1988) *Cities of Tomorrow*, Blackwell, Oxford.

Hall, P. and Taylor, R. (1996) Political science and the three institutionalisms. *Political Studies*, XLIV, 936–957.

Hall, P., Thomas, R., Gracey, H. and Drewett, R. (1973) *The Containment of Urban England*, George, Allen and Unwin, London.

Harding, A. (1997) Urban regimes in European Cities. *European Urban and Regional Studies*, 4, 291–314.

Hargreaves, M. (1995) Cambridge belt takes the strain in panel report. *Planning*, 1112.

Harris, N. and Hooper, A. (2004) Rediscovering the 'spatial' in public policy and planning: an examination of the spatial content of sectoral policy documents. *Planning Theory and Practice*, 5, 147–170.

Harvey, D. (1985) *The Urbanisation of Capital*, Blackwell, Oxford.

Harvey, D. (1989) From managerialism to entrepreneurialism: the formation of urban governance in late capitalism. *Geografisker Annaler*, 71B, 3–17.

Haugh, P. (1986) US high technology multinationals and Silicon Glen. *Regional Studies*, 20, 103–116.

Healey, P. (1983) *Local Plans in British Land Use Planning*, Pergamon, Oxford.

Healey, P. (1990) Policy processes in planning, *Policy and Politics*, 18, 91–103.

Healey, P. (1997) *Collaborative Planning: Shaping Places in Fragmented Societies*, Macmillan, London.

Healey, P. (1998a) Collaborative planning in a stakeholder society. *Town Planning Review*, 69, 1–21.

Healey, P. (1998b) Regulating property development and the capacity of the development industry. *Journal of Property Research*, 15, 211–228.

Healey, P. (1998c) Building institutional capacity through collaborative approaches to urban planning. *Environment and Planning A*, 30, 1531–1556.

Healey, P. (1999) Institutionalist analysis, communicative planning and shaping places. *Journal of Planning Education and Research*, 19, 111–122.

Healey, P. (2002) On creating the 'city' as a collective resource. *Urban Studies*, 39, 1777–1792.

Healey, P. (2004a) Creativity and urban governance. *Policy Studies*, 25, 87–102.

Healey, P. (2004b) The treatment of space and place in the new strategic spatial planning in Europe. *International Journal of Urban and Regional Research*, 28, 45–67.

Healey, P. (2006a) Transforming governance: challenges of institutional adaptation and a new politics of space. *European Planning Studies*, 14, 299–319.

Healey, P. (2006b) Territory, integration and spatial planning. In *Territory, Identity and Space*, Tewdwr-Jones, M. and Allmendinger, P. (eds), Routledge, London, pp. 64–79.

Healey, P. (2006c) Relational complexity and the imaginative power of strategic spatial planning. *European Planning Studies*, 14, 525–546.

Healey, P. (2006d) The new institutionalism and the transformative goals of planning. In *Planning and Institutions*, Verma, N. (ed.), Elsevier, Oxford, pp. 62–87.

Healey, P., de Magalhães, C., Madanipour, A. and Pendlebury, J. (2003) Place, identity and local politics: analysing partnership initiatives. In *Deliberative Policy Analysis: Understanding Governance in the Network Society*, Hajer, M. and Wagenaar, H. (eds), Cambridge University Press, Cambridge, pp. 60–87.

Healey, P., Khakee, A., Motte, A. and Needham, B. (eds) (1997) *Making Strategic Spatial Plans: Innovation in Europe*, UCL Press, London.

Healey, P., McDougall, G. and Thomas, M. (eds) (1982) *Planning Theory: Prospects for the 1980s*, Pergamon, Oxford.

Healey, P., McNamara, P., Elson, M. and Doak, J. (1988) *Land Use Planning and the Mediation of Urban Change*, Cambridge University Press, Cambridge.

Healey, P., Purdue, M. and Ennis, F. (1995) *Negotiating Development*, Spon, London.

Hillier, J. (2000) Imagined value: the poetics and politics of place. In *The Governance of Place*, Madanipour, A., Hull, A. and Healey, P. (eds), Ashgate, Aldershot, pp. 69–101.

Hillier, J. (2002) *Shadows of Power: an Allegory of Prudence in Land-Use Planning*, Routledge, London.

Hillier, J. (2007) *Stretching Beyond the Horizon: a Mutliplanar Theory of Spatial Planning and Governance*, Ashgate, Aldershot.

Hodgson, G.M. (2004) *The Evolution of Institutional Economics: Agency, Structure and Darwinianism in American institutionalism*, Routledge, New York.

Holford, W. and Wright, H.M. (1950) *Cambridge Planning Proposals: a Report to the Town and Country Planning Committee of the Cambridgeshire County Council*, Cambridge University Press, Cambridge.

Holston, J. (1998) Spaces of insurgent citizenship. In *Making the Invisible Visible: a Multicultural Planning History*, Sandercock, L. (ed.), University of California Press, Berkeley, pp. 37–56.

Hooghe, L. (ed.) (1996) *Cohesion Policy and European Integration: Building Multi-Level Governance*, Oxford University Press, Oxford.

Imrie, R. and Raco, M. (eds) (2003) *Urban Renaissance? New Labour, Community and Urban Policy*, Policy Press, Bristol.

Indovina, F. and Matassani, F. (1990) *La Citta Diffusa*, DAEST, Venezia.

Ingold, T. (2000) *The Perception of the Environment: Essays in Livelihood, Dwelling and Skill*, Routledge, London.

Ingold, T. (2005) The eye of the storm: visual perception and the weather. *Visual Studies*, 20, 97–104.

Innes, J. (1990) *Knowledge and Public Policy: the Search for Meaningful Indicators*, Transaction Books, New Brunswick.

Innes, J. (1992) Group processes and the social construction of growth management. *Journal of the American Planning Association*, 58, 440–454.

Innes, J. (2004) Consensus building: clarification for critics. *Planning Theory*, 3, 5–20.

Innes, J. and Booher, D. (1999a) Consensus-building and complex adaptive systems: a framework for evaluating collaborative planning. *Journal of the American Planning Association*, 65, 412–423.

Innes, J. and Booher, D. (1999b) Consensus-building as role-playing and bricolage. *Journal of the American Planning Association*, 65, 9–26.

Innes, J. and Booher, D. (2000) Planning institutions in the network society: theory for collaborative planning. In *The Revival of Strategic Spatial Planning*, Salet, W. and Faludi, A. (eds), Koninklijke Nederlandse Akademie van Wetenschappen, Amsterdam, pp. 175–189.

Innes, J. and Booher, D. (2001) Metropolitan development as a complex system: a new approach to sustainability. In *The Governance of Place*, Madanipour, A., Hull, A. and Healey, P. (eds), Ashgate, Aldershot, pp. 239–264.

Innes, J. and Booher, D. (2003) Collaborative policy-making: governance through dialogue. In *Deliberative Policy Analysis: Understanding Governance in the Network Society*, Hajer, M. and Wagenaar, H. (eds), Cambridge University Press, Cambridge, pp. 33–59.

Innes, J.E. and Gruber, J. (2005) Planning styles in conflict: the Metropolitan Transportation Commission. *Journal of the American Planning Association*, 71, 177–188.

Jensen, O. and Richardson, T. (2000) Discourses of mobility and polycentric development: a contested view of European spatial planning. *European Planning Studies*, 8, 503–520.

Jensen, O.B. and Richardson, T. (2004) *Making European Space: Mobility, Power and Territorial Identity*, Routledge, London.

Jessop, B. (1995) Towards a schumpeterian workfare regime in Britain? Reflections on regulation, governance and the welfare state. *Environment and Planning A*, 27, 1613–1626.

Jessop, B. (1997) Capitalism and its future; remarks on regulation, government and governance. *Review of International Political Economy*, 4, 561–581.

Jessop, B. (1998) The narrative of enterprise and the enterprise of narrative: place marketing and the entrepreneurial city. In *The Entrepreneurial City: Geographies of Politics, Regime and Representation*, Hall, T. and Hubbard, P. (eds), John Wiley, London, pp. 77–99.

Jessop, B. (2000) The crisis of the national spatio-temporal fix and the tendential ecological dominance of globalising capitalism. *International Journal of Urban and Regional Research*, 24, 323–360.

John, P. (1998) *Analysing Public Policy*, Pinter, London.

Johnstone, C. and Whitehead, M. (eds) (2004) *New Horizons in British Urban Policy: Perspectives on New Labour's Urban Renaissance*, Ashgate, Aldershot.

Jolles, A. (2005) Amsterdam's growth rings. In *Impact: Urban Planning in Amsterdam After 1986*, Buurman, M. and Kloos, M. (eds), ARCAM/Architectura and Natura Press, Amsterdam, pp. 17–38.

Jolles, A., Klusman, E. and Teunissan, B. (eds) (2003) *Planning Amsterdam: Scenarios for Urban Development 1928–2003*, NAi, Rotterdam.

Jonas, A.E.G., Gibbs, D.C. and While, A. (2005) Uneven development, sustainability and city-regionalism contested: English city-regions in the European context. In *Regionalism Contested: Institution, Society and Territorial Governance*, Sagan, I. and Halkier, H. (eds), Ashgate, Aldershot, pp. 223–246.

Katz, P. (ed.) (1994) *The New Urbanism: Towards an Architecture of Community*, McGraw Hill, New York.

Keeble, D., Lawson, C., Moore, B. and Wilkinson, F. (1999) Collective learning processes, networking and 'institutional thickness' in the Cambridge Region. *Regional Studies*, 33, 319–332.

Keeble, L. (1952) *Principles and Practice of Town and Country Planning*, Estates Gazetter, London.

Kickert, W.J.M. (2003) Beyond public management: shifting frames of reference in administrative reform in the Netherlands. *Public Management Review*, 5, 377–399.

Kickert, W.J.M., Klijn, E.-H. and Koppenjan, J.F.M. (1997) *Managing Complex Networks: Strategies for the Public Sector*, Sage, London.

Kitching, G. (2003) *Wittgenstein and Society: Essays in Conceptual Puzzlement*, Ashgate, Aldershot.

Klijn, E.-H. (1997) Policy networks: an overview. In *Managing Complex Networks: Strategies for the Public Sector*, Kickert, W.J.M., Klijn, E.-H. and Koppenjan, J.F.M. (eds), Sage, London, pp. 14–34.

Klijn, E.-H. and Teisman, G.R. (1997) Strategies and games in networks. In *Managing Complex Networks: Strategies for the Public Sector*, Kickert, W.J.M., Klijn, E.-H. and Koppenjan, J.F.M. (eds), Sage, London, pp. 98–118.

Knorr-Cetina, K. (1999) *Epistemic Cultures: How the Sciences Make Knowledge*, Harvard University Press, Cambridge MA.

Kratz, P. (1997) Cambridge in Crisis. *Planning*, 18 April.

Kreukels, A. (2003) Rotterdam and the South Wing of the *Randstad*. In *Metropolitan Governance and Spatial Planning*, Salet, W., Thornley, A. and Kreukels, A. (eds), Spon, London, pp. 189–202.

Kuhn, T.S. (1970) *The Structure of Scientific Revolutions*, University of Chicago Press, Chicago.

Kunzmann, K. (1998) Planning for spatial equity in Europe. *International Planning Studies*, 3, 101–120.

Kunzmann, K. (2001) The Ruhr in Germany: a laboratory for regional governance. In *The Changing Institutional Landscape of Planning*, Albrechts, L., Alden, J. and da Rosa Pires, A. (eds), Ashgate, Aldershot, pp. 181–208.

Lagendijk, A. and Cornford, J. (2000) Regional institutions and knowledge – tracking new forms of regional development policy. *Geoforum*, 31, 208–218.

Lascoumes, P. and Le Galès, P. (2003) Interest groups and public organisations in Europe. In *Handbook of Public Administration*, Peters, G.B. and Pierre, J. (eds), Sage, London, pp. 321–330.

Latour, B. (1987) *Science in Action*, Harvard University Press, Cambridge, MA.

Law, J. (2004) *After Method: Mess in Social Science Research*, Routledge, London.

Le Galès, P. (2002) *European Cities: Social Conflicts and Governance*, Oxford University Press, Oxford.

Lefebvre, H. (1991) *The Production of Space*, Blackwell, Oxford.

Lefèvre, C. (1998) Metropolitan government and governance in Western countries: a critical review. *International Journal of Urban and Regional Research*, 22, 9–25.

Liggett, H. (1995) City sights/sites of memories and dreams. In *Spatial Practices*, Liggett, H. and Perry, D. (eds), Sage, Thousand Oaks, CA, pp. 243–270.

Liggett, H. and Perry, D. (eds) (1995) *Spatial Practices*, Sage, Thousand Oaks, CA.

Lindblom, C.E. (1990) *Inquiry and Change: the Troubled Attempt to Understand and Shape Society*, Yale University Press, New Haven.

Logan, J. and Molotch, H. (1987) *Urban Fortunes: the Political Economy of Place*, University of California Press, Berkeley and Los Angeles.

Logie, G. (1966) *The Future Shape of Cambridge*, Cambridge City Council: Report of the City Architect and Planning Officer, Cambridge.

Lovering, J. (1999) Theory led by policy: the inadequacies of 'the New Regionalism' (illustrated from the case of Wales). *International Journal of Urban and Regional Research*, 23, 379–395.

Lowndes, V. (2001) Rescuing Aunt Sally: taking institutional theory seriously in urban politics. *Urban Studies*, 38, 1953–1972.

Lukes, S. (1974) *Power: a Radical View*, Macmillan, London.

Macchi Cassia, C., Orsini, M., Privileggio, N. and Secchi, M. (2004) *Per/To Milano*, Editore Ulriche Hoepli, Milan.

McGuirk, P. (2003) Producing the capacity to govern in global Sydney: a multiscaled account. *Journal of Urban Affairs*, 25(2), 201–223.

MacLeod, G. (1999) Place, politics and 'scale dependence': exploring the structuration of euro-regionalism. *European Urban and Regional Studies*, 6, 231–254.

MacLeod, G. (2001) Beyond soft institutionalism: accumulation, regulation, and their geographical fixes. *Environment and Planning A*, 33, 1145–1167.

Macleod, G. and Goodwin, M. (1999) Space, scale and state strategy: towards a reinterpretation of contemporary urban and regional governance. *Progress in Human Geography*, 23, 503–529.

Macnagthen, P. and Urry, J. (1998) *Contested Natures*, Sage, London.

Madanipour, A. (2003) *Public and Private spaces in the City*, Routledge, London.

Magatti, M. (2005) Logiche de Sviluppo e di governo di un nodo globale. In *Milano, nodo della rete globale*, Bassetti, P. *et al.* (eds), Bruno Mondadori, Milan.

Magnier, A. (2004) Between institutional learning and re-legitimisation: Italian mayors in the unending reform. *International Journal of Urban and Regional Research*, 28, 166–182.

Majone, G. (1987) *Evidence, Argument and Persuasion in the Policy Process*, Yale University Press, New Haven.

Mak, G. (2003) Amsterdam as the 'compleat citie'. In *Amsterdam Human Capital*, Musterd, S. and Salet, W. (eds), Amsterdam University Press, Amsterdam, pp. 31–48.

Mansuur, A. and van der Plas, G. (2003) *De Noordvleugel*, Amsterdam DRO, Gemeente Amsterdam.

Marston, S.A. and Jones III, J.P. (2005) Human geography without scale. *Transactions of the Institute of British Geographers*, 30, 416–432.

Martinelli, F. (2005) *La pianificazione strategica in Italia e in Europa: metodologie ed esiti a confronto*, Franco Angeli, Milan.

Marvin, S. and May, T. (2003) City futures: visions from the centre. *City*, 7, 213–225.

Massey, D. (1984) *Spatial Divisions of Labour*, Macmillan, London.

Massey, D. (1994) *Space, Place and Gender*, Polity Press, Cambridge.

Massey, D. (2000) Travelling thoughts. In *Without Guarantees: Essays in Honour of Stuart Hall*, Gilroy, P., Grossberg, L. and McRobbie, A. (eds), Verso, London, pp. 225–232.

Massey, D. (2004a) Geographies of responsibility. *Geografisker Annaler*, 86B, 5–18.

Massey, D. (2004b) The responsibilities of place. *Local Economy*, 19, 97–101.

Massey, D. (2005) *For Space*, Sage, London.

Massey, D., Allen, J. and Anderson, J. (eds) (1984) *Geography Matters*, Cambridge University Press, Cambridge.

Massey, D., Quintas, P. and Wield, D. (1992) *High Tech Fantasies: Science Parks in Society and Space*, Routledge, London.

Mastop, H. and Faludi, A. (1997) Evaluation of strategic plans: the performance principle. *Environment and Planning B: Planning and Design*, 24, 815–832.

Mayer, M. (2000) Social movements in European cities: transitions from the 1970s to the 1990s. In *Cities in Contemporary Europe*, Bagnasco, A. and Le Galès, P. (eds), Cambridge University Press, Cambridge, pp. 131–152.

Mazza, L. (1997) *Trasformazione del Piano*, Franco Angeli, Milan.

Mazza, L. (2001) Nuove procedure urbanistiche a Milano. *Territorio*, 16, 53–60.

Mazza, L. (2002) Flessibilita e rigidita delle argomentazioni urbanistica. *Urbanistica*, 118. Reprinted in Mazza, L. (2004c) *Prove parziali di riforma urbanistica*, Franco Angeli, Milan.

Mazza, L. (2004a) *Progettare gli Squilibri*, Franco Angeli, Milan.

Mazza, L. (2004b) *Piano, progetti, strategie*, Franco Angeli, Milan.

Mazza, L. (2004c) *Prove parziali di riforma urbanistica*, Franco Angeli, Milan.

Mehlbye, P. (2000) Global Integration Zones – neighbouring metropolitan regions in metropolitan clusters. In *Europaische Metropolregionen. Informationen zur Raumtwicklung Heft 11/12*, pp. 755–762, Bonn, Bundesamt fur Bauwesen und Raumordnung.

Melucci, A. (1989) *Nomads of the Present: Social Movements and Individual Needs in Contemporary Society*, Hutchinson, London.

Meyerson, M. and Banfield, E. (1955) *Politics, Planning and the Public Interest*, Free Press, New York.

Mintzberg, H. (1994) *The Rise and Fall of Strategic Planning*, Pearson Education Limited, Edinburgh.

Morgan, G. (1997) *Images of Organization*, Sage, London.

Morgan, K. (1997) The learning region: institutions, innovation and regional renewal. *Regional Studies*, 31, 491–503.

Morgan, K. and Murdoch, J. (2000) Organic v. conventional agriculture: knowledge, power and innovation in the food chain. *Geoforum*, 31, 159–173.

Morrison, N. (1998) The compact city: theory versus practice – the case of Cambridge. *Netherlands Journal of Housing and the Built Environment*, 3, 157–179.

Motte, A. (ed) (1995) *Scheḿa directeur et projet d'agglomeration: l'experimentation de nouvelles politiques urbaines spatialisees 1981–1993*, Les editions Juris Service, Paris.

Motte, A. (1997) Building strategic urban planning in France. In *Making Strategic Spatial Plans: Innovation in Europe*, Healey, P., Khakee, A., Motte, A. and Needham, B. (eds), UCL Press, London, pp. 59–76.

Motte, A. (2001) The influence of new institutional processes in shaping places: the cases of Lyon and Nimes (France 1981–95). In *The Governance of Place: Space and Planning Processes*, Madanipour, A., Hull, A. and Healey, P. (eds), Ashgate, Aldershot, pp. 223–238.

Motte, A (2005) Enabling the emergence of metropolitan strategic planning through network development: the French Experiment (2004–2006). In *Proceedings of the AESOP Congress*, Vienna, July.

Moulaert, F., with Delladetsima, P., Delvainquiere, J.C., Demaziere, C., Rodriguez, A., Vicari, S. and Martinez, M. (2000) *Globalisation and Integrated Area Development in European Cities*, Oxford University Press, Oxford.

Moulaert, F., Martinelli, F., Swyngedouw, E. and Gonzalez, S. (2005) Towards alternative model(s) of local innovation. *Urban Studies*, 42, 1969–1990.

Mugnano, S., Tornaghi, C. and Vicari Haddock, S. (2005) Nuove visioni del territorio: il rinnovo urbano e i nuovi spazi pubblici nel nord Milano. In *dell'Agnese, E. Bicocca e il suo territorio*, Skiria, Milan, pp. 166–193.

Murdoch, J. (1995) Actor-networks and the evolution of economic forms: combining description and explanation in theories of regulation, flexible specialisation and networks. *Environment and Planning A*, 27, 731–757.

Murdoch, J. and Abram, S. (2002) *Rationalities of Planning: Development Versus Environment in Planning for Housing*, Ashgate, Aldershot.

Natter, W. and Jones, J.P. (1997) Identity, space and other uncertainties. In *Space and Social Theory: Interpreting Modernity and Postmodernity*, Benko, G. and Strohmeyer, U. (eds), Blackwell, Oxford, pp. 141–161.

Needham, B. (2005) The new Dutch Spatial Planning Act. *Planning Practice and Research*, 20(3), 327–340.

Needham, B. and Zwanniken, T. (1997) The current urbanization policy evaluated. *Netherlands Journal of Housing and the Built Environment*, 12, 37–55.

Needham, B., Koenders, P. and Kruijt, B. (1993) *The Netherlands: Urban Land and Property Markets*, UCL Press, London.

Newman, P. and Thornley, A. (1996) *Urban Planning in Europe*, Routledge, London.

Nielsen, E.H. and Simonsen, K. (2003) Scaling from 'below': practices, strategies and urban spaces. *European Planning Studies*, 11, 911–927.

Nigro, G. and Bianchi, G. (eds) (2003) *Politiche, Programmi e Piani nel governo della Citta*, Gangemi Editore, Rome.

Nonaka, I., Toyama, R. and Konno, N. (2001) SECI, Ba and leadership: a unified model of dynamic knowledge creation. In Managing Industrial Knowledge: Creation, Transfer and Utilisation, Nonaka, I. and Teece, D. (eds), Sage, London, pp. 13–43.

Novarina, G. (ed.) (2003) *Plan et Projet: L'urbanisme en France et en Italie*, Anthropos, Paris.

Offe, C. (1977) The theory of the capitalist state and the problem of policy formation. In *Stress and Contradiction in Modern Capitalism*, Lindberg, L.N. and Alford, A. (eds), D.C. Heath, Lexington, MA, pp. 125–144.

Office of the Deputy Prime Minister (ODPM) (2003) *Sustainable Communities: Building for the Future*, ODPM, London.

Office of the Deputy Prime Minister (ODPM) (2005a) *Planning System: General Principles*, ODPM, London.

Office of the Deputy Prime Minister (ODPM) (2005b) *Planning Policy Guidance 3: Housing: supporting the delivery of new housing: (proposals for changes – August 2005)*, ODPM, London.

Oliva, F. (2002) *L'Urbanistica di Milano: Quel che resta dei piani urbanistici e nella trasformaziche della citta*, Hoepli, Milan.

Owens, S. and Cowell, R. (2002) *Land and Limits: Interpreting Sustainability in the Planning Process*, Routledge, London.

Owens, S., Rayner, T. and Bina, O. (2004) New agendas for appraisal: reflections on theory, practice and research. *Environment and Planning A*, 36, 1943–1959.

Palermo, P.-C. (2002) Osservare Milano, laboratorio sperimentale di un futuro possibile. *Urbanistica*, 119, 121–124.

Paris, C. (ed.) (1982) *Critical Readings in Planning Theory*, Pergamon, Oxford.

Parr, J. (2005) Spatial planning: too little or too much? *Scienze Regionali*, 4, 113–129.

Pasqui, G. (2002) *Confini Milanese: processi territoriale e practiche di governo*, Franco Angeli, Milan.

Perry, D. (1995) Making space: planning as a mode of thought. In *Spatial Practices*, Liggett, H. and Perry, D. (eds), Sage, Thousand Oaks, CA, pp. 209–242.

Peters, G. (1999) *Institutional Theory in Political Science: the 'New Institutionalism'*, Continuum, London.

Piccinato, L. (1956) Special issue on the *Piano Regolatore Generale di Milano. Urbanistica*, 18/19.

Pierre, J. (ed.) (1998) *Partnerships in Urban Governance: European and American Experience*, Macmillan, London.

Pierre, J. and Peters, G. (2000) *Governance, Politics and the State*, Palgrave Macmillan, London.

Ploeger, R. (2004) Regulating urban office provision. *Faculteit der Maatschappij-en Gedragswetenschappen*, Universiteit von Amsterdam, Amsterdam.

Plummer, J. (2000) *Municipalities and Community Participation: a Sourcebook for Capacity Building*, Earthscan, London.

Pomilio, F. (2001) Il Documento di Inquadrimento delle politiche urbanistiche (DdI) a Milano: un caso 'anomalo' di pianficazione locale Milano, Politecnico di Milano: unpublished course papers.

Pomilio, F. (2003) Il 'Documento di Inquadrimento delle politiche urbanistiche' di Milano: un caso anomalo di pianificazione strategica? In *Strategie per la citta: piani, politiche, azioni: Una rassegna di casi*, Pugliese, T. and Spaziente, A. (eds), Franco Angeli, Milan, pp. 187–206.

Priemus, H. (2002) Spatial-economic investment policy and urban regeneration in the Netherlands. *Environment and Planning C: Government and Policy*, 20, 775–790.

Priemus, H. and Visser, J. (1995) Infrastructure policy in the *Randstad* Holland: struggle between accessibility and sustainability. *Political Geography*, 14, 363–377.

Priemus, H. and Zonneveld, W. (2004) Regional and transnational spatial planning: problems today, perspectives for the future. *European Planning Studies*, 12, 283–297.

Pruijt, H. (2004) The impact of citizens' protest on city planning in Amsterdam. In *Cultural Heritage and the Future of the Historic Inner City of Amsterdam*, Deben, L., Salet, W. and van Thoor, M.-T. (eds), Aksant, Amsterdam, pp. 228–244.

Pugliese, T. and Spaziente, A. (eds) (2003) *Strategie per la citta: piani, politiche, azioni: Una rassegna di casi*, Franco Angeli, Milan.

Punter, J. (2003) *The Vancouver Achievement*, UBC Press, Vancouver.

Rein, M. and Schon, D. (1993) Reframing policy discourse. In *The Argumentative Turn in Policy Analysis and Planning*, Fischer, F. and Forester, J. (eds), Duke University Press, Durham, NC, pp. 145–166.

Rhodes, R.A.W. (1997) *Understanding Governance: Policy Networks, Governance, Reflexivity and Accountability*, Open University Press, Milton Keynes.

Richardson, T. (2004) Environmental assessment and planning theory: four short stories about power, multiple rationality and ethics. *Environmental Impact Assessment Review*, 25(4), 341–365.

Richardson, T. (2006) The thin simplification of European space: dangerous calculations? *Comparative European Politics*, 4(2) (in press).

Roy, A. (2003) *City Requiem, Calcutta: Gender and the Politics of Poverty*, University of Minnesota Press, Minneapolis.

Royal Commission on Environmental Pollution (RCEP) (2002) *Environmental Planning: 23rd Report*, Norwich, The Stationery Office.

Royal Town Planning Institute (RTPI) (2001) *A New Vision for Planning*, RTPI, London.

Rydin, Y. (1986) *Housing Land Policy*, Gower, Aldershot.

Rydin, Y. (2003a) *Conflict, Consensus and Rationality in Environmental Planning: an Institutional Discourse Approach*, Oxford University Press, Oxford.

Rydin, Y. (2003b) *Urban and Environmental Planning in the UK*, Palgrave, Basingstoke.

Sabatier, P.A. and Jenkins-Smith, H.C. (eds) (1993) *Policy Change and Learning: an Advocacy Coalition Approach*, Westview Press, Boulder, CO.

Salet, W. (2003) Amsterdam and the north wing of the *Randstad*. In *Metropolitan Governance and Spatial Planning*, Salet, W., Thornley, A. and Kreukels, A. (eds), Spon Press, London, pp. 175–188.

Salet, W. and Faludi, A. (eds) (2000) *The Revival of Strategic Spatial Planning*, Koninklijke Nederlandse Akademie van Wetenschappen (Royal Netherlands Academy of Arts and Sciences), Amsterdam.

Salet, W. and Gualini, E. (2003) The region of Amsterdam, Unpublished paper EU COMET Project, AME, University of Amsterdam, Amsterdam.

Salet, W. and Majoor, S. (eds) (2005) *Amsterdam Zuidas European Space*, Zuidas Reflector Foundation, Amsterdam.

Salet, W., Thornley, A. and Kreukels, A. (eds) (2003) *Metropolitan Governance and Spatial Planning: Comparative Studies of European City-Regions*, E & FN Spon, London.

Salzano, E. (2002) Il modello flessibile a Milano. *Urbanistica*, 118, 140–148.

Salzer-Morling, M. (1998) As God created the earth . . . a saga that makes sense. In *Discourse + Organisation*, Grant, D., Keenoy, T. and Oswick, C. (eds), Sage, London, pp. 104–118.

Sandercock, L. (2003a) *Mongrel Cities: Cosmopolis 11*, Continuum, London.

Sandercock, L. (2003b) Out of the closet: the importance of stories and storytelling in planning practice. *Planning Theory and Practice*, 4, 11–28.

Sanyal, B. (ed.) (2005) *Comparative Planning Cultures*, Routledge, London.

Schlosberg, D. (1999) *Environmental Justice and the New Pluralism*, Oxford University Press, Oxford.

Schmal, H. (2003) The historical roots of the daily urban system. In *Amsterdam Human Capital*, Musterd, S. and Salet, W. (eds), Amsterdam University Press, Amsterdam, pp. 67–83.

Schon, D. and Rein, M. (1994) *Frame Reflection: Towards the Resolution of Intractable Policy Controversies*, Basic Books, New York.

Scott, J.C. (1998) *Seeing Like a State: How Certain Schemes to Improve the Human Condition Have Failed*, Yale University Press, New Haven and London.

Secchi, B. (1986) Una nuova forma di piano. *Urbanistica*, 82, 6–13.

Secchi, B. (1988) Ritematizzare Milano. *Urbanistica*, 90, 89–93.

Secchi, B. (2002) Diary of a planner: projects, visions and scenarios. *Planum*, Vol. 2003, 22 July.

Secretary of State for the Environment and other departments (SoS) (1990) *This Common Inheritance: Britain's Environment Strategy*, Her Majesty's Stationery Office, London.

Segal Quince Wicksteed (1985) *The Cambridge Phenomenon*, Segal Quince Wicksteed, Cambridge.

Segal Quince Wicksteed (2000) *The Cambridge Phenomenon Revisited*, Segal Quince Wicksteed, Cambridge.

Senior, D. (1956) *A Guide to the Cambridge Plan*, Cambridgeshire County Council, Cambridge.

Shipley, R. (2002) Visioning in planning: is the practice based on sound theory? *Environment and Planning A*, 34, 7–22.

Shipley, R. and Newkirk, R. (1999) Vision and visioning in planning: what do these terms really mean? *Environment and Planning B: Planning and Design*, 26, 573–591.

Short, J.R., Fleming, S. and Witt, S. (1986) *House Building, Planning and Community Action: the Production and Negotiation of the Built Environment*, Routledge and Kegan Paul, London.

Simonsen, K. (2004) Networks, flows and fluids – reimagining spatial analysis? *Environment and Planning A*, 36, 1333–1340.

Solesbury, W. (1974) *Policy in Urban Planning*, Pergamon, Oxford.

South East Joint Planning Team (SEJPT) (1970) *Strategic Plan for the South East*, Ministry of Housing and Local Government, London.

Stoker, G. (1995) Regime theory and urban politics. In *Theories of Urban Politics*, Judge, D., Stoker, G. and Wolman, H. (eds), Sage, London, pp. 54–71.

Stoker, G. (ed.) (2000) *The New Politics of British Local Governance*, Macmillan, Houndmills, Basingstoke.

Stone, C. (1989) *Regime Politics: Governing Atlanta 1946–1988*, University of Kansas Press, Lawrence.

Stone, C.N. (2005) Rethinking the policy–politics connection. *Policy Studies*, 26, 241–260.

Storper, M. (1997) *The Regional World*, Guilford Press, New York.

Susskind, L., McKearnan, S. and Thomas-Larmer, J. (eds) (1999) *The Consensus-Building Handbook*, Sage, London and Thousand Oaks, CA.

Sutcliffe, A. (1981) *Towards the Planned City: Germany, Britain, the United States and France, 1780–1914*, Blackwell, Oxford.

Takeuchi, H. (2001) Towards a universal management of the concept of knowledge. In *Managing Industrial Knowledge: Creation, Transfer, Utilization*, Nonaka, I. and Teece, D. (eds), Sage, London, pp. 315–329.

Tarrow, S. (1994) *Power in Movement*, Cambridge University Press, Cambridge.

Taylor, M. (2003) *Public Policy in the Community*, Palgrave, Houndmills.

Taylor, P.J. (2004a) *Amsterdam in a World City Network*, AME, University of Amsterdam, Amsterdam.

Taylor, P.J. (2004b) *World City Network: a Global Urban Analysis*, Routledge, London.

Terhorst, P. and Van de Ven, J. (1995) The national growth coalition in the Netherlands. *Political Geography*, 14, 343–361.

Terhorst, P., van den Ven, J. and Deben, L. (2003) Amsterdam: it's all in the mix. In *Cities and Visitors: Regulating People, Markets and City Space*, Hoffman, L.M., Fainstein, S.F. and Judd, D.R. (eds), Blackwell, Oxford, pp. 76–90.

Tewdwr-Jones, M. (2002) *The Planning Polity: Planning, Government and the Policy Process*, Routledge, London.

Tewdwr-Jones, M. and Allmendinger, P. (eds) (2006) *Territory, Identity and Space*, Routledge, London.

Thompson, S. (2000) Diversity, difference and the multi-layered city. In *Urban Planning in a Changing World: the Twentieth Century Experience*, Freestone, R. (ed.), E & FN Spon, London, pp. 230–248.

Thornley, A. (1991) *Urban Planning Under Thatcherism: the Challenge of the Market*, Routledge, London.

Thrift, N. (1996) *Spatial Formations*, Sage, London.

Thrift, N. (1999) Steps to an ecology of place. In *Human Geography Today*, Massey, D., Allen, J. and Sarre, P. (eds), pp. 295–332.

Thrift, N. (2000) Everyday life in the city. In *A Companion to the City*, Bridge, G. and Watson, S. (eds), Blackwell, Oxford, pp. 398–409.

Throgmorton, J.A. (1996) *Planning as Persuasive Story-Telling*, University of Chicago Press, Chicago.

Throgmorton, J.A. (2004) Inventing the 'greatest': planning as persuasive storytelling in the open moral community called Louisville. In *ACSP Congress*, Portland, Oregon, USA.

Tosi, A. (ed.) (1985) *Terziario, impresa, territorio: dinamiche e politiche urbane e regionali*, Franco Angeli, Milano.

Tosi, A. (ed.) (1990) *Milano e la Lombardia: per un rete urbana policentrica*, Franco Angeli, Milano.

Ufficio Tecnico Esecutivo per la revisione del PRG (UTERP) (1975) *La variante generale al PRG di Milano. Edilizia Popolare*, Anno XXII, whole issue.

Urban, Task Force (1999) *Towards an Urban Renaissance*, E & FN Spon, London.

Urry, J. (2005) The complexity turn. *Theory, Culture and Society*, 22, 1–14.

van Duinen, L. (2004) Planning imagery: the emergence and development of new planning concepts in Dutch national spatial policy. *Faculteit der Maatschappij en Gedragswetenschappen*, Universiteit Amsterdam, Amsterdam.

Van Eeten, M. (1999) *Dialogues of the Deaf: Defining New Agencies for Environmental Deadlocks*, Eburon, Delft.

van Engelsdorp Gastelaars, R. (2003) Landscapes of power in Amsterdam? In *Amsterdam Human Capital*, Musterd, S. and Salet, W. (eds), Amsterdam University Press, Amsterdam, pp. 289–309.

Verma, N. (1998) *Similarities, Connections, Systems: the Search for a New 'Rationality' for Planning and Management*, Lexington Books, Lanham, MD.

Verma, N. (ed.) (2006) *Planning and Institutions*, Elsevier, Oxford.

Vicari Haddock, S. (ed.) (2005) *Regenerare la Citta: Practiche di innovazione sociale nelle citta europee*, Il Mulino, Bologna.

Vicari, S. and Molotch, H. (1990) Building Milan: alternative machines of growth. *International Journal of Urban and Regional Research*, 14, 602–624.

Vickers, G. (1965) *The Art of Judgement: a Study of Policy-Making*, Chapman Hall, London.

Vigar, G., Graham, S. and Healey, P. (2005) In search of the city in spatial strategies: past legacies and future imaginings. *Urban Studies*, 42, 1391–1410.

Vigar, G., Healey, P., Hull, A. and Davoudi, S. (2000) *Planning, Governance and Spatial Strategy in Britain*, Macmillan, London.

Vitale, T. (2006) Contradiction and reflexivity in social innovation: a case study from the de-institutionalisation movement. *European Urban and Regional Studies*, 13 (in press).

VROM, Ministry for Housing, Spatial Planning and the Environment (Netherlands) (1999) *Planning the Netherlands: Strategic Principles for a New Spatial Planning Policy*, Ministry of Housing, Spatial Planning and the Environment, The Hague.

VROM, Ministry for Housing, Spatial Planning and the Environment (Netherlands) (2000) *Vijfde Nota* (*Note on National Spatial Strategy*), as approved by the Netherlands Cabinet, VROM, Den Haag.

VROM, Ministry for Housing, Spatial Planning and the Environment (Netherlands) (2005) *Nota Ruimte* (*National Spatial Strategy*) VROM, Den Haag.

Wagenaar, M. (2003) Between civic pride and mass society. In *Amsterdam Human Capital*, Musterd, S. and Salet, W. (eds), Amsterdam University Press, Amsterdam.

Waide, W.L. (1955) The Cambridge Plan – retrospect and prospect. In Town and Country Planning Summer School, Cambridge, pp. 82–89.

Wannop, U. (1995) *The Regional Imperative: Regional Planning and Governance in Britain, Europe and the United States*, Jessica Kingsley, London.

Ward, S.V. (1994) *Planning and Urban Change*, Paul Chapman Publishing, London.

Webber, M. (1964) The urban place and the non-place urban realm. In *Explorations into Urban Structure*, Webber, M., Dyckman, J., Foley, D., Guttenberg, A., Wheaton, W. and Wurster, C. (eds), University of Pennsylvania Press, Philadelphia, pp. 79–153.

Weiss, R.S. (1995) *Learning from Strangers: the Art and Method of Qualitative Interview Studies*, Free Press, New York.

Wenban-Smith, A. (2002) A better future for development plans: making 'Plan, Monitor and Manage' work. *Planning Theory and Practice*, 3, 33–51.

Wenger, E. (1998) *Communities of Practice: Learning, Meaning and Identity*, Cambridge University Press, Cambridge.

While, A., Jonas, A. and Gibbs, D. (2004) Unlocking the city? Growth pressures, collective provision and the search for new spaces of governance in Greater Cambridge, England. *Environment and Planning A*, 36, 279–304.

Whittington, R. (1993) *What is Strategy and Does it Matter?* International Thompson Business Press, London.

Wilkinson, D. and Appelbee, E. (1999) *Implementing Holistic Government*, Policy Press, Bristol.

Williams, R.H.W. (1996) *European Union Spatial Policy and Planning*, Paul Chapman Publishing, London.

Woltjer, J. (2000) *Consensus Planning: the Relevance of Communicative Planning Theory in Dutch Infrastructure Networks*, Ashgate, Aldershot.

WRR (Scientific Council for Government Policy, Netherlands) (1999) *Spatial Development Policy*, SDU, The Hague.

Wynne, B. (1991) Knowledges in context. *Science, Technology and Human Values*, 16, 111–121.

Zajczyk, F., Mugnano, S. and Palvarini, P. (2004) *Large Housing Estates in Italy: Opinions and Prospects of the Inhabitants of Milan (RESTATE project)*, University of Milan-Bicocca, Department of Social Research, Milan.

Zonneveld, W. (2000) Discoursive aspects of strategic planning: a deconstruction of the 'balanced competitiveness' concept in European Spatial Planning. In *The Revival of Strategic Spatial Planning*, Salet, W. and Faludi, A. (eds), Koninklijke Nederlandse Akademie van Wetenschappen, Amsterdam, pp. 267–280.

Zonneveld, W. (2005a) In search of conceptual modernisation: the new Dutch 'national spatial strategy'. *Journal of Housing and the Built Environment*, 20, 425–443.

Zonneveld, W. (2005b) Multiple visoning: new ways of constructing transnational spatial visions. *Environment and Planning C: Government and Policy*, 23, 41–62.

Zonneveld, W. (2005c) The Europeanisation of Dutch national spatial planning: an uphill battle. DISP, 163, 4–15.

INDEX

The arrangement is letter-by-letter. References to notes are prefixed by *n*. Italic page numbers indicate illustrations, tables and figures not included in the text page range.